Risk Habitat Megacity

Dirk Heinrichs • Kerstin Krellenberg
Bernd Hansjürgens • Francisco Martínez
Editors

Risk Habitat Megacity

Editors
Dr. Dirk Heinrichs
German Aerospace Center (DLR)
Institute of Transport Research
Rutherfordstr. 2
12489 Berlin
Germany
dirk.heinrichs@dlr.de

Prof. Dr. Bernd Hansjürgens
Helmholtz Centre for
Environmental Research - UFZ
Department of Economics
Permoserstr. 15
04318 Leipzig
Germany
bernd.hansjuergens@ufz.de

Dr. Kerstin Krellenberg
Helmholtz Centre for
Environmental Research - UFZ
Department of Urban and
Environmental Sociology
Permoserstraße 15
04318 Leipzig
Germany
kerstin.krellenberg@ufz.de

Prof. Dr. Francisco Martínez
Institute for Sustainable Urban Development
Faculty of Physical and Mathematical Sciences
Universidad de Chile
Blanco Encalada 2002
Santiago
Chile
fmartine@ing.uchile.cl

ISBN 978-3-642-11543-1   e-ISBN 978-3-642-11544-8
DOI 10.1007/978-3-642-11544-8
Springer Heidelberg Dordrecht London New York

Library of Congress Control Number: 2011938468

© Springer-Verlag Berlin Heidelberg 2012

This work is subject to copyright. All rights are reserved, whether the whole or part of the material is concerned, specifically the rights of translation, reprinting, reuse of illustrations, recitation, broadcasting, reproduction on microfilm or in any other way, and storage in data banks. Duplication of this publication or parts thereof is permitted only under the provisions of the German Copyright Law of September 9, 1965, in its current version, and permission for use must always be obtained from Springer. Violations are liable to prosecution under the German Copyright Law.

The use of general descriptive names, registered names, trademarks, etc. in this publication does not imply, even in the absence of a specific statement, that such names are exempt from the relevant protective laws and regulations and therefore free for general use.

Printed on acid-free paper

Springer is part of Springer Science+Business Media (www.springer.com)

Foreword

In 2007, various research institutions of the Helmholtz Association, led by the Helmholtz Centre for Environmental Research – UFZ, came together to tackle a theme of growing urgency for the twenty-first century: the issue of megacity development and the risks for humans and the environment inherent in such large agglomerations. The project was approached jointly by the research institutions and Chilean partners.

The Helmholtz Association was not the alone in focusing on this theme. At the same time, two research programmes in Germany were and still are dedicated to similar themes, i.e. the programme 'Future Megacities', funded by the Federal German Ministry of Education and Research, and the research focus 'Megacities – Megachallenges. Informal Dynamics of Global Change', funded by the German Research Foundation DFG. Business-related research has also discovered the topic of megacities: Siemens, for instance, is working on public transport system solutions and car manufacturer BMW is busy developing vehicles and cars that might be better suited to the requirements of gigantic agglomerations.

These examples alone are proof enough that the problem field megacity is attracting growing attention in research and public awareness. In Europe and globally there are more large projects that we should mention. The seventh EU Framework Programme funds the project MEGAPOLI dedicated to the exogenous and endogenous climate effects of megacities. The Mega City Development Project, funded by the Asian Development Bank, has a regional focus on Pakistan and aims at improving the quality of life for the urban population.

This is the research background for the *Risk Habitat Megacity* research initiative of the Helmholtz Association. It differs from the other projects in its concept of focusing on only one city in a multi-thematic/integrative manner and limiting itself to that region. This approach facilitated close cooperation with local and national institutions (government organizations, NGOs, universities), complemented by the international perspective of the UN Economic Commission for Latin America (ECLAC) based in Santiago. The project theme was broad and included natural hazards (earthquakes, floods), air quality, water and energy supply, urban transport, waste management, social risks, residential issues, land-use change, and governance issues. Each individual subsystem was put into the context of the 'super-system city' and systemically integrated into the whole. The research initiative was not confined to scientific analysis of the issues but yielded some approaches to practical solutions for sustainable urban development.

The project faced considerable challenges with regard to terminology, methods and theories. Cooperation between social and natural scientists and researchers from engineering sciences holds a number of difficulties. I remember well the passionate debates about how to define risk and about the practical implementation of sustainability and good governance at the status conferences within the project. In the end, we found concepts that did not only become decisive for this project but will also provide impulses for future research. *Risk Habitat Megacity* has thus made a substantial contribution to the theoretical debate.

Thinking back on the conferences in Santiago, Berlin, and Leipzig, what I remember most vividly are the young researchers from Chile and Germany whose work was funded by this project. These young people went at their themes with great enthusiasm, accompanied and led by experienced scientists from both countries. Another special aspect of *Risk Habitat Megacity*: the young researchers set up workshops and approached interdisciplinary cooperation in a discursive manner, overcame cultural differences stemming from their countries of origin, developed and fine-tuned their ability to communicate in the languages of Humboldt, Darwin and Subercaseaux, learned from and with each other, connected theory and practice. They gained experience in international conferences and publications, presented their findings in talks, poster presentations and publications and put their names on the map of their national and the international scientific community. The project made an invaluable contribution to the training of young scientists of both countries.

As speaker of the Scientific Advisory Council, I had the enviable position of being part of this research initiative from the start. I was honoured to work with Adisa Azapagic, Paul H. Brunner, Tamara Grummt, Veronika Fuest, Marco Keiner, Joseluis Samaniego, Ben Wisner, and Wolfgang von Eckartsberg, prominent representatives of urban studies, the United Nations Economic Commission for Latin America and the Caribbean (ECLAC), business and politics. We had the task of accompanying the project for four long years with critical observation and evaluation, a task that we were happy to take on. In the course of this work we also encouraged an exchange of views and experiences between the three large megacity projects in Germany to the benefit of all.

At the end of my short foreword I want to express thanks. These are due to the project coordinators, especially Bernd Hansjürgens, Dirk Heinrichs and Kerstin Kellenberg, to the academics heading the subprojects, the scientists from Germany and Chile and, most of all, to the many young researchers working hard on their theses, who grew into a closely knit team. I thank my colleagues in the Scientific Advisory Council for their openness, their readiness to contribute pertinent criticism and for their many constructive ideas. Thanks are also due to the host institutions in Chile who provided the beautiful venues of ECLAC and the universities for our conferences.

With this volume, *Risk Habitat Megacity* presents a summary of its findings. Readers may form their own views of what this project achieved within just a few years. It is a milestone and will encourage future megacity research. We hope that this is not the end and that the Helmholtz Association or other research institutions will retain an interest in this theme which is essential for a sustainable future of mankind, an ever larger proportion of which lives in gigantic agglomerations. Only continued research can ensure that, like in the *Risk Habitat Megacity* research initiative, the experiences and findings, the knowledge of best practices and the development of regionally adapted strategies come together in terms of an intercultural and inter- as well as transdisciplinary epistemological interest.

Axel Borsdorf, Innsbruck, Speaker of the Scientific Advisory Board of the Helmholtz *Risk Habitat Megacity* research initiative.

Preface and Acknowledgement

The central motivation for this book is the observation that megacities are both places of opportunities and of risks. Regarding opportunities, megacities promise innovation, economic prosperity and liveable human conditions. The use of resources can be organized much more efficient and effective compared to rural areas. But megacities are also places of risks for humans and the built environment. These risks take many different forms and consequences. This variety, on the one hand, requires distinct risk management approaches. On the other hand, it has led different disciplines to develop a wide range of risk concepts and research methods.

This book brings together a number of different risk perspectives. It explores natural risks stemming from earthquakes, flood risks in connection with land use change, social risks from socio-spatial segregation, risks in the energy supply and the transport system, health risks associated with poor air quality, water quantity- and water quality-related risks, and risks associated with waste management. In each case the risk analysis includes underlying drivers and trends, their extent and severity, and the possibilities to adequately cope with them.

The book concentrates on one selected case study: the Metropolitan Region of Santiago de Chile. This approach offers the unique chance to bring sectoral risk analyses and sectoral policies into correspondence, revealing cross-impacts and side effects. It makes explicit how different risks overlap along different spatial and temporal scales. It sketches a comprehensive picture of the future challenges confronting the Risk Habitat Megacity. And it highlights that the governance structures are both a contributing risks factor and a key capacity to manage and mitigate risks. Overall this comprehensive approach makes the complexity of urban systems visible.

The book is a result of the *Risk Habitat Megacity* research initiative, a joint project of five German research institutes of the Helmholtz Association (German Aerospace Center, Karlsruhe Institute of Technology, GFZ German Research Centre for Geosciences, Helmholtz Centre for Infection Research, Helmholtz Centre for Environmental Research – UFZ) and five partner organizations in Latin America (Universidad de Chile, Pontificia Universidad Católica de Chile, Pontificia Universidad Católica de Valparaíso, Economic Commission for Latin

America and the Caribbean of the United Nations (ECLAC/CEPAL), Universidad Alberto Hurtado) (see www.risk-habitat-megacity.ufz.de). Nearly 60 researchers (among them 20 PhD students) contributed to this inter- and transdisciplinary project between 2007 and 2010. This book presents first comprehensive results. It systematically combines theory-based risk perspectives with concrete risk analysis and discusses the findings in the context of sustainable development and governance.

We would like to thank all those who have contributed to the *Risk Habitat Megacity* research initiative and this book. First of all our thanks go to the more than 60 contributing authors of the book: Jaime Campos, Cristián Cortés, Cristóbal Gatica, Cristián Hernán Godoy Barbieri, James McPhee, Yarko Niño, Gonzalo Paredes, Adriana Perez, Adriana Quintero, Rainer Schmitz, and Luis Vargas (Universidad de Chile); Jonathan R. Barton, Gustavo Durán, Alejandra Rasse, Sonia Reyes-Paecke, Johannes Rehner, Francisco Sabatini, Alejandra Salas, and Claudia Rodriguez Seeger (Pontificia Universidad Católica de Chile); Marcel Szanto (Pontificia Universidad Católica de Valparaíso); Ricardo Jordán, Jorge Rodriguez Vignoli, and Joseluis Samaniego (Economic Commission for Latin America and the Caribbean of the United Nations); Aldo Mascareño (Universidad Alberto Hurtado); Joachim Vogdt and Gerhard Schleenstein (Ingenería Alemana); Frank Baier, Andreas Justen, Barbara Lenz, and Sonja Simon (German Aerospace Center); Klaus-Rainer Bräutigam, Christian Büscher, Tahnee Gonzalez, Jürgen Kopfmüller, Helmut Lehn, Laura Margarete Simon, Helmut Seifert, Volker Stelzer, and Peter Suppan (Karlsruhe Institute of Technology); Marco Pilz, Stefano Parolai, and Joachim Zschau (GFZ German Research Centre for Geosciences); Ellen Banzhaf, Ulrich Franck, Carolin Höhnke, Corinna Hölzl, Sigrun Kabisch, Annegret Kindler, Michael Lukas, Karin Metz, Annemarie Müller, Henning Nuissl, Gerhard Strauch, Ulrike Weiland, and Juliane Welz (Helmholtz Centre for Environmental Research – UFZ). We are likewise grateful to the staff of the Regional Government of the Santiago Metropolitan Region, in particular Anamaria Silva and Pablo Fuentes, for their close cooperation and partnership. We further would like to thank the more than 100 representatives from public sector, private enterprises, academia and civil society organizations for contributing their knowledge and perspectives by participating in several research activities and workshops, and thereby contributing to the results presented in this book. Sunniva Greve made a tremendous effort to improve the style of the English. The production would not have been possible without the help of Katrin Barth and Kay Fiedler who assembled the text, tables, graphs, and references from the authors into a coherent manuscript. The Helmholtz Association provided financial support through its Initiative and Networking Fund.

Berlin/Leipzig/Santiago

Dirk Heinrichs, Kerstin Krellenberg, Bernd Hansjürgens, Francisco Martínez

Contents

Part I Megacities – A Challenge for Research and Implementation

1 Introduction: Megacities in Latin America as Risk Habitat 3
   Dirk Heinrichs, Kerstin Krellenberg, and Bernd Hansjürgens

2 Megacities in Latin America: Role and Challenges 19
   Ricardo Jordán, Johannes Rehner, and Joseluis Samaniego

Part II Developing the Conceptual Framework

3 Mechanisms of Systematic Risk Production 39
   Christian Büscher and Aldo Mascareño

4 Sustainable Urban Development in Santiago de Chile:
   Background – Concept – Challenges 65
   Jonathan R. Barton and Jürgen Kopfmüller

5 Megacity Governance: Concepts and Challenges 87
   Henning Nuissl, Carolin Höhnke, Michael Lukas, Gustavo Durán,
   and Claudia Rodriguez Seeger

Part III Exploring Policy Fields

6 Earthquake Risks: Hazard Assessment of the City
   of Santiago de Chile 111
   Marco Pilz, Stefano Parolai, Joachim Zschau, Adriana Perez,
   and Jaime Campos

7 Land-Use Change, Risk and Land-Use Management 127
   Ellen Banzhaf, Annegret Kindler, Annemarie Müller, Karin Metz,
   Sonia Reyes-Paecke, and Ulrike Weiland

8 Socio-spatial Differentiation: Drivers, Risks and Opportunities. . . 155
   Sigrun Kabisch, Dirk Heinrichs, Kerstin Krellenberg, Juliane Welz,
   Jorge Rodriguez Vignoli, Francisco Sabatini, and Alejandra Rasse

9 Energy Systems... 183
   Sonja Simon, Volker Stelzer, Luis Vargas, Gonzalo Paredes,
   Adriana Quintero, and Jürgen Kopfmüller

10 Santiago 2030: Perspectives on the Urban Transport System..... 207
   Andreas Justen, Francisco Martínez, Barbara Lenz, and Cristián Cortés

11 Air Quality and Health: A Hazardous Combination
   of Environmental Risks.................................. 229
   Peter Suppan, Ulrich Franck, Rainer Schmitz, and Frank Baier

12 Risks and Opportunities for Sustainable Management of Water
   Resources and Services in Santiago de Chile................. 251
   Helmut Lehn, James McPhee, Joachim Vogdt, Gerhard Schleenstein,
   Laura Simon, Gerhard Strauch, Cristian Hernàn Godoy Barbieri,
   Cristobal Gatica, and Yarko Niño

13 Municipal Solid Waste Management in Santiago de Chile:
   Challenges and Perspectives towards Sustainability............ 279
   Klaus-Rainer Bräutigam, Tahnee Gonzalez, Marcel Szanto,
   Helmut Seifert, and Joachim Vogdt

Part IV Synthesis and Perspectives

14 How Sustainable is Santiago?............................. 305
   Jürgen Kopfmüller, Jonathan R. Barton, and Alejandra Salas

15 Dealing with Risks: A Governance Perspective
   on Santiago de Chile.................................... 327
   Corinna Hölzl, Henning Nuissl, Carolin Höhnke, Michael Lukas,
   and Claudia Rodriguez Seeger

16 Synthesis: An Integrative Perspective on Risks in Megacities..... 353
   Dirk Heinrichs, Kerstin Krellenberg, Bernd Hansjürgens,
   and Francisco Martínez

List of Figures

Fig. 3.1	Attraction/exposure relation	49
Fig. 3.2	Metabolization/deterioration relation	51
Fig. 3.3	Synchronization/desynchronization relation	56
Fig. 3.4	Inclusion/exclusion relation	60
Fig. 4.1	Public institutional structure for the Santiago metropolitan region	80
Fig. 6.1	Basin of Santiago de Chile	115
Fig. 6.2	Surface geology of the city of Santiago de Chile	116
Fig. 6.3	Map of the fundamental resonance frequency in the investigated area	118
Fig. 6.4	Cross sections of the interpolated 3D shear wave velocity model within the area of investigation	120
Fig. 6.5	v_s^{30} for the area limited by a linear connection between the outermost measurement sites	121
Fig. 7.1	The components of risk exemplified for the case of floods: hazard, vulnerability, elements at risk	129
Fig. 7.2	Change in population by municipality 1992–2002 and 2002–2006 in the Metropolitan Area of Santiago de Chile	132
Fig. 7.3	Changes in built-up areas showing dynamics of urban growth	137
Fig. 7.4	Dynamics of land-use change with reference to urban built-up densities and loss of agricultural land	138
Fig. 7.5	Variables and measures relevant to flood risk generation in Santiago de Chile	142
Fig. 7.6	Land-use/land-cover changes in La Reina and Peñalolén between 1993 and 2009, changes in flood-prone areas between 1986 and 2008	143
Fig. 8.1	Sequence of analytical steps	158
Fig. 8.2	The Greater Metropolitan Area (GMAS) of Santiago and the five municipal clusters Centre, Peri-Centre, Eastern Peri-Centre, Periphery, Extra-Periphery	160

Fig. 8.3	Migration patterns differentiated by municipal cluster	162
Fig. 8.4	Average land prices (US\$/m^2) aggregated by municipal cluster	169
Fig. 8.5	Number of new social housing units in the municipal clusters (1979–2002)	172
Fig. 9.1	Electricity consumption per capita in selected municipalities of Santiago for 2002	185
Fig. 9.2	Share of final energy consumption 2007	186
Fig. 9.3	Location and data (2008) of the four power grids	188
Fig. 9.4	Share of rural households in the MRS without access to electricity	192
Fig. 9.5	Duration of interruption of electricity supply in MRS Data	193
Fig. 9.6	Amount of energy per GDP (relative, 1990 as 100) Data	193
Fig. 9.7	Share of power production from NCRE in SIC in relation to the requirements of Law 20.257 for the development of NCRE and other targets	194
Fig. 9.8	Development of CO_2 emissions in Chile under BAU assumptions	195
Fig. 9.9	Reduction of gas supply from Argentina (in percent of contracted delivery)	196
Fig. 9.10	Potential development of primary energy demand in the "Energy [R]evolution" scenario	200
Fig. 10.1	Santiago road and metro network	209
Fig. 10.2	Interaction between land-use and transport models	216
Fig. 10.3	Input and output of transport and land-use models	217
Fig. 10.4	Prioritization of transport policies in Santiago 2008	218
Fig. 10.5	Percentage increase in the number of households by municipality in the Metropolitan Area of Santiago between 2010 and 2030	221
Fig. 10.6	Saturation on major transport links 2010 (*left*) and 2030 (*right*)	222
Fig. 10.7	Increase and decrease in average travel time between 2010 and 2030 by mode and municpality	225
Fig. 11.1	Methodological approach of combining satellite information with measurements and emission data (*left*) as input for coupled meso- and micro-scale modelling tools (*centre*), a prerequisite for health impact assessment studies, stakeholder involvement and scenario development (*lower right*)	232
Fig. 11.2	Methodology of traffic emission modelling as a basic requirement for the assessment of current and future emission states	233
Fig. 11.3	Emission sources in the greater region of Santiago de Chile: NO_x emissions (*left*) and PM_{10} emissions (*right*)	237

List of Figures

Fig. 11.4	Annual mean time series of NO_2, and PM_{10} and 8 h annual maximum of O_3 based on the MACAM II monitoring network of the Metropolitan Area of Santiago de Chile	238
Fig. 11.5	Annual PM_{10} means in selected urban agglomerations worldwide	239
Fig. 11.6	Distribution of O_3 (*left*) and PM_{10} (*right*) concentrations measured in the Metropolitan Area of Santiago in 2004	240
Fig. 11.7	Distribution of PM10 (above) and NO_x (below) traffic emissions per mileage (Nogalski 2010) and traffic category for 2010 (by SECTRA) within the Metropolitan Area of Santiago de Chile	241
Fig. 11.8	Distribution of NO_x traffic emissions in the municipalities of the Metropolitan Area of Santiago de Chile (tons per year in 2010)	242
Fig. 11.9	Traffic-related annual mean NO_x concentration levels in $\mu g/m^3$ within the Metropolitan Area of Santiago (municipality border lines are included in the graphic) based on meteorological conditions in 2006	242
Fig. 11.10	Distribution of NO_x concentrations on the regional scale (*left*) and in the Greater Region of Santiago (*right*) during a 2 week period in January 2006	243
Fig. 11.11	Lagged risk increase per 10 $\mu g/m^3$ PM_{10} for hypertensive (*left*) and ischemic heart disease (*right*). Significantly increased values are indicated (CI – 95% confidence intervals)	244
Fig. 11.12	Lagged risk increase per 10 $\mu g/m^3$ PM_{10} for other forms of heart disease (*left*) and influenza and pneumonia (*right*). Significantly increased values are indicated (CI – 95% confidence intervals)	245
Fig. 11.13	Lagged risk increase per 10 $\mu g/m^3$ PM_{10} for chronic lower respiratory disease. Significantly increased values are indicated (CI – 95% confidence intervals)	245
Fig. 12.1	Distance-to-target approach	253
Fig. 12.2	The Santiago Metropolitan Region and the catchment area of the Maipo-Mapocho River System	254
Fig. 12.3	Concession areas of drinking water suppliers in Santiago de Chile	256
Fig. 12.4	Connectivity implementation between the numerical model (MOSSEM) and the management mode	257
Fig. 12.5	Long-lasting precipitation variations in Santiago – Quinta Normal Station	260
Fig. 12.6	Copper (upper value) and arsenic (lower value) concentrations in the surface waters of the Maipo/Mapocho catchment and drinking water in Providencia ($\mu g/l$)	263

Fig. 12.7	Water stress in river catchments in the year 2000 calculated as the ratio between water extraction and renewable water resources (precipitation minus evapotranspiration 1961–1990)	266
Fig. 13.1	Correlation between MSW deposited in landfills and economic growth in RMS	281
Fig. 13.2	Mass flow of waste in the Metropolitan Region of Santiago de Chile, 2007. Stock flows, emissions and residue fluxes have been omitted to simplify the diagram	283
Fig. 13.3	Recycling rates for different materials – formal and informal contributions, 2007	285
Fig. 13.4	Calculated emissions of CO_2 equivalents for "RMS" for shares of organic waste collected separately and capture rates of landfill gas	298
Fig. 14.1	Percentage of people living in poverty	312
Fig. 14.2	Infant mortality rate (per 1,000 live births)	313
Fig. 14.3	Household overcrowding	314
Fig. 14.4	Reported serious crime rate (per 100,000 people)	314
Fig. 14.5	Wage inequality by gender	315
Fig. 14.6	Gini coefficient	316
Fig. 14.7	CO_2 emissions per capita	317
Fig. 14.8	Public green area per capita	318
Fig. 14.9	Percentage of households with Internet access	319
Fig. 14.10	Unemployment rate	320
Fig. 14.11	Population aged 14–17 not attending school	320
Fig. 15.1	Actors with a major influence on governance in Santiago	330
Fig. 15.2	Evaluation of decision-making process coordination	333

List of Tables

Table 2.1	Expected population growth and urbanization	22
Table 2.2	Ageing population indicators in selected Latin American countries	23
Table 2.3	Participation in the national GDP and informal employment	25
Table 2.4	Income concentration and poverty	27
Table 3.1	Risk production mechanisms in Santiago de Chile with respect to land-use management	50
Table 3.2	Risk production mechanisms in Santiago de Chile with respect to air quality	52
Table 3.3	Risk production mechanisms in Santiago de Chile with respect to waste management	53
Table 3.4	Risk production mechanisms in Santiago de Chile with respect to transportation	56
Table 3.5	Risk production mechanisms in Santiago de Chile with respect to socio-spatial differentiation	60
Table 4.1	Sustainability rules of the Helmholtz integrative sustainability concept	75
Table 5.1	Heuristic framework for the analysis of arrangements and processes of governance	97
Table 7.1	Selected risk-related indicators	134
Table 7.2	Development of population density in the peripheral municipalities in total and in their built-up areas	139
Table 7.3	Number and surface area of green spaces in the Metropolitan Area of Santiago by size rank	140
Table 8.1	Population, intra-urban migration and net migration for GMAS 1982–2006	161
Table 8.2	Greater Metropolitan Area of Santiago (39 municipalities and five clusters): average years of education, increase in years of education in percentage and average years of education of immigrants and emigrants aged 25 or over	163

Table 8.3	Change in housing stock 1992–2002	167
Table 8.4	Number of building permits for residential dwellings aggregated by municipal cluster	168
Table 8.5	Land offered for transaction (in m^2)	169
Table 8.6	FSV I and FSV II acquisition in GMAS between 2006 and 2009	175
Table 9.1	Final energy use (in PJ)	185
Table 9.2	Import of energy resources (PJ)	186
Table 9.3	Electric system (SIC) generation by source	189
Table 9.4	Chilean dependency on imports for energy provision	195
Table 9.5	Percentage of the three most important branches in the energy sector: electricity sales, electricity purchase, and market distribution of liquid fuels	197
Table 10.1	Modal split in Santiago, 1991 and 2001	210
Table 10.2	Motorization in the metropolitan area of Santiago between 2001 and 2009	211
Table 10.3	Economic and demographic growth and expected motorisation rate	219
Table 10.4	Average velocities and travel times 2010 and 2030	223
Table 10.5	Modal split in Santiago, 2010 and 2030, morning peak hour 07.30–08.30	224
Table 11.1	Threshold values for specific pollutants in Chile compared with the air quality guidelines of the World Health Organization	236
Table 11.2	Statistics on PM_{10} concentrations of all monitoring stations in the Metropolitan Region of Santiago de Chile for 2006	238
Table 11.3	Total number of deaths in various cardiovascular and respiratory disease classes in Santiago de Chile, 2006 (ICD-10 codes are an international coding system for disease)	244
Table 11.4	Assignment of PM_{10} concentrations to scenarios	246
Table 11.5	Maximum daily death risk decreases for selected disease groups. Risk decreases are compared with the business-as-usual scenario, which defines the 100% PM_{10} associated mortality risk	246
Table 12.1	Core sustainability indicators for the water sector	259
Table 12.2	Absolute and specific amount of renewable fresh water available (supply)	260
Table 12.3	Water usage by different economic sectors in the Metropolitan Region of Santiago in 2007 and the prediction for 2017 and 2032	261
Table 12.4	Per capita drinking water consumption in the concession area of Aguas Andinas S.A for the year 2005 and predictions for the future	261

Table 12.5	Options for measures to overcome water stress and water scarcity in the RM Santiago	275
Table 13.1	Quantity of different types of waste in RMS in 2006	281
Table 13.2	Quantities of municipal solid waste in RMS in the years 1995–2007 and calculated recycling rate	282
Table 13.3	Sustainability indicators for MSW management	291
Table 13.4	Waste characteristics used for model calculations	296
Table 13.5	Calculated values for the organic carbon content of disposed waste	297
Table 14.1	General sustainability indicators	311

List of Abbreviations

AG	Aktiengesellschaft (German Public Corporation)
Agbar	Aguas de Barcelona
AMS	Área Metropolitana de Santiago (MAS - Metropolitan Area of Santiago)
APR	Agua Portable Rural
AUDP	Áreas Urbanas de Desarrollo Prioritario (Urban Development Priority Areas)
BAU	Business-as-usual
BRT	Bus rapid transit
CASEN	Caracterización Socioeconómica Nacional (National Socioeconomic Characterization)
CCHC	Cámara Chilena de la Construcción (Chilean Construction Chamber)
CDEC	Centro de Despacho Económico de Carga (Load Economic Dispatch Centre)
CDM	Clean Development Mechanism
CELADE	Centro Latinoamericano de Demografía (Latin American Centre of Demography)
CEPAL	Comisión Económica para América Latina y el Caribe (ECLAC – Economic Commission for Latin America and the Caribbean)
CESCO	Consejo Económico y Social Comunal (Economic and social Citizens'Council)
CGE	Compañía General de Electricidad (General Electricity Company)
CHP	Combined Heat Power
CI	Confidence Intervals
CNE	Comisión Nacional de Energía (National Energy Commission)
CODESUP	Corporación para el Desarrollo Sustentable de Pudahuel (Corporation for the Sustainable Development of Pudahuel)
CONAMA	Comisión Nacional de Medio Ambiente (National Environment Commission)

CONECYT	Consejo Nacional de Ciencia y Tecnología (CONACYT - National Council on Science and Technology)
CORFO	Corporación de Fomento de la Producción (Corporation of for the promotion of production)
CSM	Consorcio Santa Marta
CSD	Commission on Sustainable Development
DALY	Disability-adjusted Life Years
D.C.	Distrito Capital (Capital District)
DEIS	Departamento de Estadísticas e Información de Salud (Department of Health Statistics and Information)
D.F.	Distrito Federal (Federal District)
DGA	Dirección General de Aguas (Water Division)
DGF	Departamento de Geofísica (Department of Geophysics)
DIPRES	Dirección de Presupuestos (y Budgetary Division of the Treasury)
DLR	Deutsches Zentrum für Luft- und Raumfahrt (German Aerospace Centre)
DOH	Dirección de Obras Hidráulicas (Division for Hydraulic Construction)
DOM	Dirección de Obras Municipales (Division for Municipal Works)
DPSIR	D: Driving forces; P: Pressure; S: State; I: Impact; R: Response
D.S.	Development Studies Program
DSS	Decision Support System
EDR	Estrategia de Desarrollo Regional (Regional Development Strategy)
EG	Europäische Gemeinschaft (European Community)
EGIS	Entidad de Gestión Inmobiliaria Social (Social Housing Management Agency)
ENSO	El Ninõ – Southern Oscillation
EMOS	Empresa Metropolitana de Obras Sanitarias (Metropolitan Company of Waste Water Works)
ENAP	Empresa Nacional de Petróleo (National Petroleum Company)
EOD	Encuesta Origen Destino (Origin Destination Survey)
EREC	European Renewable Energy Council
ESTRAUS	Santiago four-stage Simulation Transport Model
FCM	Fondo Común Municipal (Common Municipal Fund)
FDI	Foreign Direct Investment
FSV	Fondo Solidario de Vivienda (Housing Solidarity Fund)
GA	Genetic Algorithm
GDP	Gross Domestic Product
G_e	Total amount of gas produced in Nm^3/ton of waste
GHG	Greenhouse Gas
GIS	Geographic Information Systems
GMAS	Greater Metropolitan Area of Santiago

List of Abbreviations

GmbH	Gesellschaft mit beschränkter Haftung (German private limited liability company)
GORE	Gobierno Regional Metropolitano de Santiago (Regional Government of Santiago)
GPS	Global Positioning System
GRAL	Grazer Lagrangian Model
GTZ	Deutsche Gesellschaft für Technische Zusammenarbeit (German development agency), since 1 January 2011 Deutsche Gesellschaft für Internationale Zusammenarbeit (GIZ)
HDI	Human Development Index
HDV	Heavy Duty Vehicles
ibid	ibidem, the same place
ICAP	Illness Costs of Air Pollution
ICD-10	International Statistical Classification of Diseases and Related Health Problems
ICLEI	International Council for Local Environmental Initiatives
IEA	International Energy Agency
IMF	International Monetary Fund
INE	Instituto Nacional de Estadísticas (National Institute of Statistics)
IPCC	Intergovernmental Panel on Climate Change
IPCC A2	Intergovernmental Panel on Climate Change Appendix 2
IISD	International Institute for Sustainable Development
ISDR	International Strategy for Disaster Reduction
LAC	Latin America and the Caribbean
LAC	Línea de Atención a Campamentos (Support Programme for Popular Settlements)
LNG	Liquefied Natural Gas
LUGC	Ley General de Urbanismo y Construcciones (General Law of Urbanism and Construction)
LULC	Land-use and Land-cover
MACAM	Red de Monitoreo Automática de Contaminantes Atmosféricos (Automatic Monitoring Network of Air Pollutants)
MAS	Metropolitan Area of Santiago de Chile
MEE	Ministerio de Economía y Energía (Ministry of Economy and Energy)
MESAP/Planet	Modular Energy System Analysis and Planning Environment/Planning Network
MIDEPLAN	Ministerio de Planificación (Ministry of Planning)
Minminería	Ministro de Minería y Metalurgia (Mining Ministry)
MinSal	Ministerio de Salud (Ministry of Health)
MINVU	Ministerio de Vivienda y Urbanismo (Ministry of Housing and Urbanism)
MITT	Ministerio de Transporte y Telecomunicaciones (Ministry of Transport and Telecommunication)

MMA	Ministerio del Medio Ambiente (Ministry of Environment)
MODEM	Modelo de Emisiones Vehiculares (Vehicle Emission Model)
MOP	Ministerio de Obras Públicas (Ministry of Public Works)
MOSSEM	Model for Sediment Transport and Morphology
MNRCH	Movimiento Nacional de Recicladores de Chile (National Movement of Chilean Recyclers)
MR	Metropolitan Region
MRS	Metropolitan Region of Santiago de Chile
MSW	Municipal Solid Waste
MUSSA	Modelo del Use de Suelo (Land-Use Equilibrium Model)
n./a.	Not available
NCh	Norma Chilena Oficial (Official Chilean Norm)
NCRE	Non-conventional Renewable Energy
NGO	Non-Governmental Organization
NIMBY	Not in my back yard
NIMTO	Not in my term of office
NMVOC	Non-methane Volatile Organic Compounds
NTAX	Non-traditional export agriculture
OECD	Organisation for Economic Co-operation and Development
OTAS	Programa de Ordenamiento Territorial Ambientalmente Sustentable (Environmentally Sustainable Spatial Planning Programme)
PAC	Programa de Aseguramiento de la Calidad (Quality Insurance Plan)
PAHO	Pan-American Health Organization
PDUC	Proyectos de Desarrollo Urbano Condicionados (Conditional Urban Development Projects)
PLADECO	Plan de Desarrollo Comunal (Muncipal Development Plan)
PM	Particulate matter
PMG	Programa de Mejoramiento de la Gestión (Management Improvement Programme)
PNUMA	Programa de las Naciones Unidas para el Medio Ambiente (UNEP - United Nations Environmental Programme)
PPDA	Plan de Prevención y Descontaminación Atmosférica de la Region Metropolitana (Plan for Atmospheric Prevention and Decontamination of the Metropolitan Region)
PPP	Public Private Partnership
PRC	Plan Regulador Comunal (Municipal Regulatory Plan)
PR China	China People's Republic of China
PRMS	Plan Regulador Metropolitano de Santiago (Metropolitan Land-Use Plan for Santiago)
PROT	Plan Regional de Ordenamiento Territorial (Regional Urban Development Plan)
PTUS	Plan de Transporte Urbano del Gran Santiago (Public Transport Plan for Santiago)

List of Abbreviations

RHM	Risk Habitat Megacity
RM	Region Metropolitana (MR - Metropolitan Region)
RMS	Region Metropolitana de Santiago (MRS - Metropolitan Region of Santiago)
S.A.	Sociedad Anónima (Chilean Public Limited Corporation)
SECPLA	Secretaría Comunal de Planificación (Municipal Secretariat of Planning)
SECTRA	Secretaria de Planificación de Transporte (Secretary of Transportation Planning)
SEIA	Sistema de Evaluación de Impacto Ambiental (System of Environmental Impact Assessment)
SENDOS	Servicio Nacional de Obras Sanitarias (National Service of Water Works)
SEREMIs	Secretarías Regionales Ministeriales (Regional Ministerial Secretariats)
SEREMITT	Secretarías Regionales Ministeriales de Transportes y Telecomunicaciones (Regional Secretariats of the Ministry of Transport and Telecommunication)
SERVIU	Servicio de Vivienda y Urbanismo (Service of Housing and Urbanism)
SING	Sistema Interconectado del Norte Grande (Northern Interconnected System)
SOA	Secondary Organic Aerosols
SUBDERE	Subsecretaría de Desarrollo Regional y Administrativo (Sub-Secretariat for Regional Development and Administration)
SWOT	Strengths-Weaknesses-Opportunities-Threats
SIC	Sistema Interconectado Central (Central Interconnected System)
SINIA	Sistema Nacional de Información Ambiental de Chile (National Environmental Information System)
SISS	Superintendencia de Servicios Sanitarios (Superintendency of Sanitary Services)
SMAPA	Servicio Municipal de Agua Potable y Alcantarillado de Maipú (Municipal Water and Sewer Services Maipú)
SUR	Corporación de Estudios Sociales y Educación
TAZ	Traffic Analysis Zones
Tcf	Trillion cubic feet
TEEB	The Economics of Ecosystems and Biodiversity
TTIK	A measurement standard for power supply interruption
UN	United Nation
ECLAC	Economic Commission for Latin America and the Caribbean (CEPAL – Comisión Económica para América Latina y el Caribe)
UCH	Universidad de Chile (University of Chile)
UCV	Universidad Católica de Valparaiso (Catholic University of Valparaíso)

UF	Unidad de Fomento (Unit of Exchange)
UNCHS	United Nations Centre for Human Settlements
UNDP	United National Development Program
UNEP	United Nations Environmental Program
UNESCO	United Nations Educational, Scientific and Cultural Organization
UNFPA	United Nations Fund for Population Activities
USD	United States Dollar
UTF	Universidad Técnica Federico Santa Maria (Technical University Federico Santa Maria)
VOC	Volatile Organic Compounds
Vs30	Average share wave velocity in the uppermost 30 m of the soil
WBGU	Wissenschaftlicher Beirat der Bundesregierung Globale Umweltveränderungen (German Advisory Council on Global Change)
WCED	United Nations World Commission on Environment and Development
WHO	World Health Organization
WRF	Weather Research and Forecasting
WRF/chem.	Meteorology/Chemistry Dispersion Model
ZODUC	Zona de Desarrollo Urbana Condicionado (Conditioned Development Zones)

Part I
Megacities – A Challenge for Research and Implementation

Chapter 1
Introduction: Megacities in Latin America as Risk Habitat

Dirk Heinrichs, Kerstin Krellenberg, and Bernd Hansjürgens

Abstract While the world has stepped into the century of cities, the emergence of the megacity is perhaps the most visible expression of the mega-trend urbanization. This introductory chapter discusses some features that characterize the megacity beyond its extraordinary population size. Furthermore, it outlines the main objectives of this book, discusses the key concepts risk, sustainability and governance and further elaborates on why and to what extent megacities are places of opportunities and risks. The chapter describes the geographical focus – Latin America – with Santiago de Chile as an *anchor city* and closes with a brief overview of the book.

Keywords Governance • Latin America • Megacity • Risk • Sustainability

1.1 The Rise of Megacities

The world has stepped into the century of cities. The so-called 'urban turn' that finds more people living in urban areas than in the countryside for the first time in history is driven by unprecedented urbanization, a trend that breeds both risks and opportunities. The expansion of urban areas transforms land cover and hydrological systems (Grimm et al. 2008), represents a major driver of habitat loss (Mc Donald et al. 2008) and exposes people to the dangers of hazardous natural and man-made events (Mitchell (Eds) 1999). At the same time, the concentration of population in cities harbours the potential to lower resource consumption. Compact urban forms, mass transit systems and walkable neighbourhoods reduce carbon dioxide (CO_2) emissions and the demand for energy (National Research Council 2009). Agglomeration effects

---

D. Heinrichs (✉) • K. Krellenberg • B. Hansjürgens
German Aerospace Center (DLR), Institute of Transport Research, Rutherfordstr. 2, 12489 Berlin, Germany
e-mail: dirk.heinrichs@dlr.de

in the economic sectors promote employment, and allow firms and workers to become more productive and creative (Quigley 1998; Bettencourt et al. 2007).

According to the United Nations (2008), the urban share of the world's population rose from around 3% in 1900 to one-third in the early 1950s, reaching 50% in 2008. Predictions suggest that this trend will continue. Urban population growth of about 1.7% per annum will bring the number of urban dwellers worldwide from currently 3.5 billion to roughly 5 billion by 2030, an increase corresponding to the creation of a city with approximately 1.5 million inhabitants every 10 days (UN Population Division 2008). Although urbanization is a global phenomenon, it does not occur as a homogeneous process across regions and countries, or even within one country. Historically rapid urbanization has its roots in today's more developed regions. The overwhelming majority of the population of Europe (72%) and North America (80%) now lives in towns and cities. Among the developing world regions, it is the region of Latin America and the Caribbean, with its exceptionally high level of urbanization (around 78%), that ranks as advanced in this context. In contrast, the majority of people in the continents of Africa and Asia still live in rural areas (approx. 60%). Urbanization levels are expected to rise in the coming decades, particularly in regions of the developing world (UN Population Division 2007, 2008).

The emergence of the megacity is perhaps the most visible expression of the mega-trend urbanization. There is some consensus on the definition of megacities as urban agglomerations of at least ten million inhabitants (UN Population Division 2002, 2004, 2006, 2008; WBGU 1996; Montgommery et al. 2003). The number increased dramatically from three megacities (Mexico City, New York, Tokyo) in 1975 to around 20 in 2007 and is projected to reach almost 30 worldwide by 2025. The group of *emerging megacities* whose populations range from five to ten million have likewise experienced a notable increase. The number is expected to increase from currently about 30 to almost 50 in 2025. Together, 'established' and emerging megacities now account for roughly 15% of the world's total urban population. This share is predicted to increase to approximately 17% by 2025 (UN Population Division 2008).

The largest urban agglomerations are not necessarily those with the fastest population growth. A third of the current megacities with more than ten million inhabitants report a population increase of less than 1.5% for the period 1975–2007. In Tokyo, for example, the growth rate was slightly less than 1% and is predicted to decline further up to 2025. On the other hand, cities like Karachi (3.5%), Dhaka (5.6%) and Bogotá (3.5%) have witnessed extremely rapid growth. Even where proportional rates seem low, the absolute population figures continue to rise significantly. The current 1.8% population growth rate for Mexico City, for example, still adds approximately 400,000 people to its urban population every year. At an approximate growth rate of 2% population doubles every 35 years, at a growth rate of 3% it doubles every 17 years (UN Population Division 2008).

Following these figures, the primary and perhaps most defining attribute of the megacity is its magnitude in terms of population. Megacities and their urban populations have reached incomparable scales. Not only are they larger in terms of population than ever before, but also in terms of physical extent, environmental impact, values and economic importance. They are 'global cities' in a global economy (e.g., Beaverstock et al. 1999; Friedmann 1986; Hall 1996; Sassen 1991) or 'primate' cities with a prominent role in their respective countries (Bronger 1996).

A second defining feature of the megacity is the speed of change. It took just 50 years for the city of Dhaka, for example, to advance from a population of less than 1 to currently almost 15 million. As populations grow, urban land expands to provide accommodation and services. While for most of history the rate of urban land expansion was low and physical boundaries shifted at moderate rates, contemporary land conversion occurs at high speed. Changes in life style, housing demands and mobility, as well as socio-economic change are key factors behind this development.

A third distinguishing characteristic of the megacity is its complexity (Heinrichs et al. 2009). The rapid change and unprecedented scale of populations and physical expansion that characterize mega-urbanization sets the stage for highly complex simultaneous and interactive processes. Megacities are, for example, embedded in complex global–local relationships. Their emergence and development is both a trend and an expression of globalization, with intense interdependencies regarding the exchange of information, goods and the use of resources (Young et al. 2006). Global dimensions are intrinsic to megacities, since the use of resources has a decisive impact on resource regimes and ecosystems worldwide. On the other hand, environmental change, and more specifically climate change, impacts back on megacities. In this sense, they are *culprits* and *victims* of global change at one and the same time (Sassen 2002; Gurjar and Lelieveld 2005; Hardoy et al. 2001).

Megacities also prove complex with regard to *governance structures*. Mexico City is an illustrative case. The metropolitan region is divided into four main administrative units: the federal district (which itself is subdivided into 16 units), the administration of the state of Mexico and Hidalgo (with 59 municipal administrations) and the national government ministries with key responsibilities in the operation of various sectors (MRC McLean Hazel and GlobeScan 2007). Interaction between government rules and spontaneous, decentralized decisions renders this complexity more acute. The provision of basic services such as water, shelter and mobility serves as an example. In many megacities it is organized by a huge variety of *decentralized* individual and *informal* institutions outside the *formal* set of rules (Webster and Lay 2003), particularly when governments fail or are unable to ensure the provision of basic goods and services.

1.2 Understanding the *Risk Habitat Megacity*: Motivation and Objectives of this Book

The objective of this book is to link the defining features of the megacity – scale, rapid change, complexity – to the study of its inherent risks and opportunities. It attempts to (1) focus systematically on the emergence of risk in mega-urban agglomerations by analysing risk elements, (2) evaluating the extent and severity of risks, and (3) developing strategies to cope with adverse risks and to guide urban development. It addresses the following questions:

- What are the risks associated with the trend towards mega-urbanization and what are the driving forces and interactions behind it?
- What research methods and approaches are appropriate to understand and provide orientation and action knowledge on the *Risk Habitat Megacity*?
- What forms and strategies of governance constitute an adequate response to these challenges?

The book explores these questions for one selected case: the Metropolitan Region of Santiago de Chile. It contains an in-depth analysis of a comprehensive collection of sectors and policy fields: land use, socio-spatial differentiation, water, energy, transportation, waste, air quality and health, and earthquake management. The examination of each policy field adopts a *risk* perspective. This firstly serves to make explicit the possible range of risk perspectives and a pronounced problem orientation (see Chap. 3), and helps to move the analysis from understanding to action by providing concrete advice for risk management in each sector. Secondly, the consideration of several sectors renders their relationships visible (e.g., transportation and air quality, energy and water, land use, risks and socio-spatial differentiation) in the attempt to adequately address complexity. Thirdly, at a more conceptual level, this multiple perspective allows for reconstruction of the *Risk Habitat Megacity* and consequently for an understanding of its complexity.

The book combines the empirical study of urban sectors with theoretical considerations of risk and explores several common risk concepts. Beyond the risk perspective, it draws on two other theory-based concepts: *sustainability* and *governance*. Sustainable development is used to introduce a normative perspective. With the formulation of indicators, it ushers in a discussion on acceptable risk levels or thresholds, for example with respect to air contamination, waste production or water consumption. In doing so, the concept is likewise exploited to define risks as a sustainability deficit. The concept of governance puts forward the action perspective, which is concerned with how best to manage risk. At the same time, the analysis of the different components acknowledges that governance structures themselves may be the source of risks, e.g., when poor planning systems lead to settlement on territory endangered by floods or landslides.

The book provides insights for urban managers, practitioners, researchers and students with a particular interest in interdisciplinary and intersectoral work. It likewise constitutes a source for scholars in various cross-cutting areas of urban research, sustainability science, hazard research, governance and planning science.

The remainder of this introductory chapter deals with the necessary definitions and sets the frame for the subsequent chapters. Section 1.3 elaborates on why and to what extent megacities are places of opportunities and risks, while Sect. 1.4 highlights the underlying definitions and concepts of risk. The geographical focus – Latin America – with Santiago de Chile as an *anchor city* is described in Sect. 1.5. Finally, Sect. 1.6 gives a brief overview of the book.

1.3 Megacities as Spaces of Opportunity and Risk

Unprecedented scale, rapid change and complexity, these are the defining characteristics that transform the urban habitat into a space of opportunity and a space of risk.

Megacities provide first and foremost *economic opportunities*. They are engines of global economic growth and contribute more than proportionately to the national output. As an illustration, approximately half of the world's megacities generate more than a third of the gross domestic product (GDP) of their respective countries. Cities like Bangkok or São Paulo are home to between 10% and 15% of the national population but contribute more than 40% to the GDP (UN 2006; see also Jordán et al. in this volume). These high levels of relative productivity are attributed to the scale advantages of large agglomerations (see, for example, Bettencourt et al. 2007) in the provision of public and private goods, and more favourable conditions for specialization. This associates with the network and cluster effects of economic branches that lead to further efficiency gains. Key to this development is the opening of national economies, promoting the mobilization of goods, capital and people, and an international division of labour (DB Research 2008). These tendencies are not expected to change in the future and may even intensify as a result of increased globalization.

Closely related to economic opportunities is the potential for *social and human capital*. In general, the provision of education (from pre-school to university) is more advanced in large agglomerations than in smaller cities and rural areas. School enrolment rates are in general higher in cities than in villages. The financial strength of the largest cities likewise opens up opportunities for diversification in the spheres of culture, the arts and science, accompanied by technological innovation. This illustrates the strong link in megacities between economic opportunities and the development of social and human capital.

A third opportunity (and another scale effect) is the potential to satisfy human energy and material requirements in a way that is both *cost-effective* and *ecologically oriented*. It points to the argument referring to economies of scale mentioned above, but takes it a step further by addressing energy and resource aspects. The concentration of population potentially reduces the per capita demand for occupied land, the cost of providing treated water or collecting solid and liquid waste. It also permits implementation of mass transit facilities, reducing the demand for private vehicles in the process. Several 'visible' examples of mobility demonstrate the potential of 'marrying economic and ecological goals'. In 2003, the city of London introduced a tax on each vehicle entering the city centre. As a result the traffic volume decreased by 15% and drivers spent 30% less time in traffic congestion (Eash et al. 2008).

Megacities, however, are likewise spaces of risk. Most of the world's largest cities are concentrated in areas where natural hazards such as earthquakes, floods and landslides are likely to happen (Mc Granahan et al. 2007). As a consequence, the amount of losses (lives and values lost) also increased (International Federation of Red Cross and Red Crescent Societies 2010). This does not imply that megacities are particularly threatened or have a monopoly on risk. Due to the concentration of people and values in hazardous locations, i.e., flood or earthquake zones, however, the extent of a potential risk event is estimated to exceed the capacity of a megacity to react, with the consequence of particularly high losses (Munich Re Group 2004).

Apart from natural hazards, there are man-made (environmental and technological) risks. Rapid change in land use often exacerbates the risk of floods, while interference with water catchments jeopardizes the quality and quantity of the water supply (UN World Water Assessment Programme 2009). Flood risk is therefore a good example of the close link between natural hazards and man-made risks. Other examples of man-made risks are uncontrolled waste disposal leading to environmental degradation and health risks to urban dwellers (in particular those living in precarious locations), system leakages, industrial and toxic waste resulting in ground-water pollution (UN HABITAT 2006). Large quantities of untreated sewage curtail the use of surface water downstream and mounting traffic contributes substantially to the worsening of air quality and health risks such as respiratory disease. Rapid increase in the consumption of energy outpaces the capacity of the energy system in place and in the long run threatens energy security. Socio-spatial segregation in cities and disparities between affluent gated communities, on the one hand, and informal neighbourhoods, on the other hand, bear risks of a widening social divide, violence and crime.

While this portfolio of risks may appear diverse, the risks themselves have two elements in common. They first of all confirm that although large agglomerations face numerous risks, on the other hand, they produce and reinforce them (Mitchell (Eds) 1999; Büscher and Mascareno, in this volume). Pelling (2003, p. 7) also states that "urbanization affects disasters just as profoundly as disasters affect urbanization". Secondly, vulnerability to the dangers associated with environmental degradation, poor housing and sanitation or the lack of access to basic services differs significantly across locations (Puente 1999; Munich Re Group 2004) and social groups (UN Habitat 2006; Wisner 1999). People in slums suffer on average more detrimental health outcomes and are more vulnerable to disasters such as floods. Women, children and the elderly are the most vulnerable of all (Hardoy et al. 2001).

The occurrence of natural and man-made risks cannot be treated as two separate entities. Rather, the natural dimension is inextricably bound up with the human or social dimension. Flood risk, for example, may depend on the likelihood of a hazardous rainfall event (as a natural cause) but is largely triggered by human action, such as settlement construction in hazard-prone areas, which can lead to functional changes in the natural system and impede the mitigation of flood extremes.

The above considerations of megacities as spaces of risk relate to the notion of risk and vulnerability. These aspects will be taken up in the following section and the underlying risk definitions and concepts briefly addressed. The intention is not to cover the entire risk debate found in the literature – this would indeed be an ambitious goal. Instead, the following section introduces basic definitions and concepts applied to the sectors and policy fields explored in subsequent chapters.

1.4 Megacities as Risk Habitat: Perspectives and Risk Concepts

There is no single over-arching concept of risk that applies to all problems. This also holds true for risks in megacities. There are, however, several risk definitions and concepts at hand, each pointing to different aspects of risk and focusing on a distinct facet of the problem under consideration. This section presents three concepts.

A first concept of risk derives from hazard research. Risks in megacities are frequently understood as an increase in *hazards* – it is the potential hazard that creates risks and the hazard itself that defines its severity. Following the classic work by Frank Knight (1921), this definition of risk is based on two factors: (1) the expected damage caused by hazard, and (2) the probability of the occurrence of damage. Following this understanding, mega-urbanization is associated primarily with an increase in hazards emerging from changes to the physical environment.

Hazards do not impact evenly on society. Instead, certain social groups, economic sectors, or specific areas are affected more than others. Vulnerability is a decisive factor in determining the degree to which a person, social group, region, or economic sector is at risk. Concepts of *vulnerability* are commonly used as analytical tools in the scientific community to explain the occurrence of disaster or crisis. Despite their diverse interpretation (Weichselgartner 2009), some key parameters are seen as crucial to the definition of vulnerability, e.g., the *exposure* to and *capacity to adapt to and/or cope with* perturbation and external stresses (cf. Adger 2006; Hansjürgens et al. 2008).

Exposure to external stresses or hazards refers to *elements at risk*. This somewhat technical expression relates to all *elements* exposed to hazard – humans, social groups, economic sectors, areas within the city or regions within a country (e.g., coastal zone areas), and the corresponding values (buildings, the urban infrastructure, etc.). In a first approximation, *elements at risk* refers to humans, physical capital and natural capital. While physical capital plays a pivotal role in mega-urban agglomerations (as cities are characterized mainly by their built environment), the natural environment (natural capital) may also have a *value*. Although frequently neglected in the urban context, these natural values deliver vital services, as seen from the ecosystem services of, for example, green spaces or parks (e.g., cooling, filtering of pollutions), all of which positively influence local water cycles and the city climate (TEEB 2010).

The working definition of vulnerability developed by Wisner et al. (2005) is helpful in understanding the *coping capacity* of affected *elements at risk*. It refers to vulnerability as the characteristics of an individual or group in terms of their ability to anticipate, cope with, resist, and recover from the impact of a natural hazard. Instead of merely emphasizing the characteristics of natural or technological hazards or the exposure (structure, building, etc.) to them, this interpretation focuses on the question of how communities and social groups deal with the impact of natural hazards. It is not so much the susceptibility of a community or a social group to a specific hazard that is of interest here, but rather the coping capacity, hence active behaviour of affected humans to improve their situation.

The focus on vulnerability emphasizes that disaster and crisis are never purely the result of natural processes but are in fact often social products (O'Keefe et al. 1976). It essentially refutes 'taken-for-granted beliefs about the essential nature of things', and shows how particular concepts and categorizations of the material world have been produced and maintained over time (Demeritt 2002). As White stated some 60 years ago: "Floods are 'acts of God', but flood losses are largely acts of man" (White 1945, p. 2).

A second concept of risk connects to sustainable development. The definition of central objectives for future megacity development requires the formulation of targets against which risks can be evaluated. The concept of *sustainable development* is one such target dimension – it can help to formulate a reference point for policy evaluation and act as a guiding principle for urban development. In line with the Brundtland Report, sustainable development is seen as a global vision for the future of human civilization. Starting with the postulation of intra- and intergenerative justice, the protection of human existence, maintaining society's productive potential, and preserving society's options for development and action are specified to be the universal objectives of sustainable development. These objectives can be defined in more detail by a set of minimum requirements, the fulfilment of which may be rightfully claimed by all members of the world's society, including future generations. Minimum requirements may serve as orientation aids for a viable city development policy, on the one hand, and as criteria for analysis of the extent to which past development was in compliance with these sustainable development objectives.

Although useful in its own right, the sustainable development perspective can also facilitate the definition and assessment of risk. The concept of sustainability allows for development of a set of *sustainability indicators*, making megacity development and policy targets explicit and concrete. This in turn allows for analysis of departures from set targets. A useful indicator to measure such departures is the *distance-to-target index*, which points to policy distance from the aspired target. At the same time it can be used to assess the severity of a problem of interest and thus serve as an indicator for risk analysis. This second perspective on risk is based on the tight link between risk and sustainability, defining risk as the distance to an aspired sustainability target.

A third definition of risk places it in the context of complexity research and *systemic risk perspectives*. The starting point is the observation that hazards, malfunctions or other *exogenous* factors are not a source of risk. Rather, the conditions that ensure the functionality of the economic, social and technical systems provide the context of risk. In this understanding risks emerge from the unintended, potentially hazardous impacts of human activity. The focus on complex social interactions and the attempts to determine the cause of harm reveal that the processes and relationships in megacities are too complex to be captured in simple models of cause and effect. This interpretation of risk is closely connected to organization theory and chaos theory, where non-linear effects, backward loops and thresholds exist.

The city is the place where social systems interrelate and deal with local, regional, national and supranational problems in material, temporal and social terms. Through the centralization, concentration and networking of various processes, urban places have gained a functional advantage over the rural areas. Megacities, as highly organized, coordinated, and synchronized systems, represent a specific form of systemic risk, where the size, concentration and dynamics of interacting processes amplify both opportunities and risks. To assume, however, that megacities are purely chaotic or random, i.e., that risks and dangers can only be described in stochastic models, would be misleading. On the contrary, millions of people in megacities find mutual orientation in living together in neighbourhoods and city districts. They follow certain rules, e.g., when they participate in traffic or build up a network of relationships. Megacities can therefore be characterized as an object that combines *complex processes with a high degree of order*.

All of this demonstrates that the three concepts outlined above explaining risk (a) as the product of hazard, exposure and coping capacity, (b) a sustainability deficit, and (c) either the negative or positive consequence of functional interaction in a complex system are suitable tools for the analysis and explanation of risk in the megacity. Chapters 6 to 13 show that the risks explored in the various sectors and policy fields fall into one or more of these three categories. The hazard perspective earthquakes (Chap. 6) and land-use change and flood risk (Chap. 7). Energy (Chap. 9), transportation (Chap. 10), related air quality and health aspects (Chap. 11), Water resources and services (Chap. 12), and waste management (Chap. 13) highlight the relevance of explaining risk as a sustainability deficit. The systemic risk perspective is used to analyse trends in socio-spatial differentiation and social exclusion (Chap. 8).

The application of this complementary set of risk concepts to the concrete problems of a megacity suggests that a single risk concept would fail to satisfy the needs of analysis and explanation of the broad range of risks concerned. As illustrated in Chap. 3, it is perhaps the systemic risk perspective that has the most potential to develop a more encompassing theoretical perspective on the *Risk Habitat Megacity*.

1.5 Geographical Focus: Latin America and Santiago de Chile

This book focuses on megacities in Latin America and the Caribbean, the most urbanized region in the world after North America, with 78% of the population living in urban areas (UN Population Division 2008). As shown in Sect. 1 above, the degree of urbanization compares with Europe and exceeds that of other world regions substantially. Beginning with the period from 1925 to 1950 (Rodriguez 2002), the intense rate of urbanization in Latin America and the Caribbean led to concentration of the region's population in a few main cities. This explains the significant number of megacities in the region. Since the peak decades in the latter half of the twentieth century, the pace of urbanization gradually slowed down at the beginning of the new century. Patterns of urban concentration have become more diversified and are now directed towards smaller cities (Rodriguez 2008). Today, however, urban dwellers are more concentrated in the megacities of this region than in the world as a whole: 14% of the world population lives in megacities (UN Population Division 2008).

The Metropolitan Region of Santiago de Chile serves as an example to explore the *Risk Habitat Megacity*. Santiago de Chile gives access to a wide range of representative problems associated with megacities (for an overview, see Jordán et al. in this volume). The Metropolitan Region is typical in terms of spatial structure and growth patterns, and a commonly used representative or *ideal* model to illustrate the stages of urban growth in Latin American cities. At the same time, urban expansion and land use in Santiago de Chile show some very recent trends in mega-urbanization and new forms of sprawl and socio-spatial transformation (Borsdorf and Hidalgo 2007). Land conversion for urban expansion is taking place at a higher rate than ever before. While the city expands outwards on an unprecedented scale, the dominance of the central area has, similar to other large cities in the region, declined sharply. Santiago de Chile is socially one of the most polarized cities in the region (De Mattos 2002). Inadequate public transport and the segregation of functions has contributed to the rise in vehicle emissions and had an immediate impact on extreme air pollution levels. The valley location of the city, between the Andes and the coastal mountains, has aggravated this condition to the extent that atmospheric contamination continues to be a major problem, despite extensive efforts to improve air quality. As shown by the recent earthquake event and its aftershocks in 2010, frequent seismic activity makes the city vulnerable to substantial risk of earthquake. Urban expansion has furthermore led to an interruption of the drainage system and consequently to an increase of flood incidences (Ebert et al. 2010). Of great concern is the standard of housing and infrastructure provided either by the government or private initiatives for low-income groups.

These trends offer a rich base for the detection of emerging risks and opportunities, and the drawing of lessons from strategies adopted in response to the associated challenges. Chile was among the first countries in Latin America to respond to globalization by shifting its macro-economic policy from import substitution to market liberalization, and promoting foreign direct investment. In terms of urbanization, it is one of the most advanced countries in Latin America (along with Argentina, Uruguay and Venezuela). Approximately 85% of the Chilean population lives in the urban centres; roughly one-third is concentrated in the capital of Santiago de Chile. Urbanization nevertheless continued to increase by 1.7% for the period 1995–2000 (ECLAC 2000). The implications of Santiago's advanced state of urbanization and the attendant demographic transformation make it an excellent illustration of the pressures and consequences of urbanization in 'mature' urban agglomerations, and the corresponding political responses. Firstly, the nature and structure of the problems involved are of growing relevance to the majority of agglomerations, where urbanization, although increasing, is less advanced, and problems are gradually being addressed. Secondly, because Santiago's large-scale urbanization took place comparatively early, the city has gained a wealth of experience. From the point of view of problem response and lessons to be learned, Santiago de Chile is a case worth studying. Significant progress was made, for example, by introducing new technologies to measure air quality and forestall critical conditions, but also by predicting drops in air quality for prompt emergency and preventive measures. This, in combination with the establishment of air quality standards, yielded positive outcomes that have led to a constant decline in air quality emergencies (ECLAC 2000). Likewise, the authorities have introduced innovative economic instruments for regulation, such as tradable particulate emission permits (ECLAC 2005) and completed *Transantiago*, a new public transport system.

1.6 Overview of the Book

Although the book is an edited volume, the chapters are not unrelated. Rather, the articles were written against the background of a common framework that seeks to tie together cross-cutting theoretical concepts (i.e., *risk, sustainable development* and *governance*) with concrete sectors and policy fields.

Part I establishes the context for the *Risk Habitat Megacity*. Following this introduction, Chap. 2 explores megacity conditions in Latin America and the Caribbean and provides a comparative overview of six megacities – Bogotá, Buenos Aires, Mexico City, Lima, Santiago de Chile, and São Paulo. Based on a set of selected indicators, this chapter gives a first impression of past developments and common trends, sustainability problems, and the risks these megacities face. The chapter also places the case of Santiago de Chile into a comparative context.

Part II consists of three chapters, all of which address the underlying conceptual framework on risk, sustainable development and governance. Chapter 3 on 'Mechanisms of Risk Production' deepens the systemic perspective of *risks* as non-linear, self-enforcing and mutually amplifying processes. It likewise presents an attempt to 'order' the diverse range of risks described in subsequent chapters into different 'risk production mechanisms'. Chapter 4, 'Sustainable Urban Development in Santiago de Chile', introduces the concept of *sustainable development*, which forms the basis for analysis of several policy fields included in the book, and discusses the extent to which this concept is suited to urban agglomerations. Chapter 5 deals with the concept of governance. It outlines the role of governance issues as a prerequisite for risk minimization but also as a factor of risk production. It furthermore defines a set of governance trends associated with decentralization, informality, participation and privatization, and shows how they relate to the different policy fields discussed in the book.

Part III comprises eight contributions, each of which explores a concrete sector or policy field. They are the result of in-depth empirical investigation and correspond closely to the concepts of *risk, sustainable development* and *governance* introduced in Part II.

Chapter 6 explores earthquake risks in Santiago de Chile. It presents a method of gaining a rough estimate of seismic hazard that combines a high resolution map of the soil's fundamental resonance frequency and a 3D shear wave velocity model. By comparing the results with mapped intensities of recent events, the chapter estimates what areas of the city are endangered and recommends more thorough investigations. It concludes with some practical recommendations such as retrofitting the existing building stock.

Chapter 7 addresses the issue of land-use change, risk and land-use management, focusing on flood risk analysis and risk prevention in Santiago de Chile. It interprets population growth and land-use change as key drivers of urban development and establishes the linkages to flood risk. It concludes with recommendations for the improvement of flood risk management.

The drivers, risks and opportunities of socio-spatial differentiation processes in Santiago are the focus of Chap. 8. The chapter analyses demographic trends and trends in the housing sector, including the strong role of social housing programmes. It explores the opportunities and risks contained in these trends with respect to social inclusion and exclusion, and identifies hotspot locations.

Chapter 9 explores the performance of the Chilean energy system in fulfilling Santiago de Chile's energy needs, and assesses future trends. It considers the risk of import dependencies on fossil fuels and increasing CO_2 emissions from energy production. The chapter concludes with a discussion of potential solutions for energy efficiency and the increased use of renewable energy sources.

Chapter 10 focuses on the urban transport system of Santiago de Chile. It summarizes recent developments in Santiago's public and private transport systems, explores likely trends up to 2030 and discusses the associated risks. Taking the results from modelling exercises, the chapter concludes with policy recommendations.

Related to the topic of transportation, Chap. 11 concentrates on air quality and health risks. It explores the links between transport-related emissions and health impacts. It presents cost-effective mitigation strategies with the aim of achieving maximum benefit for the urban population.

Risks and opportunities associated with water resources and services are the focus of Chap. 12. This chapter analyses the services that provide ground and surface water, and how human activity impacts on these services. It likewise assesses the existing water supply and waste water treatment infrastructure. Based on model calculations for trends in water availability and water demand, it identifies water scarcity as a major risk.

Finally, Chap. 13 concentrates on the challenges and perspectives for sustainability in the area of municipal solid waste management. As shown, population growth and rising living standards in Santiago de Chile in recent decades have led to the production of increasing quantities of municipal solid waste. The chapter discusses the appropriate handling of waste and its disposal as a preventive measure against negative impacts on health and the environment.

Part IV contains two chapters and delivers a summary and a synthesis of findings. Chapter 14 provides an integrated view of the sustainability performance of Santiago de Chile. Chapter 15 presents a perspective on governance challenges and options. Chapter 16 returns to the key questions outlined in this introduction. It summarizes the findings on risk associated with the trend towards mega-urbanization and the interaction of the driving forces behind this phenomenon. It discusses research methods and approaches considered appropriate for the understanding and provision of orientation and action knowledge. The book concludes with forms of governance and strategies that seem a fitting response to these challenges of the *Risk Habitat Megacity.*

References

Adger, N. W. (2006). Vulnerability. *Global Environmental Change, 16,* 268–281.
Beaverstock, J. V., Smith, R. G., & Taylor, P. J. (1999). A roster of world cities. *Cities, 16*(6), 445–458.
Bettencourt, L. M. A., Lobo, J., Helbing, D., Kuhnert, C., & West, G. B. (2007). Growth, innovation, scaling and the pace of life in cities. Proceedings of the National Academy of Sciences USA (PNAS), *104,* 7301–7306.
Borsdorf, A., & Hidalgo, R. (2007). New dimensions of social exclusion in Latin America: from gated communities to gated cities, the case of Santiago de Chile. *Land Use Policy, 25,* 153–160.
Bronger, D. (1996). Megastädte. *Geographische Rundschau, 48*(5), 74–81.
De Mattos, C. (2002). Mercado metropolitano de trabajo y desigualdades socials en el Gran Santiago. Una ciudad dual? *EURE, 28*(85), 51–70.
Demeritt, D. (2002). What is the 'social construction of nature'? A typology and a sympathetic critique. *Progress in Human Geography, 26*(6), 767–790.
Eash, C., Jasny, P. R., Roberts, L., Stone, R., & Sugden, A. M. (2008). Reimagining cities. *Science, 319,* 339–355.
Ebert, A., Welz, J., Heinrichs, D., Krellenberg, K., & Hansjürgens, B. (2010). Socio-environmental change and flood risks: the case of Santiago de Chile. *Erdkunde, 64*(4), 303–313.
ECLAC (2000). *From rapid urbanization to the consolidation of human settlements in Latin America and the Caribbean. A territorial perspective.* Santiago de Chile: ECLAC.
ECLAC (2005). *Social panorama of Latin America.* Santiago de Chile: ECLAC/CEPAL.
Friedmann, J. (1986). The world city hypothesis. *Development and Change, 17,* 69–83.
Grimm, N. B., Faeth, S. H., Golubiewski, N. E., Redman, C. L., & Wu, J. G. (2008). Global change and the ecology of cities. *Science, 319,* 756–760.
Gurjar, B. R., & Lelieveld, J. (2005). New directions. Megacities and global change. *Atmospheric Environment, 39,* 391–393.
Hall, P. (1996). *The world cities.* London: World University Press.
Hansjürgens, B., Heinrichs, D., & Kuhlicke, C. (2008). Mega-urbanization, risk and social vulnerability. In K. Warner (Ed.), *Megacities: Social vulnerability and resilience building* (pp. 20–28). Summer Academy for Social Vulnerability. United Nations University UNU EHS and Munich Re Foundation, UNU Press
Hardoy, J., Mitlin, D., & Satterswaite, D. (2001). *Environmental problems in third world cities.* London: Routledge.

Heinrichs, D., Kuhlicke, C., Meyer, V., & Hansjürgens, B. (2009). Mehr als nur Bevölkerung: Größe, Geschwindigkeit und Komplexität als Herausforderung für die Steuerung Megastädten. In U. Altrock, R. Kunze, E. Pahl-Weber, & D. Schubert (Eds.), *Jahrbuch Stadterneuerung* (pp.47–57). Arbeitskreis Stadterneuerung an deutschsprachigen Hochschulen und dem Institut für Stadt- und Regionalplanung der Technischen Universität Berlin.

International Federation of Red Cross and Red Crescent Societies (2010). *World disaster report.* Geneva: Focus on urban risk.

Knight, F. (1921). *Risk, uncertainty, and profit.* New York: Houghton Mifftlin.

Mc Donald, R. I., Kareira, P., & Formana, R. T. T. (2008). The implication of current and future urbanization for global protected areas and biodiversity conservation. *Biological Conservation, 141*, 1695–1703.

Mc Granahan, G., Balk, D., & Anderson, B. (2007). The rising tide: assessing the risks of climate change and human settlements in low elevation coastal zones. *Environment and Urbanization, 19*(1), 17–37.

Mitchell, J. K. (Ed.) (1999). *Crucibles of hazard. Megacities and disasters in transition.* Tokyo: UN University.

Montgommery, M. R., Stren, R., Cohen, B., & Reed, H. E. (Eds.) (2003). *Cities transformed – Demographic change and its implications in the developing world.* London: National Academic Press.

MRC Mc Lean Hazel and GlobeScan (2007). *Megacities und ihre Herausforderungen.* München.

Munich Re Group (2004). *Megacities – Megarisks. Trends and challenges for insurance and risk management.* München: Munich Re Group.

National Research Council (2009). *Driving and the built environment: The effects on of compact developments on motorized travel, energy use, and CO2 emissions.* Washington, DC: Transportation Research Board.

O'Keefe, P., Westgate, K., & Wisner, B. (1976). Taking the naturalness out of natural disasters. *Nature, 260*, 566–567.

Pelling, M. (2003). *The vulnerability of cities – Natural disasters and social resilience.* London: Earthscan.

Puente, S. (1999). Social vulnerability to disasters in Mexico City – An assessment method. In J. K. Mitchel (Ed.), *Crucibles of hazards. Mega-cities and disasters in transition* (pp. 295–334). Tokyo/New York/Paris: United Nations University Press.

Quigley, J. M. (1998). Urban diversity and economic growth. *Journal of Economic Perspectives, 12*(2), 127–138.

Research, D. B. (2008). Megacitys: Wachstum ohne Grenzen? *Aktuelle Themen, 412*, 1–18.

Rodriguez, J. (2002). Distribución territorial de la población de América Latina y el Caribe: tendencias, interpretaciones y desafíos para las políticas públicas. Población y Desarrollo del Centro Latinoamericano y Caribeño de Demografía (CELADE) – División de Población. Serie 32, Santiago de Chile.

Rodriguez, J. (2008). Spatial distribution of the population, internal migration and development in Latin America and the Caribbean. *Proceedings of the United Nations expert group meeting on population distribution, urbanization, internal migration and development,* New York, 21–23 January 2008.

Sassen, S. (1991). *The global city.* New York/London/Tokio/Princeton: Princeton University Press.

Sassen, S. (2002). Locating cities on global circuits. *Environment and Urbanization, 14*(1), 13–30.

TEEB (2010). *The Economics of Ecosystems and Biodiversity for local and regional policy makers.* Edited by Wittmer H. and H. Gundimeda. www.teebweb.org

UN HABITAT (2006). State of the world's cities 2006/2007. The millennium development goals and urban sustainability. London.

United Nations Population Division (2002). *World urbanization prospects. The 2001 Revision.* New York.

United Nations Population Division (2004). *World urbanization prospects. The 2003 Revision.* New York.
United Nations Population Division (2006). *World Urbanization Prospects. The 2005 Revision.* New York.
United Nations Population Division (2007). *World urbanization prospects. The 2006 Revision.* New York.
United Nations Population Division (2008). *World urbanization prospects. The 2007 Revision.* New York.
Webster, Ch, & Lay, L. W. Ch. (2003). *Property rights, planning and markets. Managing spontaneous cities.* Cheltenham/Northampton: Edward.
Weichselgartner, J. (2001). Disaster mitigation: The concept of vulnerability revisited. *Disaster Prevention and Management, 10*(2), 85–94.
White, G. F. (1945). *Human adjustment to floods: A geographical approach to the flood problem in the United States.* Chicago: University of Chicago.
Wisner, B. (1999). There are worse things than earthquakes: Hazard vulnerability and mitigation capacity in Greater Los Angeles. In J. K. Mitchell (Ed.), *Crucibles of hazards. Mega-cities and disasters in transition* (pp. 375–427). Tokyo/New York/Paris: United Nations University Press.
Wisner, B., Blaikie, P., & Cannon, T. (2005). *At risk. Natural hazards, people's vulnerability, and disasters.* London/New York: Routledge.
Wissenschaftlicher Beirat der Bundesregierung globale Umweltveränderungen (WBGU) (1996). Welt im Wandel: Wege zur Lösung globaler Umweltprobleme – Jahresgutachten 1995. Berlin, Heidelberg, New York.
World Water Assessment Programme (2009). *The United Nations world water development report 3: Water in a changing world.* Paris: UNESCO Publishing, and London: Earthscan.
Young, O. R., Berkhout, F., Gallophin, G. C., Janssen, M. A., Ostrom, E., & Van der Leeuw, S. (2006). The globalization of socio-ecological systems: An agenda for scientific research. *Global Environmental Change, 16*(3), 304–316.

Chapter 2
Megacities in Latin America: Role and Challenges

Ricardo Jordán, Johannes Rehner, and Joseluis Samaniego

Abstract This chapter discusses the primary issues and challenges of urban sustainability that have come to light in six Latin American metropolises. It identifies leading urbanization trends in Latin America and their relevance to the selected city regions. Furthermore, it presents stylized facts from each city under review and summarizes the work of the recently developed Regional Panorama Latina America – Megacities and Sustainability, which evaluates urban risks along sustainability criteria.

Keywords Latin American megacities • Urban sustainability • Urbanization trends

2.1 Introduction

For the purposes of policy design and the definition of research agendas, and assuming the identification of common urban trends in the region, this chapter takes a closer look at the challenges to metropolitan areas in Latin America. It presents a set of dominant trends in and characteristics of Latin American cities, focusing on metropolitan cities.

The chapter serves three purposes: Firstly, to make urbanization processes and common urbanization trends in Latin America and the Caribbean (LAC) explicit, and to point to the role of megacities in Latin America and the major challenges they face. Secondly, by comparing megacities in Latin America, to present an overview of their individual specifics. And thirdly, to put the particular case of Santiago de Chile into context. Although there are distinctions between these

---

R. Jordán (✉) • J. Rehner • J. Samaniego
Economic Commission for Latin America and the Caribbean ECLAC/CEPAL, Sustainable Development and Human Settlements Division, Av. Dag Hammarskjöld, 3477 Vitacura/Santiago de Chile, Chile
e-mail: Ricardo.Jordán@cepal.org

megacities, the case of Santiago – with its characteristic economic, environmental and societal features – demonstrates that in many respects it is typical of other megacities in Latin America. Common trends, characteristic features and several differences between Santiago, on the one hand, and other mega-urban agglomerations in the region, on the other, are rendered visible.

Information is based on the Regional Panorama Latina America – Megacities and Sustainability (Jordán et al. 2011). The Panorama objective is to understand the leading challenges that have emerged in six Latin American metropolitan cities. Although metropolitan cities and regions in LAC are tackling similar problems, specific aspects related to sustainability differ in intensity from one city to another. The Panorama study was conducted by officials of the UN Economic Commission for Latin America and the Caribbean (UN ECLAC), from the Division of Sustainable Development and Human Settlements, and external consultants in cooperation with the Helmholtz Association and with the support of the Deutsche Gesellschaft fûr Technische Zusammenarbeit (GTZ) GmbH. It contains a review of the comprehensive indicator set in the six selected metropolitan cities. Information was gathered primarily via indicator-based data provided by national statistic offices, local governmental institutions, and international organization and scientific research publications. As the report depended on available data, the comparability of detailed information was limited.

The case studies include Latin American cities that have already become megacities or will do so in the near future. For pragmatic reasons the number of case studies was confined to a maximum of six metropolitan cities, which were chosen for their socio-economic conditions, structural characteristics and local environmental contexts, and their urban planning system frameworks. In accordance with these criteria and data availability, we selected the following megacities as case studies: *Bogotá, Buenos Aires, Lima, Mexico City, Santiago de Chile and São Paulo*. Guatemala City, Porto Alegre and Rio de Janeiro, all of them likewise Latin American megacities, are not discussed in the Panorama document.

References are made in Sect. 2.2 below to major processes and common urbanization trends in Latin American cities. We focus on typical aspects such as demographic change, economic growth and environmental issues, but we also include socio-cultural aspects related to urban life and observations on governance structures. In Sect. 2.3 we concentrate on the six major megacities in the region, mentioned above. Section 2.4 brings the chapter to a close and offers some concluding remarks.

2.2 Urbanization Trends and Characteristics in Latin America

This section discusses common trends in Latin American cities (see Hoornweg et al. 2007, pp. 5–7 and Jordán et al. 2010, pp. 13–14.)

2.2.1 Megacities: Growth and Primacy

Latin America is a continent of urban primacy. Urban primacy refers to the predominance of the main city in the national context in terms of population, which goes beyond the relation between it and other cities, e.g., by the rank-size rule that states the existence of a mathematical relation between a city's rank position and its population as a percentage of the largest city. Such dominance of a single urban centre is considered typical of developing countries, particularly in its initial stage of rapid urbanization (UN Habitat 2009, p. 36). Urban structures in Latin America and the Caribbean are characterized by a high degree of urban primacy, based on their historical legacy. A single city gravitates to a position of outstanding political and economic importance as a result of the colonization process, and particularly post-colonial industrial and commodity-based growth (Cuervo González 2004). This structure was reinforced by the migration waves of the twentieth century, which led to the creation of several urban megacities in the region Williams Montoya (2009).

'Megacities', that is, cities with more than ten million inhabitants, are becoming more common worldwide and tend to concentrate in the developing world. By 2020, all but four of the world's largest cities will be in developing countries. Latin America and the Caribbean are among the most urbanized regions of the developing world and characterized by a high degree of metropolization. In the year 2000, approximately 20% of the total population of Latin America, i.e., more than of any other world region, lived in cities with populations exceeding five million inhabitants (UN Habitat 2008, p. 22). In Table 2.1 below, we present a population projection as reported by UN Habitat and the urbanization rate of the countries concerned. The table clearly shows the conspicuous role of the urban areas in Latin America and the expected sustained growth of megacities.

Patterns of urban growth are shifting towards small and intermediate cities. Despite the above-mentioned prevalence of megacities and their sustained growth, population growth in recent years has been registered higher for large rather than primate cities. The current trend in Latin America is towards growing small and intermediate cities. Almost 50% of its urban population lives in these cities (less than 500,000 inhabitants), whose economies are gaining momentum.[1] An estimated 200 new small and intermediate cities accounting for more than 30 million inhabitants were established throughout Latin America, most of them in Brazil, Mexico and Colombia (UN Habitat 2009, p. 36). This trend reflects not only a process of decentralization, as in Mexico, for example, with the growing importance of the northern frontier area and the so-called Bajío,[2] but also of peri-urban

---

[1] The corresponding percentage in East Asia is 40 per cent; in Western Europe 72%; in North Africa 58% (UN Habitat 2009, pp. 231–232).

[2] Located to the north of Mexico City, this region includes parts of the Guanajuato, Jalisco, Querétaro and Michoacán states and has recently been particularly successful in attracting economic activities.

Table 2.1 Expected population growth and urbanization

	Population 2000	Population 2020	Urbanization rate (national level 2000)	Urbanization rate (national level 2020)
Bogotá/Colombia	6.4	9.3	72.1%	78.0%
Buenos Aires/Argentina	11.8	13.7	90.1%	93.8%
Lima/Peru	7.1	9.3	70.7%	73.6%
Mexico/Mexico	18.0	20.7	74.7%	80.7%
Santiago/Chile	5.3	6.2	85.9%	91.0%
São Paulo/Brazil	17.1	21.1	81.2%	89.5%
Comparison				
Shanghai/PR China	13.2	18.5	35.8%	53.2%
Bangkok (Krung Thep)/ Thailand	6.3	7.8	31.1%	38.9%
Cairo/Egypt	10.5	14,5	42.6%	45.0%
Johannesburg/South Africa	2.7	3.9	56.9%	66.6%
New York/USA	17.8	20.4	79.1%	84.9%
Berlin/Germany	3.4	3.4	73.1%	75.6%

Data source: UN Habitat 2008: 238–247 and UN Habitat 2009: 242–244

growth and the progressive significance of secondary centres in the environs of megacities. With the slow-down in population growth in most Latin-American megacities, attention has shifted to aspects of structure, inner contrasts, consumption patterns and the functional linkages of cities as key elements of the sustainable development endeavour.

2.2.2 Socio-demographic Aspects

Changing demographics impact significantly on the region's megacities. In demographic terms the shift in population numbers is not the only decisive factor in urban development. The changing age structure of the population is progressively becoming a challenge to Latin American societies, since it increases dependency ratios (the number of children and elderly people per 100 population of working age). Mexico, Colombia and Peru still have a high dependency ratio of young people, but anticipate a sharp decline of this indicator in the long run (see Table 2.2).

Although recent decades have seen the dominance of economically active groups in the Latin America population, this 'demographic bonus' is likely to disappear soon in the most powerful countries of the region. Chile, for example, will be one of the first Latin American countries to have this bonus capped (by 2015); Brazil, Colombia and Mexico will reach this point around the year 2020 (CELADE and UNFPA 2005, pp. 12–14). Table 2.2 gives an overview of the expected dramatic increase in the percentage of elderly people in Latin American countries (2000–2050). Chile and Argentina are considered to have a moderately advanced ageing population, while the ageing of other countries is at a moderate stage, i.e., they still anticipate a long

Table 2.2 Ageing population indicators in selected Latin American countries

	Population >60 (2000)	Population >60 (2050)	Dependency ratio young (2000)	Dependency ratio young (2050)	Dependency ratio elderly (2050)
Argentina	13.5%	24.8%	47.0	30.9	41.9
Brazil	8.1%	29.4%	45.4	30.7	45.0
Chile	10.2%	28.2%	46.4	31.7	41.8
Colombia	6.7%	23.9%	54.2	32.0	37.1
Mexico	7.5%	27.4%	55.3	30.6	43.8
Peru	7.1%	21.8%	59.2	30.7	36.0

Source: The authors, based on CEPAL and UNFPA 2009, p. 14

period of growth in the elderly population. In some countries, especially Argentina, the latter is concentrated in urban rather than rural areas.

This predicted shift in the population profile is likely to impact on health insurance and retirement at national level, and suggests an inevitable adjustment of urban planning instruments. In terms of direct impacts, it will produce a greater need for age-related infrastructure, an adjustment of the existing infrastructure and in certain instances, a slight reduction in the current youth-related infrastructure. As an indirect impact, the prime social challenge may shift to the issue of vulnerability of the elderly poor. Today, securing an income for the elderly takes place under very different conditions in each of the six countries under review. In Mexico, for instance, more than 40% and in Columbia more than 50% of people over the age of 60 have no income at all. Peru, Mexico and Colombia register a high percentage of elderly people below the poverty line (CEPAL and UNFPA 2009, p. 28–32).

The traditional city-rural dichotomy is gradually disappearing. In spite of its designation as highly urbanized, Latin America is still characterized by substantial rural areas. In territorial, economic and socio-cultural terms, the rural areas are of crucial importance, not least for the urban centres. However, recent trends indicate a shift towards economic specialization of the rural areas and their integration in global production chains, thereby tightening their functional links to the dominant cities. This scenario is complemented by characteristic changes within the wider rural context of the main metropolis. New spatial patterns of population and economic growth, conceptualized as 'peri-urban growth', have been observed in these areas. They are marked by a fractal and discontinuous form, and by contrasting life styles, working patterns and income levels. This recent phenomenon is highly relevant to urban planning, challenging as it does transport systems and access to services of all kinds. In the case of Santiago de Chile, for instance, 'peri-urban' areas became a prime location for new housing settlements, transforming originally rural areas, changing socio-spatial structure and increasing substantially non-agricultural employment in these areas (Hidalgo et al. 2005). This metamorphosis notwithstanding, these areas are still important in terms of agricultural production (Salazar 2010). Due to the extent of peri-urban growth, the unmistakable boundaries between cities and their rural 'hinterland' became fuzzy and finally disappeared.

Furthermore, metropolitan cities are still growing, albeit mostly outside their traditional or administrative limits. In terms of governance this means that new actors are now involved in metropolitan issues and new challenges arise in terms of land use and transportation. As an illustration, the metropolitan area of Mexico expanded in the 1990s at a rate of approximately 4.400 ha/year. Similarly, urban expansion in Santiago reached about 1.400 ha each year, indicating an even higher speed, given the very much smaller population of Santiago (Jordán et al. 2010, p. 60).

2.2.3 Economic Importance and Centrality of Metropolitan Cities in Latin America

Despite the significance of extractive activities, economic centrality prevails to a high degree. The megacities of the region constitute the economic core in their respective national context. Mexico City and São Paolo are the Latin American megacities of the greatest economic importance and, classified as global cities, with the highest level of interconnectivity (Taylor 2004). In the case of Chile and Peru, almost half of the economic activities are concentrated in the capital city, and around a quarter in the capital cities of Argentina and Colombia (see Table 2.3). By comparison, Brazil and Mexico's historic legacies are more polycentric, evidenced in decentralized gateway cities and industrial activities. In spite of foreign trade and investment-driven decentralization processes, more than 57% of foreign direct investment (FDI) inflows in Mexico are concentrated in the D.F. (Distrito Federal) and a further 6% in Estado de México, (between 1996 and 2005; Delgadillo Macias 2008, p. 86).

Metropolitan cities focus on the service and creative industries. In recent decades, advanced services (e.g., financial services, consultancy, Research & Development) have shown evidence of more rapid growth than manufacturing businesses and industries. Despite the exponential growth of information and communication technologies and the impact of globalization, location has not lost its significance. On the contrary, place now matters more than ever and is crucial when it comes to creative activities (Florida 2008). Advanced producer service providers tend to concentrate their activities in national and international centres. The latter are conceptualized as gateway cities and global cities, and partially disrupt the primacy structure, rendering them simultaneously elements of both agglomeration and diffusion (Sassen 2007; Consoni Rossi and Taylor 2007). Advanced producer services are seen as a vital link between global cities and globally organized production chains. They are particularly important in Mexico City, São Paulo and Santiago de Chile, where the command and control functions considered essential for economic globalization are concentrated (Parnreiter et al. 2007). For Mexico City and Santiago, Parnreiter et al. (2007) see increased participation in global city networks. Since their national economies are heavily involved in globally oriented production chains, these are managed to a certain extent by the two capitals, which in turn participate in flows (information, capital) involving other global cities. The outstanding level of locally represented global financial institutions and their concentration in Mexico City and Santiago de Chile is the

Table 2.3 Participation in the national GDP and informal employment

	Participation in the national GDP	Informal employment
Bogotá	25%	–
Buenos Aires	24%	44.0%
Lima	47%	53.1%
Mexico	29%	45.7%
Santiago	43%	34.0%
São Paulo	12%	40.8%

Source: Jordán et al. 2010, p. 72

logical consequence of this trend, and considered the missing link between the decentralized production chains and city networks.

Increasing informal sector employment and the working poor. In the major metropolitan cities in Latin America the informal economy is vital – between a third and more than half of the labour market is considered to be informal (see Table 2.3). This economic sector is likely to continue growing and is nowadays bound to a chain of services and products reaching beyond the city limits. Since informal labour is generally characterized by lack of social and legal protection and low wage levels, it constitutes a major challenge specifically to metropolitan cities with low income levels, such as Lima. As a direct consequence of the prevailing disparities in levels of income, and economic and financial crises such as in Buenos Aires, the issue of the working poor has gathered momentum in Latin American cities.

Growing informality means greater challenges for local governments and governance structures, since informal activities increase the complexity of megacities, making them more vulnerable.

2.2.4 Sustainability and Environmental Trends

Despite remarkable progress towards sustainability and the reduction of urban externalities, huge challenges persist. Urban sustainability is an issue of major relevance to the future of society (McGranahan et al. 2001), and all the more in Latin America, where urbanization is more advanced than that of other regions. In most regions with large urban areas, the coincidence of economic and population growth, the frequently inefficient use of natural resources and energy, and high levels of green house gas and polluting agent emissions have generated a mix of negative externalities such as congestion, and air and water pollution. Although the 1990s witnessed vast reductions in the level of emissions in several metropolitan areas, addressing sustainability issues in terms of planning requires a strong regional perspective. This not only covers ongoing expansion in urbanized areas, but also the broader regional context of, for example, water supply catchment areas and the rural areas as environmental service provider to the metropolis. The metropolitan areas absorb resources from their regional context, using it as a sink for the massive accumulation of polluting agents. In the case of water supplies, the principal metropolitan areas have depended for several decades on water resources from areas of growing distance from the metropolis itself. São Paulo receives only a small percentage of its water consumption from the Rio Tiete or Mexico City, which in turn is becoming more and more dependent on drinking water from the distant catchment areas.

This is of particular relevance to metropolitan regions such as Santiago de Chile, where the speed of urbanization exceeds that of population growth. Furthermore, in terms of greenhouse gas (GHG) emissions and environmental externalities in general, all societies (urban and rural) produce negative impacts on distant areas through consumption of products made and externalities produced in spaces far from the confines of the city or the village.

The linkages between urban sustainability and climate change are multiple. Metropolitan cities have been in the focus of recent discussions on sustainability and climate change, since they are in a position to play a key role in emission mitigation (Barton 2009, p. 5). Contrary to common perception, population density in large Latin American cities is high (albeit lower than those in Asia), and speaks for the economically and environmentally efficient provision of piped water, water treatment and waste management, public transport, and waste water treatment (Satterthwaite 1999; UN Habitat 2006). The field of urban transport in Latin American metropolitan cities has vast potential for mitigation, some of which has been realized in Bogotá, Mexico City and Santiago de Chile. Kyoto instruments like Clean Development Mechanisms are important signposts on the way, for example, to abolishing deficiencies in terms of carbon intensity. The particularly high potential for GHG emission reduction in the area of building insulation and the residential use of heat and energy has not yet been widely harnessed in terms of policy, although progress has been made in the implementation and promotion of low-consumption household appliances.

The impact of climate change, on the other hand, is highly relevant to urban systems, as it represents an outstanding risk to urban infrastructure and such critical resources as water, posing therefore a threat to the functionality of the urban system as a whole. Of the utmost importance from a sustainability point of view is the extreme vulnerability of low-income sectors and their habitual location in areas with high hazard levels. The development of new infrastructure frequently plays a major role in adaptation and mitigation strategies. Explorative case studies of public policies on climate change in various cities worldwide identify climate action mostly in traditional policy fields and the failure to address consumption patterns (Heinrichs et al. 2009).

2.2.5 Socio-cultural Aspects and Quality of Urban Life

Latin American cities are considered to be among the most unequal and the most segregated cities in the world. All major megacities in the region far exceed the

Table 2.4 Income concentration and poverty

	Gini coefficient	Poverty
Bogotá	0.61	23.8
Buenos Aires	0.52	21.8
Lima	–	24.2
Mexico	0.56	39.2
Santiago	0.55	10.4
São Paulo	0.61	19.6

Source: UN Habitat 2008

international alert line of the Gini income concentration index of 0.4 (UN Habitat 2008) (see Table 2.4). This overall image of unequal urban societies has a series of impacts on sustainable development in terms of unequal living conditions and future options (e.g., arising from unequal access to education), seriously challenging social cohesion. Furthermore, income inequalities in Latin American cities are generally manifested territorially by high segregation levels. By comparison, an extremely high level is registered for Johannesburg (0.75) and low levels for Shanghai (0.32), while Bangkok's Gini coefficient is similar to the level of Latin American cities (0.48) (UN Habitat 2008). Although socio-spatial segregation can be seen as a characteristic element of the Latin American metropolis, critical voices of a 'naturalistic view' are now coming to the conclusion that "there are no cultural, sociological or economic obstacles to reducing segregation" (Sabatini and Brain 2008, p. 5). The "naturalistic view" sees urban socio-spatial inequalities as a natural process that cannot be controlled.

The segregation process is nonetheless ubiquitous in large Latin American cities, with the old 'dual city' concept gradually being replaced by the fragmentation concept, characterized by a more polycentric and diverse structure – albeit still highly segregated. Characteristic features are individual types and scales of real estate and urban development projects of various types and scales, the emergence of new centralities, and so-called 'artifacts of globalization' such as shopping malls and airports (de Mattos 2004; Pereira and Hidalgo 2008). This trend towards fragmentation is associated to some extent with a loss of territorial solidarity (Prevot-Schapira and Cattaneo 2008). In the same discursive context, the multi-facetted connection between cities and violence or crime is discussed (Briceño-León 2002). Most Latin American metropolises now show high crime levels, which is discussed as a result of fragmentation processes that include increased poverty, lack of prospects to improve income or social contexts, and a loss of identity. The imaginary and the sheer materiality of fragmentation, particularly stigmatization, and the perceived need for protection, gated communities and other forms of protected territory are seen as further reasons for this development.

In spite of economic growth and increased welfare, urban poverty and habitat precariousness are widespread and call for new solutions (Winchester 2008). Poverty is still a central issue in Latin American cities and frequently related to specific characteristics of the labour market and the informal sector. In addition, poverty in Latin America is directly reflected in the precariousness of the habitat. Hence some authors prefer to speak of vulnerability rather than poverty, introducing a broader concept that takes the ability to overcome negative impacts into account. Vulnerability combines aspects of income, education and living conditions. Outstanding asymmetries with regard to living conditions of various socio-economic groups are characteristic for the region's urban areas and include quality of housing and basic services, e.g., drinking water supply, sewage, electricity supply and public transport. On a global level, social inclusion objectives and policies, and specific urban interventions to overcome poverty and slums are still among the main issues to be dealt with (World Bank 2009). The urban poor and their economic vulnerability have been the subject of numerous governmental initiatives, but criticism of the lack of integrated public policy to address extreme poverty still persists (Winchester 2008, p. 27).

2.2.6 Urban and Territorial Governance Issues

Cities and urban regions are acquiring new roles and responsibilities. As the economic and cultural centres of their respective national context, metropolitan regions in Latin America are the places where investments (public as well as private) are concentrated. Financially in a better position than other local entities, they are nevertheless dependent on the national level when it comes to financial resources and decision-making. Most of the cities under analysis are distinguished by a strong central government that is both directly involved and a key player in local decisions. The crucial point, however, is the absence of a powerful regional institution that would link local and national levels but above all constitute the proper scale to address sustainability issues. In general terms the overlapping competencies of national, regional and local levels lead to numerous difficulties, e.g., the regulation of the upper level could contradict activities at local level and vice versa.

2.3 Selected Sustainability Issues in Latin American Metropolitan Cities

This section gives a brief summary of the six city profiles in terms of the sustainability issues elaborated in the Panorama (Jordán et al. 2010).[3] The city profiles allow for a discussion of the specific situations in each of the metropolitan cities, the systemic characteristics of sustainability and the relevant public policy elements contained in each of the case studies. The latter are based on statistical and qualitative information sourced in detailed studies and official reports; it is further

---

[3] This chapter presents summaries on each of the six megacities as presented in the Regional Panorama: Latin America - Megacities and sustainability, based on the work of the experts involved and the literature quoted in the Panorama (Jordán et al. 2010).

complemented by input from local consultants and the collaborative Panorama authors in each of the metropolitan regions concerned.

2.3.1 Bogotá

An exceptional feature of Colombia's capital in terms of land use is its compact form, indicating a high population density, both in the central area and on the periphery. This is the result of public housing policies, on the one hand, and expanding informal settlements, on the other. A series of public housing programs to combat urgent issues have been implemented in recent decades. Nonetheless poverty and precarious housing conditions continue to be urgent challenges. In socio-economic terms, high spatial segregation persists and informal settlements are characterized by low levels of education and, to a certain extent, by the lack of access to water, sewage and health services. Public policies have addressed the topics of segregation and the accessibility to basic services. Examples are water supply prices in accordance with income levels or the implementation of principles of social integration in the design of public transport. The poor quality of housing in some areas continues to lead to social conflict and urban violence. In terms of the labour market and equal opportunities, Bogotá shows evidence of vast gender disparities in terms of income and employment opportunities for women.

Planning issues and competencies are concentrated at city scale (Bogotá D.C.) although – as in other cases discussed – the influence of national government is strong. The development and implementation of the Bus Rapid Transit (BRT) System *Transmilenio* is considered one of the most successful examples of public transport system reforms in Latin America. Furthermore, in Colombia social and civil society participation is generally high. Sustainable development, however, requires the concerted action of various entities involved in environmental protection and territorial planning, and a strengthening of institutionalization on a regional scale. Criticism can still be heard on centralized decision-making, such as the isolation of local environmental policies and their implementation from policies at the national level.

With regard to the natural environment, the pollution of water bodies, including wetlands and underground water, can be seen as the principal challenge. The sewage treatment facilities introduced several years ago are inadequate. Although water supplies and water management have been reformed and pricing policies differentiated, the pollution charge of the main water bodies remains extremely high. The subsequent and dramatic degradation of the Bogotá River affects the rural areas in the river basin far beyond the city. Bogotá, on the other hand, contains a large number of green areas, primarily due to the presence of the mountain ranges in the city's immediate surroundings, but also as a result of long-term public land-use policies and urban planning. This notwithstanding, incentives are low for the conservation of vulnerable strategic ecosystems at the regional level, e.g., the paramos, which provide the metropolitan city with natural resources and environmental services.

2.3.2 Buenos Aires

Although Argentinian society is considered on the whole to be egalitarian, with a broad educational base and substantial welfare levels, the metropolitan area of Buenos Aires displays rapidly increasing levels of poverty and poverty-related risk, due first and foremost to the impact of the economic and financial crisis in 2001/2002. The partial loss of dollar-based private capital and the dramatic devaluation of the Argentinian peso in 2002 were responsible for this surge in poverty levels across different social classes. In particular, the loss of the traditional middle class has been discussed as the most important structural change in the society of Buenos Aires. Confronting unemployment and the new urban poverty that soared during the 1990s, as well as the negative spin-offs of the above-mentioned crisis constitutes the main socio-economic challenge. The growth of informal and precarious housing areas is perceived today as a high risk factor. The export-driven economic growth that began in 2003, however, was accompanied by a sharp decline in the incidence of poverty and extreme poverty.

The almost complete lack of regional planning coordination is the major obstacle to integrated governance in the Buenos Aires metropolitan area, which is made up of the autonomous city of Buenos Aires and the province of Buenos Aires, with overlapping functions and responsibilities. There is also a lack of vision with regard to the city and the implementation of participatory instruments. In the last 20 years Buenos Aires has shown strong suburban and peri-urban growth characterized by high levels of segregation, where precarious housing conditions and exclusive gated communities co-exist within a short distance of each other. As a result of this expansion and investment deficits, the public transport system has suffered decay.

The most threatening environmental risk is the degradation of the main water bodies, especially the Riachuelo River, which still boasts conspicuous pollution levels. Moreover, the pollution issue presents a serious social risk as most of the inhabitants in the immediate vicinity of the river belong to the low-income sector. Air pollution levels are high in terms of emissions but not of concentration. This is due to climate conditions in the metropolitan area. In terms of sustainable development challenges, strategies to overcome deficiencies in drinking water access, the sewage water system and waste water treatment need to be defined. The growing number of flood events and the expected effects of climate change magnify the risk, especially in vulnerable areas.

2.3.3 Lima

The primary challenges of the metropolitan city of Lima are rapid urban growth and an unsatisfied demand for such basic services as drinking water supply. High pollution levels of the main river pose a major health risk to low-income sectors due to the inadequate drinking water supply systems and informal wells. Lima is now distinguished by growing socio-economic differentiation and a highly segregated metropolitan region with vast low-income sectors, which in turn elevates the levels of environmental risk. Growing social conflict is based on unmet basic needs and unequal environmental conditions. The metropolitan area furthermore still lacks an efficient public transport system. As a result, traffic volumes and the motorization rate are on the increase, posing not least a health risk.

A major obstacle to regional planning processes is the existence of two departmental governments (Lima and Callao), the segmentation of the city into 49 districts, and the absence of an integrated development strategy of economic, social and environmental dimensions. Difficult tasks in terms of spatial flows and the coordination of land-use policies need to be tackled, and high levels of clustering and specialization in goods and services in the individual districts make demands on spatial and economic planning processes.

Extreme poverty has to a great extent been reduced, especially with respect to the satisfaction of basic needs, although extreme poverty indicators for the six metropolitan cities still register high values in Lima. Moreover, as a result of underemployment, informality and a decline in wage levels for low qualified jobs, income differentiation is rising. Informal employment should be regarded as the principal challenge in socio-economic terms, as informal labour is usually characterized by lack of social protection and low wages. The issue of the 'working poor' is thus highly relevant in Lima.

2.3.4 Mexico City

Mexico City is considered to be among the places with the highest level of air pollution in the world. On the other hand, it represents a positive example of governance output, i.e., the successful application of policies for sustained air pollution reduction – albeit these policies need updating.

In terms of environmental issues, water management is the most conflictive area, as the city requires fresh water from other river basins, having suffered for more than 50 years from water distribution infrastructure of inferior quality. Climate change has altered the frequency and intensity of the rainfall, causing floods in the lower areas of the city. As a result of urban growth and poor land-use planning, serious environmental degradation has occurred in rural areas in the vicinity of the city.

The obvious limitations of the transport system, congestion and high travel times in Mexico City challenge urban mobility and economic growth, despite the existence of the largest and most efficient subway system of the six cities under study. Strengthening the metropolitan public transport system is therefore an outstanding planning issue.

Territorial planning in the metropolitan area of Ciudad de México is marked by a major obstacle: The area includes the Distrito Federal and a large number of localities in the Estado de Mexico with low capacity to cope with megacity growth and the scarcity of natural resources. The absence of coordination of public institutions at the various governmental levels makes the development of an institutional framework for planning and management at the metropolitan level a matter of urgency. Coordination between different governmental levels and greater recognition of the urban-regional context is likewise vital.

2.3.5 Santiago de Chile

Santiago has a high level of socio-spatial segregation and the lowest level of poverty and extreme poverty. Decades of social housing policies have reduced informal housing to almost zero but contributed at the same time to extending the above-mentioned patterns of segregation. This has led in part to criminality and a weakening of social cohesion.

There is little evidence of regional planning in the metropolitan area of Santiago de Chile, since the most powerful institutions are ministries at the national level. The management of environmental issues still suffers from institutional fragmentation and the interference of different scales. The ongoing expansion of the urbanized area of Santiago de Chile generates new transport needs and contributes to the loss of agricultural areas and green space. Urban expansion is hence a key constraint on sustainable development. Another land-use issue related to sustainable development is the scarcity of public space, which is considered crucial to social inclusion and of importance to ecological cycles.

Urban expansion, highly segregated structures and clustered economic development have put pressure on the efficiency of the transport system. Investment in a vast automotive transport infrastructure and the complete remodelling of the public transport system Transantiago are some of the projects implemented in recent years. Air pollution, once the greatest environmental risk in the metropolitan area, has been reduced significantly as a result of legal and technical measures.

2.3.6 São Paulo

Like most cities in Latin America, São Paulo is characterized by a high level of socio-spatial segregation and has one of the highest slum populations and level of inner-city degradation. This landscape produces contrasting levels of access to education, health, public assistance and community centres. Of the six metropolitan cities under review, São Paulo has the highest level of violence and delinquency, due in part to the low presence of state institutions in certain neighbourhoods. Reducing rates of criminality and violence have highest priority on the political agenda of the city.

A major obstacle to sustainable development in São Paulo is ensuring access to mobility or the risk of non-mobility: Traffic congestion is an almost constant concern no longer confined to rush hours. The city shows the highest level of private car participation in a modal split of daily commuting in the six metropolitan cities, a direct impact of the inadequate public transport system. Consequently automotive transportation is the main source of air pollution.

Concerning environmental issues, the most urgent tasks are to reduce air and water pollution, and to ensure access to treated water supplies and sanitation services. Water is a burning issue in most metropolitan cities in Latin America. Striking in the São Paulo case is, however, that only half of the region's water demand is covered by its own water basin, despite a climate context of water abundance.

2.4 Concluding Remarks

In this chapter based on the Regional Panorama Latina America – Megacities and Sustainability, we began with a brief review of predominant urban trends in and stylized facts on the six Latin American metropolitan cities of Bogotá, Buenos Aires, Lima, Mexico, Santiago de Chile and São Paulo. Section 2.2 gave an overview of general trends in Latin America, identified as follows: Although a continent of urban primacy, Latin American patterns of urban growth are shifting towards small and intermediate cities. The remarkable level of centrality also prevails in economic terms, with service and creative industries gaining new currency. At the same time, however, informal sector employment and the challenge of the so-called 'working poor' have gathered momentum. Latin American cities are considered to be worldwide among the most unequal in terms of income and the most segregated in spatial terms. In spite of economic success and enhanced welfare, urban poverty and habitat precariousness are still highly relevant and call for new solutions.

As for environmental issues, the progress registered in some of the metropolitan cities in terms of pollution reduction and improved treatment infrastructure should be emphasized. On the other hand, total emission and resource use levels have soared as a result of the link between welfare and consumption levels.

Section 2.3 was devoted to a brief characterization of the six megacities selected in the region. It emerged that numerous trends, features and problems specific to Santiago de Chile were typical of Latin American cities in general, although many of the indicators in the case of Santiago showed major improvements, positioning the Chilean capital in a more positive light.

Acknowledgments We are grateful to the editor for comments on a previous version of this paper and to Katherine Phelan for her revision of this version.

References

Barton, J. (2009). Adaptación al cambio climático en la planificación de ciudades-regiones. *Revista de geografía Norte Grande, 43*, 5–30.
Briceño-León, R. (2002). La nueva violencia urbana en América Latina. In R. Briceño-León (Ed.), *Violencia, sociedad y justicia en América Latina* (pp. 13–26). Buenos Aires: CLACSO.
CELADE and UNFPA (2005). *Dinámica demográfica y desarrollo en América Latina y el Caribe* (CEPAL serie población y desarrollo, Vol. 58). Santiago de Chile: CEPAL.
CEPAL and UNFPA (2009). *El envejecimiento y las personas de edad. Indicadores sociodemográficos para América Latina y el Caribe*. Santiago de Chile: CEPAL.
Consoni Rossi, E., & Taylor, P. J. (2007). Gateway cities: Circulos bancarios, concentration y dispersión en el ambiente urbano brasileño. *EURE, 33*(100), 115–133.
Cuervo González, L. M. (2004). Desarrollo económico y primacía urbana en América Latina. Una visión histórico-comparativa. In A. C. Torres Ribeiro (Ed.), *El rostro urbano de América Latina. O rostro urbano da América Latina* (pp. 77–114). Buenos Aires: CLACSO.
De Mattos, C. (2004). Santiago de Chile. Metamorfosis bajo un nuevo impulso de modernización capitalista. In C. De Mattos, M. E. Ducci, A. Rodriguez, & G. Yañez (Eds.), *Santiago en la Globalización ¿Una nueva ciudad?* Santiago de Chile: EURE Libros.
Delgadillo Macias, J. (2008). Desigualdades territoriales en México derivadas del tratado de libre comercio de América del Norte. *EURE, 34*(101), 71–98.
Florida, R. (2008). *Who's Your City?: How the creative economy is making where to live the most important decision of your life*. New York: Basic Books.
Heinrichs, D., Aggrawal, R., Barton, J., Bharucha, E., Butsch, C. Fragkias, M., Johnston, P., Kraas, F., Krellenberg, K., Lampis, A., Ling, O.G. (2011). Adapting Cities to Climate Change: Opportunities and Constraints. In: Hoornweg. D. et al (Eds), Cities and Climate Change. Responding to an urgent Agenda. The World Bank. Washington. 193–224.
Hidalgo, R., Salazar, A., Lazcano, R., Roa, F., Álvarez, L., & Calderón, M. (2005). Transformaciones socioterritoriales asociadas a Proyectos Residenciales de Condominios en comunas de la periferia del área metropolitana de Santiago. *Revista INVI, 20*(1), 104–133.
Hoornweg, D., Ruiz Nunez, F., Freire, M., Palugyai, N., Villaveces, M. & Herrera, E. W. (2007). *City indicators: Now to Nanjing* (World Bank Policy Research Working Paper 4114).
Jordán, R., Rehner, J., & Samaniego, J. (2011). *Regional panorama Latin America. Megacities and sustainability*. Santiago de Chile: CEPAL.
McGranahan, G., Jacobi, P., Songsore, J., Surjadi, C., & Kjellén, M. (2001). *The citizens at risk: From urban sanitation to sustainable cities*. London: Earthscan.
Parnreiter, C., Fischer, K., & Imhof, K. (2007). El enlace faltante entre cadenas globales de producción y ciudades globales: El servicio financiero en Ciudad de México y Santiago de Chile. *EURE, 33*(100), 135–148.
Pereira, P. C. X., & Hidalgo, R. (2008). Producción inmobiliaria y reestructuración metropolitana en América Latina. In P. C. X. Pereira, & R. Hidalgo (Ed.), *Producción inmobiliaria y reestructuración metropolitana en América Latina* (pp. 7–22). Santiago: Serie GEO Libros 11.
Prévot-Schapira, M.-F., & Cattaneo Pineda, R. (2008). Buenos Aires: La fragmentación en los intersticios de una sociedad polarizada. *EURE, 34*(103), 73–92.
Sabatini, F., & Brain, I. (2008). La segregación, los guetos y la integración social urbana: Mitos y claves. *EURE, 34*(103), 5–26.
Salazar, A. (2010). Transformaciones socio-territoriales en la periferia metropolitana: La ciudad periurbana, estrategias locales y gobernanza en Santiago de Chile. *Scripta Nova Revista electrónica de geografía y ciencias sociales, XIV*, 331–347. http://www.ub.edu/geocrit/sn/sn-331/sn-331-47.htm. Accessed Oct 2010.
Sassen, S. (2007). El reposicionamiento de las ciudades y regiones urbanas en una economía global: Empujando las opciones de políticas y gobernanza. *EURE, 33*(100), 9–34.
Satterthwaite, D. (Ed.). (1999). *The Earthscan reader in sustainable cities*. London: Earthscan.
Taylor, P. J. (2004). *World city network: A global urban analysis*. London: Routledge.

UN Habitat – United Nations Human Settlements Programme (2006). *The state of the world's cities. The millennium development goals and urban sustainability*. London: Earthscan.
UN Habitat – United Nations Human Settlements Programme (2008). *State of the world cities 2008–2009*. London: Earthscan.
UN Habitat – United Nations Human Settlements Programme (2009). *Planning sustainable cities: Global report on human settlements 2009*. London: Earthscan.
Williams Montoya, J. (2009). Globalización, dependencia y urbanización: La transformación reciente de la red de ciudades de América Latina. *Revista de geografía Norte Grande, 44*, 5–27.
Winchester, L. (2008). La dimensión económica de la pobreza y precariedad urbana en las ciudades latinoamericanas.Implicaciones para las políticas del hábitat. *EURE, 34*(103), 27–47.
World Bank (2009). Reshaping economic geography. World development report 2009. Washington, DC: World Bank.

Part II
Developing the Conceptual Framework

Chapter 3
Mechanisms of Systematic Risk Production

Christian Büscher and Aldo Mascareño

Abstract The concentration and densification of social processes is the quintessential feature of cities in general. This offers manifold opportunities: in a material/factual dimension they sustain functional processes for the provision of basic human needs such as energy, food, water and housing; in a temporal dimension they organize and coordinate the numerous municipal processes required to achieve synchronization; in a social dimension they implement measures to include the population at structural and normative levels with regard to participation in functional system services. We argue that while reproducing these functions, megacities, in particular, simultaneously create conditions that jeopardize them with what we call systematic risk production mechanisms. Using four distinctions (Attraction/Exposure, Metabolization/Deterioration, Synchronization/Desynchronization, Inclusion/Exclusion) we analyse their development in order to exemplify the non-linear dynamics and self-enforcing, mutually amplifying processes of such complex research objects as megacities. Our ultimate goal is to produce a heuristic for interdisciplinary research and create a scientific approach with a common frame of reference for the different research disciplines involved.

Keywords Danger • Functions • Hazard • Interdisciplinary research • Risk • Social mechanisms • Systemic risk

C. Büscher (✉) • A. Mascareño
Karlsruhe Institute of Technology (KIT), Institute for Technology Assessment and Systems Analysis (ITAS), Hermann-von-Helmholtz-Platz 1, 76344 Eggenstein-Leopoldshafen, Germany
e-mail: christian.buescher@kit.edu

3.1 Introduction: Megacities as Risk Habitat?

Two phenomena have been in the limelight in recent years: anthropogenic climate change and urbanization processes of hitherto unknown dimensions. With regard to the latter, megacities in several world regions have become the subject of film, television, literature and newspaper articles, mostly accompanied by a fascination bordering on revulsion at their outrageous population figures. The accounts almost constitute a canon, in which cities such as Mumbai, Teheran and Lagos are portrayed as the 'last stations before hell' and a rapidly approaching crisis.[1] Suketu Mehta (2005, p. 537) experienced a city in its 'last stage' when he returned home to Mumbai, having spent most of his life in Western metropolises.

Numerous research projects have likewise assessed megacities as 'risk areas'. In summary these studies postulate that megacities are spaces where people and objects are subjected to processes that cause injury and damage: megacities are dangerous places to be. Research of this kind has been dominated by the natural sciences and the assessment of physical, or in the case of the Munich Re study, economic variables (Wenzel et al. 2007; Kraas 2003; Pelling 2003; Greiving 2002; Berz 2004).

The characterization of megacities as risk habitats refers to these observations and indicates the ambition to address the notion of risk in current mega-urbanization trends. This raises the question of whether such a generalization is justified. To what extent can we genuinely characterize megacities as risk habitats, risk areas or dangerous places? What are the typical risks and dangers associated with megacities? What are the driving forces behind them? Where are the differences to non-megacities or rural regions? A response to these questions poses first and foremost a challenge to research. This article will therefore focus on the theoretical and methodological questions of interdisciplinary risk research.

Theoretical problems arise because it is not feasible to analyse the myriad events that occur in everyday urban life associated with some form of danger. Our attempt at characterization is confined to a valid generalization, one that describes what floodings, car accidents, power failures or violent riots might have in common. We put forward the hypothesis that all of these events possess a similar underlying risk production mechanism. It is widely accepted that the centralization and densification of natural, technical and social processes within a limited space is a major trend in mega-urbanization.[2] This not only magnifies the degree of exposure of people and protected goods to dangerous events but also intensifies the degree of complex system coupling required to generate critical interdependence (OECD 2003; Hellström 2009, p. 327). To analyse these developments research requires theoretical approaches capable of dealing with complex systems.

---

[1] Amir Hassan Cheheltan describes his home city of Teheran; see the *Süddeutsche Zeitung* of April 10, 2007.

[2] For research on the effects of *densification* on social relations, see Mayhew and Levinger 1977.

Methodological problems arise because the emergence of risks and dangers cannot clearly be attributed to natural, technical or social causes. The disastrous events of the recent past (e.g., the earthquake in Haiti, Hurricane Katrina in Middle America, the tsunami in East Asia) have shown that the magnitude of loss and destruction varies with the vulnerability of technical facilities – such as buildings or infrastructure – and of the affected social groups or individual human beings. Consequently, inter- and transdisciplinary research is of vital importance if the theoretical and methodological knowledge derived from the different natural and social science disciplines is to be harnessed. This is where the trouble starts. Without dwelling on the demands of interdisciplinary research, the obvious challenge is to provide a shared perspective on the development of megacities in order to achieve the organizational but foremost cognitive integration of several disciplines. This calls for a cross-cutting concept that elaborates a problem (here risk production) to which other disciplines can relate and contribute in line with their original disciplinary approaches. Our proposition is to utilize theories dealing with the notion of self-organization, self-enforcement or mutual amplification processes. Our ultimate aim is to provide a useful heuristic that will bring the diverse aspects of risk inherent in megacities into a common perspective, and designate this as *systematic risk production mechanisms*.

In the following we discuss a number of interdisciplinary risk research approaches in an effort to extract useful arguments and insights for the task in hand. We hypothesize that *risk* is not merely a synonym for injury, damage or hazard, but a *two-sided constellation* Sect. 3.2. Several authors share the opinion that cities constitute a hub where risk-taking (by sizing opportunities) is centralized and densified in political or economic terms and in formal or informal networks. This leads to an ongoing distinction between centres and peripheries, and the emergence of the 'functional primacy' of certain cities in their respective regions Sect. 3.3. Additionally we discuss arguments from various disciplines (organizational, financial, climate and cybernetic research) on the feasibility of analysing complex systems. We assume that risk production is – in many cases – a consequence of the development and reproduction of social systems. In this sense, risks and dangers are the result of a *regular* (or *normal*) *mode*, to which potential opportunities and problems are for the most part simultaneously attached Sect. 3.4. These arguments are employed to distinguish three crucial functional problems of city sustainment in factual, temporal and social dimensions, with the aim of deducing risk production mechanisms. Our ultimate goal is to develop an analytical tool that can – in principle – be applied to all cities and will expose the specific problems of megacity development. This allows for analysis of an individual case (Santiago de Chile) as seen in relation to other cities, adopting an approach that integrates several disciplines and leads to inter- and transdisciplinary risk research Sect. 3.5. Finally, we discuss the merits and drawbacks of this approach Sect. 3.6.

3.2 Risk Research: From Individual Perception to Social Complexity

The field of risk research is highly diverse in terms of definitions, theories and methods, and no single approach applies easily to the task at hand. While the natural and engineering sciences tend to favour a probabilistic approach to risk assessment and the analysis of natural and technical processes and their potential consequences (see, for example, Rowe 1977), psychology has explored the perception and acceptance of risk by individuals (Kahneman 1982) and its 'subjective probability' (Kahneman and Tversky 1982). The issue under review was the comprehensive trade-off between technological risks and benefits, and the problem of an overall valid, objectively calculable, level of 'acceptable risks' (Starr 1993), even when the 'human factor' is considered (Fischhoff et al. 1981). The idea of a dissent-free level of risk in modern society has been thoroughly discarded with regard to the 'cultural bias' of the individual (Douglas and Wildavsky 1983) or the distinctive attribution of possible damage to the phenomenon of risk – as the negative consequences of decision-making – or of danger – as affected by those consequences (Luhmann 2005).

Also, a look at the state of (natural) hazard research reveals that the dominance of probabilistic risk assessment has been targeted by research groups dealing with hazards and the notion of *social* vulnerability. Subsequent concepts introduced social variables into the equation by determining not only hazards (as natural occurrences) but also *vulnerability* as the physical, economic, political or social susceptibility or predisposition of a community to risk. As a convolution of hazard and vulnerability, risk represents potential loss to exposed subjects or systems. In order to describe the situation of individuals and social groups at risk, Ben Wisner et al. (2004) use a set of variables distinguished as *root causes*, *dynamic pressures* and *unsafe conditions*. In this context the authors produce a generalized account of "affected" individuals or social groups (e.g., families, neighbourhoods). They describe harmful processes (such as earthquakes, floodings) in megacities, as well as the individual and social conditions of those exposed to such processes. *Yet there is a need for a better understanding of the causes and dynamic pressures associated with social processes that lead – not so much to individual actions – but to the immense exposure of both persons and goods.*

International risk research has focused more and more on the emergence of *systemic risks*. With regard to our initial hypothesis, the term indicates a change in the perspective of risk research: the key topic is not construction failures, malfunctions, erroneous actions or hazardous exogenous processes but the analysis of functional aspects of technological or social systems, and the assessment of unintended, potentially hazardous, consequences that are produced systematically. This opens up a new perspective for interdisciplinary research on megacities. If we look at a city like Los Angeles, for example, the metropolitan area of a region that represents a major economic force in the world (USA), and if this city has severe difficulties in maintaining its basic functions (e.g., water and energy provision, traffic, education, health care), sustaining social life becomes a matter of immediate concern (Wisner 1999).

The above observation refers to a separate research level, namely, complex social entities. Attempts to determine the causes of the observed harm rapidly reveal that processes and relationships in megacities are too complex to be captured in simple cause and effect models. The price for the predictive accuracy of causal models – a commonplace in hazard research – is that identification of (dangerous) processes and their predominant features is based on extremely minute and detailed factors. Research has produced a greater number of variables to describe physical and social vulnerabilities, which have subsequently been used in risk calculations. Hazard research models now consider effects that are distant from the triggering events or represent the continuation of a chain of events that causes damage, making the models more and more complex (Turner et al. 2003).

At the same time, processes and relationships in megacities are highly organized, coordinated and synchronized. They are not chaotic or purely random, i.e., risks and dangers cannot be adequately described in stochastic models (in analogy to the daily weather report). Although they may be total strangers, millions of people in megacities find mutual orientation living together in neighbourhoods and participating in traffic or work interaction. In other words, we are dealing here with an object that combines complex processes with a high degree of order. A holistic approach to risk poses innovative theoretical requirements: "It should be founded on a theoretic basis of complexity that takes into account not only geological and structural variables, but also those of an economic, social, political and cultural nature. An approach of this type could assess, in a more consistent manner, the non-linear relations of the contextual parameters and the complexity and dynamics of social systems" (Cardona 2003).

3.3 Cities as Spaces of Opportunity and Risk

By realigning our research focus we also want to clarify the term *risk*. A closer look at the term reveals the denotation of risk as a modern form of reflection on how striving to take advantage of *opportunities* can produce negative consequences that appear to someone else, at another location or another time as *danger*. Taking a risk in the face of an uncertain future is the only mean that holds the promise of achieving what cannot be achieved any other way. Not to take a risk would mean to pass up an opportunity. Taking a chance is a calculated risk and embraces the possibility of loss. Hence opportunity signifies the preferred value of an expectation linked to action, while risk marks the possible disappointment of expectations. Historically the term risk was used to characterize a 'venture' and the calculation of the possible loss of invested capital, goods or people. Early sea trade is the prime example (Conze 2004, p. 848). Risk is therefore inherently connected to opportunity and subsequently to the fixation of an action beyond the range of other possible actions (in the sense of a commitment as decision).

This leads to the question of whether – and if so, to what degree – entire cities can be characterized by this two-sided concept of risk, considering the infinite number of events and multiple consequences that have taken place and are taking place now. Cities in particular serve as laboratories of modernity (Nassehi 2002, p. 212), where new forms of socialization and social differentiation are tested and a vast number of opportunities, risks and dangers emerge. The city is apparently the place where social systems relate to one another and deal with local, regional, national and supranational issues in material, temporal and social terms. Urban centres have gained a functional advantage over rural areas through the centralization, densification and networking of various processes. Today, the mass movement of people to the cities creates technical and organizational needs, i.e., at the very least the production and distribution of goods, the provision of water and energy, the disposal of waste, the creation of living space and travel infrastructure, and the means of treating the sick. These (and many other) functional spheres enable and sustain life in the city. In this context, the city depends on the performance of differentiated social systems and the corresponding outputs and services that must be made available in a more or less synchronized fashion (Germani 1973, p. 28). This points to functional necessity in a temporal dimension, namely, the simultaneous presence of diverse processes via coordination and organization. In a social dimension, cities face the functional necessity of including individuals as 'persons' in the functional spheres of society and allowing them to engage in the city's performance. It can be concluded that in the course of the development of modern cities, functional advantages become functional necessities when it comes to the survival of the city (and its inhabitants).

Cities gain functional primacy because they centralize and densify economic or political opportunities and serve as economic hubs for global economic trade – while the hinterland pales into insignificance. The concentration of expertise, capital and contacts is part of a global network. Although the specific rationalities of functional systems appear to free themselves of local ties, they are also dependent on cities, which now serve in certain business quarters as the 'grounding' for such processes (Marcuse 2006, p. 207). Cities are therefore both spaces of opportunity and of risk. On the one hand, they supply the prerequisites for social systems, with risk-taking as a key component of their reproduction, foremost in economic trade (Knight 1921). If, on the other hand, risk-taking is embedded in the reproduction of the cities' vital functions, problems may occur: "The problem becomes important when structures assume this function and encourage, force, and normalize the taking of risks, or even absorb the risks invisibly present in numerous individual decisions" (Luhmann 2005, p. 71). The prerequisite for social system reproduction (risk-taking) can also lead to possible dysfunctional effects, due to lack of inherent control of 'sufficiency'.

Megacities – as we argue in the following – are mega-amplifiers of economic, political, scientific, educational, media, legal, religious, artistic and entertainment interaction and thus constitute a mega-arena for risk-taking.

3.4 Complex Systems and Risk Production

Megacities, as emphasized above, can be characterized as complex entities that combine non-linear dynamics with a high degree of order. To reach a better understanding of their development, we address some of the original research arguments on the subject of complex organizational systems. We refer first of all to Todd La Porte (1975, p. 6) on the degree of complexity of organized social systems as a function of the number of system components (elements), their relative differentiation or variety, and their degree of interdependence. Secondly, we refer to Charles Perrow (1984) – following La Porte – on high-technology and normal accidents: the main cause of accidents lies not in human errors in construction or equipment, or a disregard for safety regulations but rather in the inherent structure of the respective technology itself. Perrow makes use of two independent dimensions for classification: linear or complex interaction and the tight and loose coupling of technical elements. Complex interaction refers to the non-linearity of the processes concerned and accordingly to lack of determinism in the systems, whose consequences cannot be anticipated or their relationships visualized in causal schematics. Tight coupling is a close arrangement in space and time of interdependent elements in technical systems. These elements are arranged without any slack or buffer. Whatever happens to one unit immediately affects another in the same system. By combining both dimensions Perrow gains a crosstab that points to dangerous technological systems in the field of close coupling and complex interaction,, e.g., nuclear energy systems, genetics and chemical industries. Coming back to the initial hypothesis, both La Porte and Perrow argue that complexity is a prerequisite for high performance and rationality (opportunities) but *also* gives rise to undesired consequences such as uncontrollability due to non-linear dynamics, which in turn lead to error propagation and self-enforcing effects (risks). Perrow therefore predicts normal accidents in light of social and technological system designs.

Similar arguments can be found in other disciplines, e.g., economics or climate research:

1. The notion of 'systemic risks' is employed regularly in sociological and economic research on the *financial system*. Since the end of the 1940s, there has been a strong correlation between the emergence of a financial market and the growth of the economy. The highly symbolic money-based economy, where money itself is the commodity (the 'monetary sector', as economists put it), on the one hand, and the production, goods and services economy, on the other (the 'real' sector), share a functional relationship. Financial markets have progressed to the 'real economy', albeit with different standards of success: short-term maximizing of investment returns versus long-term reproduction of an economic system in society (Windolf 2009). Systemic risks arise from the presence of increasing carriers of wealth in search of new investment opportunities, an emerging class of investment professionals (i.e., consultants motivated to find investment opportunities and take risks); the division of 'property' as company

value and its subsequent investment, borne by individual investors rather than investment professionals associated with the financial incentives of the 'shareholder value' system; the highly symbolic value of money (for the most part transferred electronically), all of which shift money-based economic communication even further away from real life and the natural experience of real people: "Under the banner of financial innovation the modern financial system is constantly producing new symbolic forms leading to an exceedingly intransparent architecture of financial instruments, models and business processes" (Willke 2007, p. 146).

The consequence for any supply and demand market (e.g., asset 'mortgage-backed securities') is the potential vicious circles that drive asset markets to a low-level equilibrium and the virtuous circles that drive them to a high-level equilibrium (locally and temporally stable) – as witnessed in the last 10 years. A shock to trust in the value of these assets generates a self-enforcing process of asset price decline, which is reinforced globally, as Krugman (2008) argues, by the 'international finance multiplier'. There is a real possibility of global contagion via the balance sheets of *financial intermediaries* (e.g., hedge funds), whereby financial losses in one global region lead to credit cut-offs in others due to balance sheet contraction (Krugman 2008).

It is important to recognize that a system (real economy and financial markets) is reproduced through transaction networks and builds up its own state of affairs (specific levels of demand and supply) as self-regulation – interfered with up to a point by political and legal regulations – but *simultaneously produces crisis conditions* (lack of credit and credibility, trust) and a possible catastrophe (interruption of financial flows) as a state of self-endangerment.

2. In climate research similar arguments come from a discipline entitled 'geophysiology', which is concerned with an 'earth system theory' (Lovelock 2009). The main hypothesis is that organisms on the earth evolve in an environment that is the product of their predecessors and not merely a consequence of the earth's geological history. The oxygen in the atmosphere is almost wholly the product of photosynthetic organisms, which adapt to a dynamic rather than a static world, generated by the organisms themselves. The earth must be seen as a dynamic interactive system. This has consequences for climate models, since life on the planet is linked to climate. Lovelock uses a simple model planet with specific atmospheric conditions (earth-sun-distance) and is inhabited by two principal ecosystems: algae in the ocean and plants on land. Interlinkages between variables (temperature, carbon dioxide abundance, plant and algae growth, atmospheric clouds, ocean water temperature) lead to non-linear differential equations. Strong negative or positive feedback links the biosphere to the atmosphere. In terms of stress testing, such as the infusion of a large amount of carbon dioxide, the system reacts with a process of self-enforcement: "Dynamic self-regulating systems ... if sufficiently stressed, change from stabilizing negative feedback to destabilizing positive feedback. When this happens they become amplifiers of change" (Lovelock 2009, p. 52). Lovelock argues for the

planet as a self-regulatory system that converts itself – as a symbiosis of the different spheres (bio-, atmo-, geo-, etc.) – into a cooler or warmer state with unforeseeable consequences for life on earth.

Both examples deliver arguments for the observation that 'systematic risk production mechanisms' exist. This perspective of risk production becomes keener when Magoroh Maruyama's (1963) idea of 'Deviation-Amplifying Mutual Causal Processes', referring to cybernetic research, is added. Maruyama describes processes of mutual causal relationship that amplify the original insignificant or accidental kick, the build-up of deviation and the divergence from the initial condition. He argues that compared to the initial kick, the subsequent development is disproportionally large. Differentiation, stabilization and the development of complexity are possible outcomes when deviation-amplifying mutual positive feedback results in networks. The process then assumes greater significance than the original occurrence.[3] One of his examples (of social processes) is the development of human settlements, where the difference between centre/periphery (city/hinterland) was – and still is – a major driving force behind the emergence of what are known today as megacities.

Summing up and combining the above-mentioned arguments,

- The need for a more holistic approach,
- The conceptualization of the notion of risk as a distinction of opportunity/risk/danger,
- The focus on complex systems and
- The focus on the process of risk production,

we introduce an approach to the assessment of systematic risk production mechanisms associated with the development of megacities. By identifying and analysing these mechanisms, we seek to describe ongoing processes that have the potential to generate hazards, crisis situations or catastrophes, all of which can be found in various forms and degrees of intensity in every major city.

3.5 City Functions and Risk Production Mechanisms

As both the product and the bearer of social differentiation, cities must solve problems in factual, temporal and social dimensions, as stated above. We are currently witnessing the development of political, economic, scientific, artistic

---

[3] Mutual causal relationships are thought of as loops, whereby individual elements influence each other mutually, either directly or indirectly. There is no hierarchical causal priority attached to any one element. As a possible outcome: "Not only are there deviation-amplifying loops and deviation-counteracting loops in the society and in the organism, but also under certain conditions a deviation-amplifying loop may become deviation-counteracting, and a deviation-counteracting loop may become deviation-amplifying" (Maruyama 1963: 13).

and religious centres in regions of global society. These centres are obliged to provide certain services if they are to survive as anthroposphere (referring to the physical existence of their inhabitants) and social cosmos (referring to the shared life orientation of their inhabitants). Their common problems can be identified in three dimensions:

1. *Material/Factual Dimension*: Regarding the physical necessities of life, cities as systems are required to sustain functional processes in order to provide, e.g., energy, food, water and housing. They have to deal with the relationship between functional necessities and services/outputs, and to implement the organizational and technical means to provide these services/outputs.
2. *Temporal Dimension*: Regarding the need for synchronization, cities have to coordinate and organize manifold local (municipal), national and supranational processes in order to provide vital services/outputs when required.
3. *Social Dimension*: Regarding the need for participation in functional system services, cities are compelled to implement measures to include their inhabitants at structural and normative levels (e.g., at the structural level, to grant access to political and legal systems, education, health care, social security; at the normative level, to accord fundamental rights such as freedom, dignity and equality).

In order to cope with the complexity of interrelated risks and opportunities, we propose four mechanisms derived from a combination of the above-mentioned functional problems and several research disciplines. These mechanisms are explained in the following in more detail.

3.5.1 Attraction/Exposure

Whereas several cities in Europe are in the process of shrinking, cities in other world regions are rapidly expanding. On the whole, the ratio of urban to rural population indicates a steady increase worldwide in favour of the former. This contradictory observation is undoubtedly the result of an uneven differentiation of centres and their peripheries in various parts of the world. On the one hand, urban research describes a blurring of the differences between cities and rural areas as cities turn into regions (metropolitan areas). This process, referred to as *suburbanization*, frequently leads to several centres evolving into one region (Stichweh 1998, p. 355). In the course of this process, competition between a number of metropolitan areas in one region may cause certain cities to lose their attraction in favour of others. Hence urbanization does not automatically foster growth in cities. In Asia, Africa and Latin America, on the other hand, distinctions between urban and rural areas gain in significance when the relevant social attractors (e.g., employment, education, health care) are relocated to the city. The urban/rural divide is heavily accentuated and the draw of the city has intensified (Feldbauer and Parnreiter 1997, p. 13). Those who move to an urban centre or already live there harbour the expectation of being able to live, work and enjoy their leisure time with complete strangers. In this sense the city is a social institution with stable expectations of social interaction (Baecker 2004). *Attraction* thus designates the expectation that the chances of survival, of some sort of life, business or pleasure, are higher in the city than anywhere else. Whether these expectations correspond to reality remains a secondary issue as long as they motivate migratory streams.

The *exposure* of megacity regions to diverse natural processes and technically induced catastrophes has been described thoroughly in multidisciplinary research and can be seen as the original motivation for conferring the title of 'risk habitat' on today's megacities. Exposure, however, is only one of many aspects – albeit an important one. We emphasize the relationship between attraction and exposure in such a way that it is comparable with other risk aspects related to megacity development. If we regard processes of growth in population and/or space – which certainly cannot be called a natural occurrence – as the endogenous variable *attraction* and relate them to the endogenous variable *exposure* as a result of socio-spatial differences, a range of contingent developments is created and can be used to describe current situations in megacities and trends in urban development (see Fig. 3.1). Low attraction combined with low exposure is the most favourable situation/development (e.g., *Berlin*) and leads to functional integrity. High attraction combined with high exposure, on the other hand, is highly unfavourable and crisis-laden (e.g., *Tokyo*; Berz 2004), and can be termed functional endangerment. Likewise, two situations/developments arise from an adaptation to one trend and are tailored in each case to only one unfavourable feature. These two fields indicate contingent developments. The task of interdisciplinary research is to gather evidence on the direction a particular city is taking.

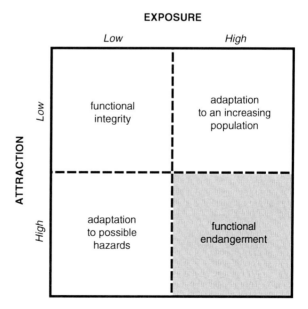

Fig. 3.1 Attraction/exposure relation (Source: The authors)

Table 3.1 Risk production mechanisms in Santiago de Chile with respect to land-use management

	Research field: land-use management[a]
Possible negative occurrence	Damage to people and protected goods
Cause, trigger, condition	Demographics and land-use dynamics as driving factors of exposure to: 1. Flooding (caused by heavy rainfall) 2. Tectonic movements leading to earthquakes
Exposed elements	1. People and social groups as socially (exclusion) and physically (organism) vulnerable: defenceless, lack of means to cope with stress and shock 2. Physical elements with certain characteristics as physical vulnerability
Methods	Probabilistic risk assessment: risk = f (hazard, elements at risk, vulnerability)

Source: The authors
[a]See in detail Chaps. 6 and 7 of this volume

A risk production mechanism in this context is present when the solution to functional problems in megacities produces a surge in expectations of opportunity – with reference to individuals and families as well as to large-scale economic investments. This is accompanied by increased attraction, which again leads to further migration flows. Urban population growth has the potential to aggravate functional problems, the solution to which, in turn, intensfies the attraction. At the same time, such feedback loops or mutually amplifying effects increase the presence of human beings and protected goods but also of vital infrastructures and organizations (e.g., administration) in a space potentially exposed to natural hazards and technical accidents.[4] As a rule, however, people tend to be in danger because they are forced to live in the exposed areas of megacities, often excluded from vital services and deprived of the capacity to move to safer places that are less exposed.

As in the case of all risk production mechanisms, we refer in the following to preliminary observations in other chapters of this volume that address concrete research areas with specific risk aspects. The aim is to exemplify the many contributions to a particular risk production problem. We seek to standardize these observations with the help of four dimensions relevant to risk research: (1) the observation of possible negative occurrences, (2) the cause, trigger or condition of these occurrences, (3) the exposed elements, and (4) the applied research methods in the respective case studies. The notion of attraction/exposure in the following table refers to risks stemming from land-use management (Table 3.1).

---

[4] According to Charles Perrow (2007: 14) this is the general trend in modern society: "Societies put their people in harm's way. Modern societies do so with especial vengeance because their technology and resources encourage risk."

3.5.2 Metabolization/Deterioration

The radical differences between urban and non-urban life lead to extreme asymmetry in the demand for and consumption of resources (e.g., energy, water, construction materials). Materials brought into cities are necessarily discharged into the environment as metabolized matter/materials. Resource import and emission in terms of a system of material flow are basic processes of cities (Brunner and Rechberger 2004). Hence the exploitation of natural resources from the surrounding areas is indispensable to their existence. At the same time, this creates conditions for a possible deterioration of the urban environment, thereby putting basic living conditions in cities at risk. If the high metabolization of resources and the high emission of unusable or hazardous substances concur, key conditions for the reproduction of municipal functions will be jeopardized for the future. On the other hand, if the import of resources can be reduced (e.g., cities shrink, efficient resource management) and the release of pollutants in the environment controlled (e.g., filtering, recycling), vital municipal functions will remain intact (see Fig. 3.2). Again, interdisciplinary research needs to gather evidence on the direction a city appears to be taking.

We refer to a general mechanism rather than a dimension that specifies a critical metabolization and/or deterioration threshold ex ante: each increase in efficiency in the exploitation of resources has led to a rise in production and consumption (Dauvergne 2008). This applies particularly to cities where, for example, greater traffic opportunities lead to more traffic, more and cheaper water supplies lead to higher water consumption per person, and more and cheaper energy leads to greater usage of electronic devices. Thus the crucial issue for cities is whether urban

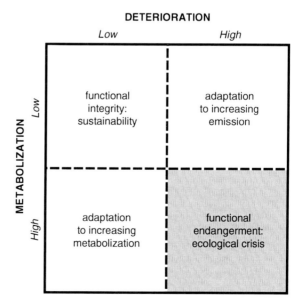

Fig. 3.2 Metabolization/deterioration relation (Source: The authors)

Table 3.2 Risk production mechanisms in Santiago de Chile with respect to air quality

	Research field: air quality[a]
Possible negative occurrence	High emission and high exposure of air pollutants to the ecosystem
	Deterioration of air quality
	Severe health impacts: cardiovascular and respiratory diseases, premature death
Cause, trigger, condition	Emission of NO_2, O_3, PM_{10}, $PM_{2.5}$, NMVOC from traffic, domestic heating and industry
Exposed elements	Human organisms
Methods	Dose–response analysis, probabilistic risk assessment
	Chain of models: transportation, traffic emissions, air quality simulations, health impact assessment (epidemiological modelling)

Source: The authors
[a]See in detail Chap. 11

metabolization endangers basic living conditions provided by the hinterland (e.g., water resources).

This aspect of urban development is primarily linked to the idea of sustainability. Options for the current satisfaction of needs (as opportunity) and the potential destruction of conditions for the future satisfaction of these needs are weighed against each other (as calculated risks and decisions in the present). The outcome will affect future generations (in danger of being affected by decisions made in the past). Only cut the wood that will grow back in the long term was the original dictum of the timber trade and, in short, became the global mandate of "Meeting the needs of society while sustaining the life support systems of the planet" (Turner et al. 2003, p. 8074). Cities (and particularly megacities) face the issue of whether and to what extent they are in a position to recognize and dictate their limits to resource imports and emissions or whether in the course of their development they reach thresholds that, if crossed, imply the irreversible and comprehensive reshaping of the urban anthroposphere.

Concrete examples of metabolization/deterioration trends are shown in Tables 3.2 and 3.3. They are taken from the fields of air pollution and waste management (Chaps. 11 and 13 of this volume).

3.5.3 Synchronization/Desynchronization

A further research problem arises in the context of functional necessities in a temporal dimension. Here the function of the city, as stated above, is to synchronize the manifold social processes. A key feature of modern society is the division of labour, i.e., the differentiation of functional spheres such as politics, law, economics, science, religion, art and the mass media. Numerous authors see cities as nodes for global processes (Castells 1996; Marcuse 2006), as laboratories of modernity, centralizing and densifying multiple social activities in one place (Nassehi 2002).

Table 3.3 Risk production mechanisms in Santiago de Chile with respect to waste management

	Research field: waste management[a]
Possible negative occurrence	Functional failure of technical system: costs, hazards (explosions)
	Externalization of undesired consequences: deterioration of water resources, land, climate system
Cause, trigger, condition	High use of resources, high amount of Municipal Solid Waste (MSW)
	Most critical issue: amount of pre-treated waste sent to adequate landfills in relation to total waste produced
	MSW is transported to one of three existing sanitary landfills without pre-treatment
Exposed elements	Human organisms: health risks to employees, residents
	Environment: ground and drinking water, land consumption, global climate
	Technical system: reduced landfill life span, explosions on landfills, need for long-term landfill monitoring
Methods	Sustainability deficits: distance-to-target analysis

Source: The authors
[a]See in detail Chap. 13

In this sense, they represent the ecological precondition for simultaneous and synchronized inclusion in numerous functional systems that call for spatial proximity and reciprocity. Cities create visible spaces in which diversity co-exists in *loosely coupled* systems (Nassehi 2002, pp. 223f). Due to the need for organization and coordination, cities develop strongly condensed technical and organizational processes that are highly dependent on each other. But is the argument of *loose coupling* still valid for megacities? And what consequences can be expected if the networking, centralization and densification of municipal processes intensifies?

From organizational research we are familiar with the argument that technical and organizational processes grow more dangerous when non-linear interactions, which cannot be predicted or controlled, and closely coupled elements magnify the impact of initially minor events (Perrow 1984, p. 387). The question is whether there is a conceivable degree of complexity and density to produce a crisis, where the maintenance and synchronization of municipal functions are at stake. These issues are discussed as 'systemic risks' (Hellström 2009) and reveal the difference between the factual and the temporal dimensions of systemic events. Systemic risks occur when several mutually influenced events take place simultaneously. In the financial sector, systemic risks are the result of mechanisms inherent in the economic rationale, as seen above, and of the network character of the financial system, which allows a financial crisis to gain global magnitude. Another example is the possibility of systemic events in infrastructure systems, e.g., energy provision, the corresponding complex interactions of technical facilities (power plants, electricity networks, information technology), and the organization of development, implementation, operation and regulation. As for a comparison between financial and energy systems, Bartle and Laperrouza (2008) see differences and similarities in the amount of damage, the time of error propagation and the degree of uncertainty involved in systemic risks. Whereas damage to the financial system could reach global magnitude, potential damage to socio-technical energy supply systems would be confined to a regional, sporadically national and in rare cases continental magnitude. The spread of failure or undesirable events through the respective networks could occur rapidly in both cases. A crisis in the financial system has the potential to endanger the entire economic system and, therefore, society as a whole. A catastrophe of this magnitude in the sense of an irreversible change in the state of the financial sector could lead to a reorganization of the coordination mechanisms of financial transactions and governance structures. The consequences of a system-wide damaging event to existing energy supply systems would be grave but normally bounded and constrained, with a 'repair' rather than reorganizational effect. These issues are relevant to risk research given the mutual factual dependencies of vital urban infrastructures e.g., energy, water or waste management.

Municipal processes also require temporal synchronization. One of many issues in megacities in this context is transportation. Here, the traffic jam is the perfect image for a state of crisis, since traffic blocks traffic. The transportation of people and goods is a fundamental necessity of social system reproduction, notwithstanding the wide dissemination of electronic media. Apart from the transportation of goods, organized systems are also forced to rely on the presence of blue- and white-collar workers in need of transportation, whose working hours must be organized and synchronized. Traffic that stands still precisely when movement is urgently required not only loses the function attributed to it but also produces a disturbance of numerous processes at other locations. Traffic depends on the simultaneous interaction of technical systems (tracks and streets, railways and cars, traffic lights and points), rule systems (driving on the left or on the right), and – above all – systems of social interaction, in which total strangers encounter one another. The list of elements that affect the functioning of traffic is endless.

When we refer to temporal synchronization, the basal function of ongoing actions despite an indefinite future comes into focus, and with it the issue of uncertainty and its reduction. In terms of traffic, we confine ourselves to statements on the *expectations* placed in functioning technology, the validity of traffic rules and that others will behave in traffic in a calculable manner. Although this list is incomplete, we should bear in mind the need to stabilize the behavioural expectations that enable traffic to be maintained and, in particular, allow individuals to participate in a traffic system whose proper functioning is uncertain (possibility of arriving late for work, having to schedule the whole day to visit a doctor) or can lead to serious consequences (e.g., death, injury, damage). Cities have to organize transportation (individual or public) to ensure a certain reliability. Not unlike everything else related to professional work, unreliable services affect many if not all other functional spheres. Nowadays cities adapt to regular traffic congestion and people calculate the effects (the loss of time) of this situation with individual choices. The unreliability of public transport services can lead to an increase in individual travel and a decline in public travel. This in turn gives rise to a mutual amplifying effect that clearly aggravates the situation. If car-based travel becomes more and more attractive, the result will be even more congestion: a vicious circle.

Formal organizations reduce uncertainties in society, stabilizing expectations in large cities that the synchronization of countless processes is under control, which again makes this synchronization possible. In other words, when people expect synchronization to take place and assume that operations will function normally, they hold these expectations even in the case of occasional frustration in reality. A collapse of these expectations would endanger the integrity of such systems. If disturbances occur arbitrarily, expectations are shattered, causing further aggravation expressed in abnormal behaviour (extreme cases are *mass panic* or acts of *looting* after a hazardous event). Achievements associated with organization and synchronization, similar to the interplay of social and technical systems, are also given for other urban functional spheres, such as water and energy supplies or waste collection. The common denominator in all cases is that individual services are available when required and that people expect as much. Solutions to problems in these areas are as numerous as cities on the planet. Each city develops its own social solutions, and in some instances regional traditions ensure a convergence of different methods, techniques and forms of organization. Waste disposal methods in the city of Karlsruhe differ from those of Dusseldorf, for example, while in São Paulo the solutions are again entirely different.

Under highly organized, complex conditions (in the form of non-linear interaction between system elements) and tight coupling (of various technical and organizational systems), disturbances and crises are more and more likely (Perrow's 'normal accidents'). At least one mechanism will aggravate the situation. It is common knowledge that attempts to solve security problems have a tendency to render a situation even more precarious. Organizations commonly react to problems arising from tightly coupled and complex interactions with an increased use of technology (a technology fix), e.g., of redundant systems, on the one hand, and further implementation of decision-making routines, on the other. This creates additional conditions for possible disturbances, since events become more sophisticated and more complex (Bechmann 1990, pp. 126f; Sagan 1994).

Because networking, centralization, densification – and hence complexity – are predominant features of megacities, observation and analysis of the potentially precarious coupling of technical and organizational systems is a major issue when it comes to assessing risks in these cities. The task is to gather evidence on whether and, if so, to what extent cities develop these perilous structures over time (see Fig. 3.3).

For the megacity of Santiago de Chile we refer to the above-mentioned transportation issue as an example of this type of problem. The dimensions of risk production are explained in Table 3.4 below.

3.5.4 Inclusion/Exclusion

With regard to the consequences of urbanization for the social fabric of society, the United Nations Economic Commission for Latin America stated as early as the 1960s that new, segregated urban nuclei (barriadas, favelas, villa miserias, shanty

Fig. 3.3 Synchronization/desynchronization relation (Source: The authors)

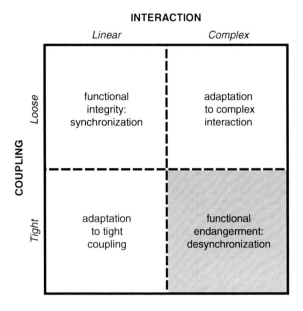

Table 3.4 Risk production mechanisms in Santiago de Chile with respect to transportation

	Research field: transportation[a]
Possible negative occurrence	Unreliability of public transport services leads to more individual and less public travel → causing greater congestion
	Increased emission and land use
	System breakdown (temporary)
Cause, trigger, condition	Lack of trust in public transport services → change in decision modes → increasing attractiveness of car-based travel
Exposed elements	Infrastructure system, air quality, fertile grounds in the hinterland etc.
Methods	Models of transportation, land use and individual choice

Source: The authors
[a]See in detail Chap. 10

towns) had been emerging over the previous decades. Housed in dilapidated structures, a vast section of the urban population lived in economically, socially and politically marginalized conditions (UN ECLA 1973, p. 159). The problem was analysed as a severe lack of integration: people live in conditions of extreme marginality and are cut off from the rest of the city (UN ECLA 1973, p. 160). Likewise in Europe, expectations are high – and have been disappointed – in terms of municipal powers to provide integration. Some authors argue that integrative processes in the social city are in a state of crisis (Häußermann 2006, p. 312).

A common feature of urban life is the 'impersonal cultural pattern', whereby the issue of individual participation in the benefits of social institutions is no longer coupled with belonging to a particular social group, stratum or class. It is beyond dispute that the family into which a person is born today still makes a difference to their weal or woe, and there is adequate empirical evidence to support this claim.

Our purpose, however, calls for systematic and thus comparative access to the multitude of problems arising from unevenly distributed wealth, legal security, political participation, education and medical care in society in general, and in megacities in particular. We thus follow Luhmann's reasoning that inclusion and exclusion denote the extent to which individuals are relevant to the communication of social systems (Luhmann 1995, p. 241). The term social inclusion conveys the expectation that in principle each individual is treated as a legal entity and may assert his personal rights in court; that everyone is entitled to their own property and can trade with it; that everyone has the right to attend educational institutions and take part in democratic elections, to name but a few examples.

We argue that the conditions for the inclusion of individuals are a result of decision processes in the autonomous, functional areas of modern society: the legal system decides on the premises of civil rights (rights of municipal/state citizens); the economy decides what constitutes an adequate income as a prerequisite for participation in economic transactions or consumption; the educational system decides the individual levels of intelligence, attentiveness and hard work required for participation in schooling, vocational training or higher education; finally, politics decides who can vote at what age and how often. On the whole, our point of reference is a functionally differentiated society. It is also the question of people's inclusion in functional systems – not their integration in society – and, in terms of risk research, the threat of their exclusion.

The megacities of today are confronted in particular with the systematic and extensive effects of exclusion. We are currently seeing a steady increase in the number of segregated nuclei; the partial isolation of these areas from any form of administrative influence; the frequent outbreaks of spontaneous violence when the underprivileged, i.e., the largely excluded who live in close contact with those who visibly own much of what they covet; the disastrous events when those living in unsafe conditions die as a result of natural hazards, and much more. These phenomena cannot simply be explained by referring to observations of integration/segregation or poverty/wealth. We point to a systematic risk production mechanism in terms of the effects of mutual amplification on global society and their regional specifics. These stem from the distinction between centres and peripheries, on the one hand, and the simultaneous existence of formally organized functional systems and informal networks based on influence, cohesion and violence, on the other.

The main source of the deep-seated dynamics of exclusion in most Latin American cities lies in how formal organizations are co-opted by informal networks. Systems or organizations, however, have no inbuilt normative motivation to deny access to benefits and goods to any specific human group or community. On the contrary, the functional and normative institutionalization of fundamental rights must take care of the fact that "subsystems of society remain available for each other, for only then is reciprocal interdependence possible" (Luhmann 1999, p. 35). Indeed, functional differentiation operates on the principle of full inclusion (Stichweh 2005); normatively it runs against any kind of informal constraint on its outcomes. Problems arise when certain communities establish specific normative conditions and mechanisms that restrict this access. The gap subsequently widens between the principle of full inclusion and the prescriptions of particularistic communities. This theoretical perspective leads us to the illustration of some mechanisms behind the threat of exclusion, with reference to the persistence of informal structures as a result of self-enforcing and mutually amplifying processes. We refer to three mechanisms:

1. *Informal networks undermine formal institutions.* As Klaus Japp states: "In this case we see networks that simultaneously use functional (administration, NGOs) and local institutions (tribal and kinship structures, personal networks). These networks have no boundaries and can therefore contain several functional communications, which also reduces their efficiency" (Japp 2007, p. 188; translation by the authors). This likewise applies to Latin American cities. As an example, Veronika Deffner argues that informal networks in Brazilian favelas constitute a separate authority in the city. Due to lack of power and order, control of the favelas and the prerogatives of interpreting security and protection rest on the parallel power of paramilitary groups (Deffner 2007, p. 215). From this, Deffner infers that informal networks replace formal power. It can be concluded that mechanisms such as violence, corruption and coercion play a significant role in governance issues when formal procedures and the rule of law fail to provide universal conditions of inclusion. In such cases, clientelism, corruption and violence via informal networks assume responsibility for particularistic and uneven social inclusion.
2. *Informal networks substitute formal institution outputs.* In the case of political clientelism in Argentina, as Javier Auyero illustrates, political networks function "as a problem-solving network that institutes a web of material and symbolic resource distribution" (Auyero 2000, p. 57). They operate as a source of goods and services, and as a protection network against the risks of everyday life. Not only does the well-known Latin American political practice of rewards for votes prove highly useful during election periods, but also the fulfilment of formal non-secure inclusion expectations through informal mechanisms in a complex governance network of particularistic reciprocity: "A broker is related to the members of his or her inner circle through strong ties of long-lasting friendship, parentage, or fictive kinship [...] Members of the outer circle (the potential beneficiaries of the brokers' distributive capacities) are related to brokers by weak ties. They contact the broker when problems arise or when a special favor is needed (a food package, some medicine, a driver's license, the water truck, a friend in jail). But those in the outer circle do not develop ties of friendship or fictive kinship with brokers. [...] While the brokers' ties to their inner circle are dense and intense, their ties to the outer circles are more sparse and intermittent" (Auyero 2000, pp. 64, 66, 67).
3. *Informal networks reiterate the distinction inclusion/exclusion.* The examples of Brazil and Argentina clearly demonstrate how networks function under conditions of high exclusion. In the example of Santiago de Chile, we see how they operate in the upper, middle and lower classes, and how they simultaneously combine procedures of inclusion and exclusion. Referring to the

democratic process in Chile in the 1990s, Peter Siavelis (2006) identifies three informal institutions that collaborated with democratic governance: the *cuoteo político* (i.e., the distribution of government positions in accordance with the clientelistic structures of political parties), the *partido transversal* (i.e., the non-ideological coalition of government elites), and the *democracia de acuerdos* (i.e., informal governance agreements between the government and the military). Without these informal mechanisms, the history of the democratic process in Chile might have been quite different. The very existence of these mechanisms, however, exercises restraint on the evolution of democracy, since they achieve informally what democracy seeks to formalize, namely, public, transparent decision-making processes.

Social scientists tend to justify this type of networking as the last resort against economic exclusion. Undoubtedly there are good reasons for this attitude. The question, however, is why formal mechanisms of inclusion are unable to do what they are supposed to do. Once again, the answer is that the existence of informal mechanisms bridles the development of formal inclusion: if inclusion can be achieved informally, the motivation for individuals, networks or indeed the government to improve governance practices automatically falls by the wayside. Leaving a 'secure' situation already in operation for the promise of an 'insecure' formal situation where trust still has to be gained is a fragile 'leap of faith'. The first hint of possible exclusion from participation in the services demanded increases the likelihood of a fall-back on proven patterns, i.e., activation of established informal networks. We argue that formal/informal inclusion represents a path dependency of self-enforcement in either one direction or another. A breakthrough in the sense of an immediate switch from one mode to the other is highly unlikely: "Personal trust (in networks) takes priority over trust in systems" (Japp 2007, p. 188, translated by the authors).

In summary we argue that inclusion produces exclusion. It seems highly probable that in the course of increased sophistication in functional areas, barriers to people's communicative relevance will be even higher. The threat of being 'left behind' in modern society is structural, particularly in megacities. Problems arise systematically in numerous megacities due to a concatenation of exclusion effects. Many immigrants are denied citizens' rights. They live in illegal settlements and therefore have no official abode or address. As a result, they cannot take on regular work (if at all available). Likewise, children cannot be registered for school and are thus denied the opportunity of education. This type of scenario shows how a legal obstacle to inclusion can lead to further exclusion effects. The continued existence of social networks for the informal and frequently criminal provision of services, whose benefits (e.g., money, work, education) are available only to those prepared to integrate into these networks (via someone who knows someone with access to benefits and who can provide this access in return for future favours), reinforces the failure of social systems and the concomitant uneven inclusion.

In this context we describe the relation between the variables of inclusion thresholds and social transfers. Barriers to inclusion in urban society are the

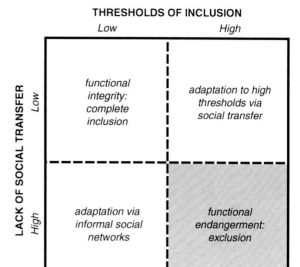

Fig. 3.4 Inclusion/exclusion relation (Source: The authors)

Table 3.5 Risk production mechanisms in Santiago de Chile with respect to socio-spatial differentiation

	Research field: socio-spatial differentiation[a]
Possible negative occurrence	Stigmatization, isolation of individuals, families, groups etc. Increase in violence and criminality, unemployment etc.
Cause, trigger, condition	Socio-spatial differentiation: homogenization as spatial isolation
Exposed elements	People and social groups as socially (exclusion) and physically (organism) vulnerable: defenceless, without the means to cope with stress and shock
Methods	Social vulnerability analysis, policy analysis

Source: The authors
[a]See in detail Chap. 8

outcome of rationalizing functional spheres (on which cities, in turn, are dependent). Their features can differ, e.g., the price of drinking water in relation to income or the opportunity to transfer wealth or social benefits – as in the example of Santiago de Chile government subsidies for a specific volume of water for low-income urban residents. If low inclusion barriers coexist with a high number of transfer opportunities, we speak of a favourable inclusion/exclusion ratio. In contradistinction, a combination of high inclusion barriers and low transfer opportunities leads to unfavourable, i.e., crisis-laden, circumstances (see Fig. 3.4).

With respect to the effects of inclusion and exclusion, we refer to Chap. 8, where the impact of socio-spatial segregation on risk production is analysed further. Table 3.5 gives a brief summary of the dimensions of risk production.

3.6 Conclusion

In this chapter we argued for directing megacity research towards specific mechanisms of systematic risk production. This model allows us to compare the different aspects of risk and danger using a common point of reference, i.e., the jeopardizing of municipal functions by self-enforcing, mutually amplifying dynamics. It is an approach that simultaneously carries the hope of examining entirely different cities from a common perspective. Rather than taking inevitable developments as read, we assume contingent possibilities characterized by the terms opportunity, risk and danger.

The perspective adopted here is a sociological one and must refer back to the specialized knowledge of other research fields to analyse problems of a functional, temporal and social nature. This applies, for instance, to research on natural disasters, on the flow of goods in cities, on traffic and refuse systems and climate research, and on the specialized disciplines of municipal planning and development. Research on concrete aspects of risk and danger is of necessity a multidisciplinary concern.

We illustrated our arguments with case studies from Santiago de Chile, supplemented by brief preliminary observations from some of the research fields addressed in other chapters of this volume. In the following chapters we find abundant evidence of the systematic character of risk production in today's megacities. Further interdisciplinary research faces the task of homogenizing terms and definitions, and operationalizing the above-described processes in order to implement complex models of megacity development.

References

Auyero, J. (2000). The logic of clientelism in Latin America: An ethnographic approach. *Latin American Research Review, 35*(3), 55–81.

Baecker, D. (2004). Miteinander leben, ohne sich zu kennen: Die Ökologie der Stadt. *Soziale Systeme, 10*(2), 257–272.

Bartle, I., Laperrouza, M. (2008). Systemic risk in the network industries: is there a governance gap?, 5th ECPR general conference, Potsdam University, September 10th–12th, 2009, Potsdam; published by Centre for the Study of Regulated Industries, School of Management, University of Bath, available online at: http://infoscience.epfl.ch/record/142565/files/Bartle%20Laperrouza%20ECPR%20Sept09%20systemic%20risk.pdf, last access at 2010-09-29.

Bechmann, G. (1990). Großtechnische Systeme, Risiko und gesellschaftliche Unsicherheit. In J. Halfmann & K. P. Japp (Eds). *Riskante Entscheidungen und Katastrophenpotentiale: Elemente einer soziologischen Risikoforschung* (pp. 123–149). Opladen: Westdeutscher Verlag.

Berz, G. (Ed.) (2004). *Megacities – Megarisks. Trends and challenges for insurance and risk management.* München: MunichRe.

Brunner, P. H., & Rechberger, H. (2004). *Practical handbook of material flow analysis.* Boca Raton: Lewis.

Cardona, O. D. (2003). The need for rethinking the concepts of vulnerability and risk from a holistic perspective: A necessary review and criticism for effective risk management. http://www.desenredando.org. Accessed 22 April 2008.

Castells, M. (1996). *The rise of the network society – The information age: Society, economy, and culture* (Vol. 1). Cambridge: Blackwell.

Conze, W. (2004). Sicherheit, Schutz. In O. Brunner, W. Conze, & R. Koselleck (Eds.), *Geschichtliche Grundbegriffe: historisches Lexikon zur politisch-sozialen Sprache in Deutschland - Band 5* (pp. 831–862). Stuttgart: Klett-Cotta.

Dauvergne, P. (2008). *The shadows of consumption. Consequences for the global environment.* Cambridge: MIT Press.

Deffner, V. (2007). Soziale Verwundbarkeit im 'Risikoraum Favela' – Eine Analyse des sozialen Raumes auf der Grundlage von Bourdieus "Theorie der Praxis". In R. Wehrhahn (Ed.), *Risiko und Vulnerabilität in Lateinamerika* (Vol. 117, pp. 207–232). Kiel: Selbstverlag des Geographischen Instituts der Universität Kiel.

Douglas, M., & Wildavsky, A. (1983). *Risk and culture.* Berkeley: University of California Press.

Feldbauer, P., & Parnreiter, C. (1997). Megastädte – Weltstädte – Global cities. In D. Bronger, P. Feldbauer, K. Husa, & E. Pilz (Eds.), *Mega-cities: die Metropolen des Südens zwischen Globalisierung und Fragmentierung (clone)* (Vol. 12, pp. 9–19). Frankfurt am Main: Brandes Apsel.

Fischhoff, B., Lichtenstein, S., Slovic, P., Derby, S. L., & Keeney, R. L. (1981). *Acceptable risk.* Cambridge: Cambridge University Press.

Germani, G. (1973). Urbanization, social change, and the great transformation. In G. Germani (Ed.), *Modernization, urbanization, and the urban crisis* (pp. 3–58). Boston: Little, Brown.

Greiving, S. (2002). *Räumliche Planung und Risiko.* München: Gerling Akad. Verlag.

Häußermann, H. (2006). Die Krise der sozialen Stadt. Warum der sozialräumliche Wandel der Städte eine eigenständige Ursache für Ausgrenzung ist. In H. Bude & A. Willisch (Eds.), *Das Problem der Exklusion: Ausgegrenzte, Entbehrliche, Überflüssige* (pp. 295–313). Hamburg: Hamburger.

Hellström, T. (2009). New vistas for technology and risk assessment? The OECD programme on emerging systemic risks and beyond. *Technology in Society, 31,* 325–331.

Japp, K. P. (2007). Regionen und Differenzierung. *Soziale Systeme, 13*(1 & 2), 185–195.

Kahneman, D. (1982). *Judgment under uncertainty: Heuristics and biases.* Cambridge: Cambridge University Press.

Kahneman, D., & Tversky, A. (1982). Subjective probability: A judgement of representativeness. In D. Kahneman (Ed.), *Judgment under uncertainty: Heuristics and biases* (pp. 32–47). Cambridge: Cambridge University Press.

Knight, F. H. (1921). Risk, uncertainty and profit. http://www.econlib.org/library/Knight/knRUP.html. Accessed 03 May 2011.

Kraas, F. (2003). Megacities as global risk areas. *Petermanns Geographische Mitteilungen, 147*(4), 6–15.

Krugman, P. (2008). The international finance multiplier. http://www.princeton.edu/~pkrugman/finmult.pdf. Accessed 17 May 2009.

La Porte, T. R. (1975). Organized social complexity: Explication of a concept. In T. R. La Porte (Ed.), *Organized social complexity: Challenge to politics and policy* (pp. 3–39). Princeton: Princeton University Press.

Lovelock, J. (2009). *The vanishing face of gaia: A final warning.* New York: Basic Books.

Luhmann, N. (1995). Inklusion und Exklusion. In N. Luhmann (Ed.), *Soziologische Aufklärung: Aufsätze zur Theorie der Gesellschaft* (pp. 237–264). Opladen: Westdeutscher Verlag.

Luhmann, N. (1999). *Grundrechte als Institution: ein Beitrag zur politischen Soziologie* (4th ed.). Berlin: Duncker Humblot.

Luhmann, N. (2005). *Risk: A sociological theory.* New Brunswick/New Jersey: Transaction Publishers.

Marcuse, P. (2006). Die Stadt – Begriff und Bedeutung. In H. Berking (Ed.), *Die Macht des Lokalen in einer Welt ohne Grenzen* (pp. 201–215). Frankfurt/Main: Campus-Verlag.
Maruyama, M. (1963). The second cybernetics. Deviation-amplifying mutual causal Processes. *American Scientist, 51*, 164–179.
Mayhew, B. H., & Levinger, R. L. (1977). Size and density of interaction in human aggregates. *The American Journal of Sociology, 82*(1), 86–110.
Mehta, S. (2005). *Maximum city: Bombay lost and found*. New York: Knopf.
Nassehi, A. (2002). Dichte Räume. Städte als Synchronisations- und Inklusionsmaschinen. In M. Löw (Ed.), *Differenzierungen des Städtischen* (Vol. 15, pp. 211–232). Opladen: Leske + Budrich.
OECD. (2003). *Emerging risks in the 21st century. An agenda for action*. Paris: Organisation for Economic Co-operation and Development.
Pelling, M. (2003). *The vulnerability of cities: Natural disasters and social resilience*. London: Earthscan.
Perrow, C. (1984). *Normal accidents. Living with high-risk technologies*. New York: Basic Books.
Perrow, C. (2007). *The next catastrophe: Reducing our vulnerabilities to natural, industrial, and terrorist disasters*. Princeton: Princeton University Press.
Rowe, W. D. (1977). *An anatomy of risk*. New York: Wiley.
Sagan, Scott D. (1994). Toward a Political Theory of Organizational Safety; in: Journal of Contingencies and Crisis Management, Band 2, Nr. 4, S. 228–240.
Siavelis, P. (2006). Accommodating informal institutions and Chilean democracy. In G. Helmke & S. Levitsky (Eds.), *Informal institutions and democracy: Lessons from Latin America* (pp. 33–55). Baltimore: Johns Hopkins University Press.
Starr, C. (1993). Sozialer Nutzen versus technisches Risiko. In G. Bechmann (Ed.), *Risiko und Gesellschaft: Grundlagen und Ergebnisse interdisziplinärer Risikoforschung* (pp. 3–24). Opladen: Westdeutscher Verlag.
Stichweh, R. (1998). Raum, Region und Stadt in der Systemtheorie. *Soziale Systeme, 4*(2), 341–358.
Stichweh, R. (2005). *Inklusion und Exklusion: Studien zur Gesellschaftstheorie*. Bielefeld: Transcript.
Turner, B. L., Kasperson, R. E., Matson, P. A., McCarthy, J. J., Corel, R. W., Christensen, L., Eckley, N., Kasperson, J. X., Luers, A., Martello, M. L., Polsky, C., Pulsipher, A., & Schiller, A. (2003). A framework for vulnerability analysis in sustainability science. *PNAS, 100*(14), 8074–8079.
Un, E. C. L. A. (1973). Some consequences of urbanization for the total social structure (United Nations Economic Commission for Latin America). In G. Germani (Ed.), *Modernization, urbanization, and the urban crisis* (pp. 151–167). Boston: Little, Brown.
Wenzel, F., Bendimerad, F., & Sinha, R. (2007). Megacities – Megarisks. *Natural Hazards, 42*, 481–491.
Willke, H. (2007). *Smart governance: Governing the global knowledge society*. Frankfurt/Main: Campus-Verlag.
Windolf, P. (2009). Zehn Thesen zur Finanzmarktkrise. *Leviathan, 37*, 187–196.
Wisner, B. (1999). There are worse things than earthquakes: Hazard vulnerability and mitigation capacity in Greater Los Angeles. In J. K. Mitchell (Ed.), *Crucibles of hazard: Mega-cities and disasters in transition* (pp. 375–427). Tokyo: United Nations University Press.
Wisner, B., Blaikie, P., Cannon, T., & Davies, I. (2004). *At risk: Natural hazards, people's vulnerability and disasters*. London: Routledge.

Chapter 4
Sustainable Urban Development in Santiago de Chile: Background – Concept – Challenges

Jonathan R. Barton and Jürgen Kopfmüller

Abstract The main objective of this chapter is to reflect on one element of the conceptual frame for urban development analysis – the goal dimension of the sustainability vision – and its application to the case of Santiago de Chile. The chapter provides essential insights into the sustainability concept in general and the current situation, debates and controversies in Santiago de Chile in particular. Basic sustainability documents are discussed in terms of their local applicability and potential for associated programmes and activities. For the case of Santiago, political and institutional characteristics and current thematic priorities are outlined. The Helmholtz Integrative Sustainability Concept is tendered as an appropriate tool for sustainability analysis. Using indicators as a basic tool, application of the concept to the Santiago case within a broader conceptual landscape provides orientation for a variety of decision-makers. Initial findings on the translation of the concept into indicators and its application to several thematic fields are presented and the most urgent sustainability performance deficits, defined as risks for future development, are highlighted. Based on an overview of the current sustainability policy in the Santiago Metropolitan Region, future challenges are identified and practical recommendations put forward.

Keywords Santiago de Chile • Strategic planning • Sustainability analysis • Sustainability performance deficits • Sustainability policy • Sustainable development

J.R. Barton (✉) • J. Kopfmüller
Pontificia Universidad Católica de Chile (UC), Instituto de Estudios Urbanos y Territoriales, El Comendador, Santiago, Chile
e-mail: jbarton@uc.cl

4.1 Introduction

Sustainable development is a conceptual framework that emerged in the 1970s and was refined in the early 1990s, particularly after the Rio de Janeiro Conference on Environment and Development. Its application in urban and regional settings was instrumental in shaping thinking about metropolitan development patterns, via the framing of Agenda 21, for example. By building on an integrated, future-oriented framework within which equity and participation play central roles, urban sustainability policies, plans and programmes promote the need for information exchange, collaboration and cooperation in planning exercises at all spatial scales.

Also central to sustainable development thinking is the role of praxis. While there has been much discussion on the conceptual heterogeneity of sustainable development and sustainability perspectives, emphasis on action and practical implementation is paramount. It is with this practical, action-oriented focus in mind that the chapter communicates a structured framework of sustainable development according to the Helmholtz Concept, and successive, logical steps that translate it into practical tools for the measurement and design of interventions and investments. The chapter notes the importance of being able to move between this conceptual development and practical proposals, transcending some of the conventional limitations of sustainability discourse trapped at the level of meaning.

In the case of the Santiago Metropolitan Region, the challenges are made evident through topical research, and their operationalizing through the Helmholtz Concept enables the grounding of sustainable development in policy and practice across public, private and civil society sectors simultaneously. For the metropolitan socio-ecological system, sustainable development is both the orientation and – by means of collaboration and decision-making processes – the glue that binds it together. The argument flowing through the chapter is one that highlights the need to promote clarity in what is understood by urban and regional development, to focus on performance and targets, and to ensure that integrated approaches are employed.

4.2 Towards Stronger Urban Sustainability: An Historical Overview

Urban development as a process whereby urban spaces are transformed in terms of their multiple, interrelated structures has been a source of academic and policy interest from the earliest origins of human settlements. For much of this time different paradigms have dominated the thinking related to desirable urban development processes. Sustainable urban development is one of these paradigms and characterizes the type of development perceived as desirable in a normative way. Many of the traits understood to be core to sustainability thinking, or sustainability science, are by no means new and reflect more traditional understandings of urban transformations. Nevertheless, the application of the sustainability guiding vision as a method of understanding urban processes and shaping policy agendas is both relevant to and significant for contemporary analysis.

While the 1972 Stockholm conference on human settlements and the environment marked an early starting point for the process of implementing the guiding vision, the formulation of an urban sustainability agenda is an evolutionary process still 'in progress'. The agenda is distinguished by reflection on the dominant trends in human development, internationalization and the changing resource base since the 1970s. From Garrett Hardin's 'tragedy of the commons' in 1968, Paul Ehrlich's *The Population Bomb* (1968), the apocalyptic *Limits to Growth* report (Meadows et al. 1972), through to the Vancouver conference of 1976 and the creation of the UN urban agency Habitat, a significant process of reflection was set in motion. This culminated decades later in the Rio Summit on Environment and Development (1992), the Istanbul Habitat II conference (1996), the meeting of Latin American environment ministers at the Santa Cruz de la Sierra Summit (1996, and again in 2006), and the Johannesburg Summit on Sustainable Development (2002). A primary objective of such diverse events was to translate these reflections into more precise declarations, guidelines and agenda.

Central to the different steps taken towards formulating a sustainable development agenda and the corresponding strategies was the prominence given to the multiple forms in which humans impact on local and global ecosystems. It revealed the ways in which societies depend on natural resource transformation for production and consumption, and suffer the consequences of poor environmental management, leading to low air and water quality, and soil degradation at different geographic scales. Considering crucial development issues such as poverty, hunger and mass unemployment, which despite decades of multilateral, bilateral and non-governmental interventions have persisted since the UN Development Decade of the 1960s, the pivotal Brundtland report (WCED 1987) focused on the interrelation between social development and the rapid deterioration of the human environment and natural resource stocks. On the one hand, it accentuated the economic roots of the socio-ecological system changes that had occurred under modernity and clearly imposed limitations on bio-physical systems, with repercussions for human subsistence and quality of life. On the other hand, the core idea of sustainable development introduced in the report is to promote justice with respect to the living conditions of current and future generations, and to institutionalize the associated responsibilities.

The fundamental ideas in the Brundtland report, including its clear global perspective, provided a key basis for the United Nations Conference on Environment and Development in 1992 in Rio de Janeiro and its core documents. This is particularly evident in Agenda 21, the blueprint for international development flagged up in the Rio Summit and signed up to by most of the participating countries. In 40 chapters across 800 pages, Agenda 21 set forth a long list of social, economic and environmental fields of action, objectives, means of implementation, and financing issues to be implemented at international, national, regional and local levels. Meanwhile, the principles outlined in the Rio Declaration highlighted the 'essential code' of sustainability (focusing on the state of the socio-ecological systems and on relations between elements within and between these systems) and sustainable development (focusing on the process of development and its normative content).

Agenda 21 identified the local level as the scale where sustainable development is likely to be implemented most effectively, since it is closest to societal groups and individuals. Metropolitan areas and their political and administrational fragments, e.g. municipalities, are obvious examples of this local scale. This is due to the historical role of urban settlements and to the estimated higher number of urban as distinct from rural inhabitants around the globe since the mid-2000s. These urban dwellers are also characterized by more intensive per capita activity levels in terms of production and consumption, which in turn draw on resources from both surrounding and remote spaces. While the Stockholm conference in 1972 increased awareness of the substantial influence of settlements beyond their designated administrative areas, indicators such as the ecological footprint (Wackernagel and Rees 1996) reveal beyond doubt the true extent of their impact on non-urban socio-ecological systems. The importance of local level responses to these phenomena is also embedded in the notion of subsidiarity, since local political and administrative actors and institutions are in immediate contact with the population. This facilitates awareness-raising and the potential mobilization of actors around sustainability issues. It also provides suitable preconditions to involve these actors in different learning and decision-making processes.

Chapter 7 of Agenda 21 is specifically dedicated to the sustainable development of human settlements, highlighting eight different fields of action that can be constituted as focal areas: (a) Providing adequate shelter for all; (b) Improving human settlement management; (c) Promoting sustainable land-use planning and management; (d) Promoting the integrated provision of environmental infrastructure: water, sanitation, drainage and solid waste management; (e) Promoting sustainable energy and transport systems in human settlements; (f) Promoting human settlement planning and management in disaster-prone areas; (g) Promoting sustainable construction industry activities; (h) Promoting human resource development and capacity-building for human settlement development. Additionally, Chap. 28 emphasizes the leading role of local authorities as regulating and planning actors in fulfilling the Agenda objectives.

Consequently, setting up local Agenda 21 initiatives has to date been a crucial element of the so-called Rio implementation process. The United Nations Commission on Sustainable Development (CSD) and the International Council for Local Environmental Initiatives (ICLEI) define them as "a participative and multi-stakeholder process to reach the objectives of Agenda 21 at the local level through the preparation and implementation of a long-term strategic plan that addresses priority local sustainable development concerns" (CSD 2002). Central to this strategic (urban) planning orientation are the following core elements: development visions and practical targets, SWOT (strengths-weaknesses-opportunities-threats) analyses, appropriate implementation measures, evaluation and monitoring systems, and complementary temporal (short- to long-term) perspectives (see, for example, de la Espriella 2007; Steinberg 2005). While strategic planning had been a dominant feature of urban policy and development during the 1960s and 1970s, e.g., in Brazil, embedded in the emblematic 1965 Master Plan of Curitiba and the new capital of Brasilia, it lost ground during the 1980s. This was a consequence of the emerging 'urban management' approach that centred on development facilitation and more short-term, project-led priorities. The return to strategic planning as a tool for sustainability analysis and proposals illustrates its potential for dealing with inter- and intragenerational sustainable development challenges (Williams 1999).

Several international programmes were established to support municipalities in their Agenda 21 activities following the Rio Conference. Examples are the Sustainable Cities Programme of the United Nations Centre for Human Settlements (UNCHS) and the United Nations Environment Programme (UNEP), as well as the Local Agenda 21 Initiative of the International Council for Local Environmental Initiatives (ICLEI). The urban sustainability agenda that these programmes pursued (among others) has two principal strands of interest or concern: the first is the subject of poverty, housing, segregation and basic living conditions; the second is that of resource use and contamination. The two are, of course, interrelated, but are often managed separately, which neither meets specific requirements nor facilitates a comprehensive understanding of sustainable development as presented in the Brundtland report. While the report defined sustainable development in terms of satisfying needs and enhancing capacities for people to address future challenges, there is still a strong tendency in policy formulation and implementation to concentrate on contamination and resource issues (water and energy supply, for example), almost to the exclusion of the first topic.

A specific and more comprehensive engagement with urban areas can be discerned in the range of initiatives that complement Agenda 21 and the work of UN Habitat following the Istanbul Conference. Likewise, the adaptation of sustainable development principles is evident in the Nuñoa Charter of 2002 (Latin American municipal representatives), the Melbourne Principles for Sustainable Cities (2002), the Gauteng Principles of Regional Sustainability (2002), the Hong Kong Declaration of Asian Cities (2004) and the Leipzig Charter of Sustainable European Cities (2007). In the USA, it was the Smart Growth movement that put forward similar urban planning principles. Although common strands align these with the Brundtland sustainability prerequisites, with the Bellagio Principles and similar initiatives, the strongest links are to the 1994 Aalborg Charter of European Towns and Cities Towards Sustainability. However, the synthetic breakdown of basic elements is best offered by the Melbourne Principles, which frame the fields of action of Agenda 21 (Chap. 7) in terms of ten basic criteria (see UNEP 2002):

1. Provide a long-term vision for cities based on sustainability, on intergenerational, social, economic and political equity, and their individuality;
2. Achieve long-term economic and social security;
3. Recognize the intrinsic value of biodiversity and natural ecosystems, and protect and restore them;
4. Enable communities to minimize their ecological footprint;

5. Build on the characteristics of ecosystems in the development and nurturing of healthy and sustainable cities;
6. Recognize and build on the distinctive characteristics of cities, including their human and cultural values, history and natural systems;
7. Empower people and foster participation;
8. Expand and enable cooperative networks to work towards a common, sustainable future;
9. Promote sustainable production and consumption through appropriate use of environmentally sound technologies and effective demand management, and
10. Enable continual improvement, based on accountability, transparency and good governance.

These principles should not, however, obscure the fact that considerable debate remains on the 'what, where, how and who' of sustainable development. Even in the urban setting, Henri Acselrad (1999) points to three contrasting views on sustainability that compete for policy space: the technical-material representation of cities, the city as 'quality of life' space, and the city as a space for the legitimization of urban policy. Each of these strains or perspectives creates different priorities, discourses and social constructions of what urban sustainability is and where it should lead (see also Naredo 1996; Jimenez 2000). It is within this complexity that case studies play an increasingly vital role in sustainability thinking. With the specific application of sustainability thinking and principles, it is possible to move beyond rhetoric into the policy-relevant realm. The case of Santiago de Chile provides one such challenge.

4.3 Sustainability Issues in Santiago de Chile: Challenges and Responses

Understanding urban areas as socio-ecological systems is a contribution made by Gilberto Gallopin (2003), resonating with the integrated systems thinking of von Bertalanffy (1968) and Fritjof Capra (1982). It also facilitates an understanding of the urban area and its relationship with surrounding systems, such as those within a watershed. While system models can be constructed to demonstrate components and flows, it is application to specific cases that generates policy-relevant analyses. Selected findings of such analyses are presented in the various chapters of this book, for which the Helmholtz Integrative Sustainability Concept provides a common overall framework.

The case of Santiago de Chile offers a metropolitan complex of over six million people, with a highly concentrated Metropolitan Area in a region characterized by the topography of the Maipo water catchment. As with all metropolitan areas, common themes such as housing, transport and waste are highlighted in Chap. 7 of Agenda 21. Nevertheless, the ways in which the relevant sectors interact are prioritized differently, and how decisions are taken varies considerably in each metropolitan context.

In Latin American terms, Santiago de Chile performs well in certain areas (see Chap. 2 of this volume). Throughout the region, the growing urbanization of poverty, and with it the rise in crime and violence, has taken centre stage on the urban policy agenda. In Santiago, however, similar to Chile as a whole, absolute and relative poverty levels are far lower than the regional average. Likewise, although perceptions of crime are fairly significant, in real terms crime statistics are lower than for other Latin American metropolitan regions, particularly in the case of homicide. The comparatively strong performance in these social areas has led to a focus in Santiago de Chile slightly different issues than those prioritized in most large cities across the continent.

Of particular concern are the basic supply systems of energy, transportation and water (see Chaps. 9, 10 and 12). These issues are no less relevant to housing and social equality. As identified in the GEO Santiago report (2004), using the UNEP pressure-state-impact-response methodology, urban expansion and rising demands for private transportation and housing have put increasing pressure on the ecological systems in the Maipo catchment. Air contamination (see Chap. 11) has been a matter of singular concern since the early 1990s, given the topography of the basin and the atmospheric inversion layer, and has led to the development and implementation of the 1996 PPDA (Plan for Atmospheric Prevention and Decontamination of the Metropolitan Region), following the declaration of the area as a saturated zone in PM_{10}, CO and ozone. Since the tighter control of industrial pollutants over time, much of the atmospheric contamination now lies with urban transportation sources. However, urban expansion in Santiago de Chile in recent decades has, for instance, given rise to an increase in the use of private rather than public transport. This process was facilitated by the concessioned urban motorway system, whose logic dates back to the strategic planning of the Metropolitan Area in the 1960s. Only in the late 2000s did the Transantiago initiative (a master plan for urban public transport) make efforts to redress this growing privatization of urban mobility, with a shift towards integrated metro and bus routes. The issue of atmospheric contamination is a clear indication of the integrated nature of urban planning, with its links between demand (for housing on the urban periphery, for greater connectivity with private vehicle use, and air quality) and supply (e.g., the gradual privatization of housing and transportation), and how these move ahead of public policy and instrument responses.

The multidimensional nature of this particular feature of the Santiago socio-ecological system is also a characteristic of other priority topics, i.e., energy and water. While energy is managed nationally through a highly privatized and integrated grid system dominated by the SIC network (Sistema Interconectado Central), accounting for about 70% of national energy generation, Santiago provides a particularly strong demand point due to its concentration of over 40% of the national population. In the absence of regional energy planning, the country can be seen as an isotropic plane across which energy supply and demand is harmonized with national logic. In the light of energy issues associated with restrictions on gas supplies from Argentina, a national energy efficiency programme, and public debates around new energy generation facilities (both thermoelectric and hydropower), this dominant, unitary approach has recently shown signs of change. The implications of gas shortages for the Santiago Metropolitan Region were significant, as many industries had to switch fuels, resorting for the most part to diesel back-up systems, or reduce their output. The switch to diesel, including at the city's large Renca energy generation facility, impacted negatively on air quality, although it did put a slight brake on the energy crisis.

The national eco-efficiency plan that emerged from the crisis, not only for industry but for wider society, has a regional profile, with 2010 marking the introduction of a regional policy in this regard, to be implemented by the regional government. Nevertheless, public debates on energy needs remain locked in the national logic and large-scale generation, and fail to discuss complementary contributions from housing and urban design, building technologies, and renewable energy in the form of sun and wind in urban space or local catchments. Although cities in general, especially large cities, are unlikely to become energy independent, their potential to meet some of the demand locally is underexploited to date. As a more recent phenomenon, provoked by the Argentinian gas crisis but highlighted in terms of plans to increase hydroelectric generation in the Aysén region and the environmental impact assessment approvals for coal-based plants, the stage is gradually being set for the role of energy and its composition in metropolitan sustainability in the twenty-first century.

Perhaps the most specific long-term issue in the Santiago case is the water balance in the Maipo catchment. Rising population numbers and increased per capita water consumption have led to growing concerns about water availability. The Chilean water code of 1981 (modified 2005) is strongly based on water markets and the privatization of water through user right provisions. However, due to rising extraction rates as well as climate change, no new surface water rights are being made available by the regional water directorate, and concerns about the replenishment rates of underground sources are growing. Given the climate change projections for the region to 2070 (based on the IPCC A2 scenario in particular), this challenge will become more urgent and require changing consumption patterns, new technological applications, and changes in land use, albeit with a significant impact on the region, a Mediterranean biodiversity 'hotspot'.

Although air quality, energy and water are merely mentioned to highlight priorities, they are useful in revealing the interconnected nature of several issues and the ramifications for planning and policy generation. It is here that a shift from sectoral thinking on urban development to more integrated, problem-oriented and system-based sustainability thinking is required. This change in the public sector perspective must be allied to broader governance shifts, with civil society and the private sector becoming more active in urban decision-making. The creation of new decision-making spaces within the scope of contemporary urban governance regimes is vital if the demands for increased participation – so crucial to sustainable development – are to be met. In the Santiago case, urban planning and sectoral planning remain fixed on resolving specific issues demarcated within well-defined areas. Although small cracks have begun to show in this logic, as seen in the more participatory approach to the *Quiero mi Barrio* local development initiatives, the Vitacura plebiscite on the local regulatory plan, the Las Condes community plebiscite on a neighbourhood mall, and the Inter-Ministerial Committee on City and Territory, there are considerable obstacles to more integrated planning within the existing structure. Consequently, urban governance represents a major challenge to more sustainable metropolitan development, restricting the possibility of conceptualizing the issues concerned, generating responses, and enhancing legitimacy (see Chap. 5).

The focus on governance also emphasizes that these phenomena are essentially social in nature. Although objective concerns about water and energy availability are tangible, they are also embedded in social constructions of these very problems, and possible responses. The same holds true for ongoing concerns regarding segregation (see Chap. 8), for example, and health and education, as well as equity and the redistribution of benefits through parallel private and public systems of transport. Despite low rates of poverty, crime and violence in Santiago compared to most other Latin American cities, in relative terms, this does not reduce concerns in these areas. Perceptions of delinquency, mounting unease about drug trafficking and drug consumption in certain areas of the city, and the lack of opportunities for social advancement are cause for ongoing concern for a variety of social groups. These matters demonstrate the importance of case study analysis. Rather than producing crude data on different fields of action, it seems more apt to temper the data with perceptions and priorities that emerge from wider society in the case study context, and to present more sophisticated data on intra-urban distributions and opportunities.

Although there are numerous mechanisms for responding to the sustainable development priorities flagged up by governmental and non-governmental organizations, current institutional instruments lack the strength to reshape the underlying structures. Planning instruments such as Regional Development Strategies, the Metropolitan Regulatory Plan, the Regional Urban Development Plan, Local Regulatory Plans and Local Development Plans all have a role to play, but in most instances are reduced to technical tools that complement sectorally generated programmes and plans, e.g., in terms of housing, transport or economic development.

The adjustments called for in the case of Santiago are deep-seated rather than superficial, as in the case of other cities, and entail more than problem identification and immediate response (mostly from a sectoral perspective). They are more structural in origin and require major transformation in terms of how metropolitan development is conceptualized, who is involved in urban decision-making, how issues are framed and responses generated, and what time frame is planned for these issues. It is here that evaluation and analysis of urban sustainability in a specific case plays a significant role and can equip decision-makers with insights relevant to planning ahead. As a cross-cutting concept relevant to metropolitan transformations, sustainable development demands unambiguous conceptualization if the framework for action is to be logical and consistent.

The Helmholtz Integrative Sustainability Concept (Kopfmüller et al. 2001), which will be addressed in the subsequent section, builds on the different initiatives mentioned in Sect. 4.1 to ground sustainable development thinking in specific cases, such as the Santiago Metropolitan Region.

4.4 The Helmholtz Integrative Sustainability Concept as an Analytical Tool

The analysis of development processes of systems, such as a city or a metropolitan region, and their assessment calls for elaboration of the direction future development is to take. Since the publication of the Brundtland report (WCED 1987) and subsequent international conferences held in Rio de Janeiro (1992) and Johannesburg (2002), the concept of sustainability has gained broad political and scientific appeal, despite the fact that controversies surrounding its conceptual foundation and a precise definition as a basis for further operationalizing steps still prevail (Acselrad 1999; Kates et al. 2005).

This section refers to a sustainability concept that provides normative goal orientation and responds to the question of what direction future (metropolitan) development should take. We refer to the Helmholtz Integrative Sustainability Concept developed by Kopfmüller et al. (2001). Contrary to most sustainability concepts, the basic idea here is to avoid defining sustainable development along 'classic' economic, ecological and social lines. Instead, the Helmholtz Concept begins with the constitutive elements of the sustainability *Leitbild*, derived from key documents such as the Brundtland report, the Rio Declaration and Agenda 21: (1) inter- and intragenerational justice, (2) a global perspective and (3) an anthropocentric approach.

These three elements were operationalized in two steps (see Table 4.1):

- Firstly, they were translated into three general sustainability goals (1) *To secure human existence*, (2) *To maintain society's productive potential* and (3) *To preserve society's options for development and action*. In terms of sustainable development, these elements refer to (1) the individual precondition, (2) the material basis and (3) the non-material basis.
- Secondly, these goals were made concrete by a set of sustainability rules.

These rules constitute the very core of the Concept. The substantial rules describe the minimum conditions for sustainable development to be guaranteed for all human beings of current and future generations. They give credence to the sustainability concept with respect to the various dimensions of social development, such as the handling of natural resources or equality of opportunity. Complementarily, the instrumental rules specify ways of implementing the substantial rules in terms of basic framework conditions for sustainable development processes.

Table 4.1 Sustainability rules of the Helmholtz integrative sustainability concept

Substantial rules		
General sustainability goals		
To secure human existence	To maintain society's productive potential	To preserve society's options for development and action
Protection of human health	Sustainable use of renewable resources	Equal access for all to information, education and occupation
Ensure satisfaction of basic needs (e.g., nutrition, housing, medical care)	Sustainable use of non-renewable resources	Participation in social decision-making processes
Autonomous subsistence based on income from own work	Sustainable use of the environment as a sink for waste and emissions	Conservation of cultural heritage and cultural diversity
Just distribution of opportunities to use natural resources	Avoidance of technical risks with potentially catastrophic impacts	Conservation of the cultural function of nature
Reduction of extreme income or wealth inequality	Sustainable development of man-made, human and knowledge capital	Conservation of social resources (e.g., tolerance, solidarity or adequate conflict solution mechanisms)

Instrumental rules
– Internalization of external environmental and social costs
– Adequate discounting
– Limitation of public indebtedness
– Fair international economic framework conditions
– International cooperation
– Capacity for social response
– Capacity for reflexivity
– Capacity for self-management
– Capacity for self-organization
– Balance of power between social actors

Source: Kopfmüller et al. (2001)

With its theoretical foundation and operational approach as outlined above, the Helmholtz Integrative Sustainability Concept was developed as a well-founded answer to some fundamental criticisms put forward in the sustainability discourse:

– To those who consider the sustainability idea diluted and imprecise, and who suspect arbitrariness behind its content – by providing a palpable, positive definition of sustainable development and applying it to different contexts;
– To those who emphasize over-complexity with respect to the multi-dimensional integrative approach – by determining the rules as minimum conditions, and
– To those who criticize sustainability debates that still begin at the conceptual level as obstacles to the required action – by designing a concept that is consistent and employing it systematically in analysis as a basis for reflected action.

The concept claims, and has indeed so far proved, to be a comprehensive, theoretically valid and feasible tool for the analysis and assessment of sustainability.

Since its development and publication, it has been applied to a number of research project and consultancy contexts within and outside the Helmholtz Association, both in Germany and in other countries dealing with various topics at municipal and other spatial scales (for an overview, see Kopfmüller 2006).

As a sustainability concept it sets forth basic preconditions for the analysis of specific topics, e.g., energy, water, air quality management, and at the more general level, of the entire Metropolitan Region of Santiago, covering all sectors and the challenges the city is now facing. As a comprehensive analytical tool it has a two-level approach and can be applied first of all to the case of Santiago and, secondly, to other Latin American megacities (see Chap. 2). This necessitates the appropriate contextualization of the Concept and its adjustment to the conditions in question.

The ultimate goal of applying this Concept to the analysis and assessment of urban development processes in Santiago is to work out an appropriate orientation for political and societal decision-makers active in the current process of designing a sustainable development strategy. The analytical framework is based for the most part on the systematic design and use of sustainability indicators.

Without a coherent framework to conceptualize what sustainable development means in a specific context, concerns about a persistent plurality of perspectives remain, as explained earlier (see Acselrad 1999). It is only by offering a clearly defined conceptual framework that the particular perspective for sustainable development proposed can be made transparent and concrete. This in itself should prevent confusion about the ambiguity of sustainability and sustainable development concepts (Lele 1991; Naredo 1996). Particularly vital, however, is the praxis of sustainable development. As a paradigm of development, its focus on action has been a key driver, particularly at local levels. This action-oriented perspective means that the grounding of concepts in action and local level transformation is fundamental to the process.

The core dimension of sustainability praxis calls for appropriate tools to measure and evaluate existing conditions of sustainability, and for normatively constructed ways forward for policy design, investment and intervention decisions based on these evaluations. Indicators and scenarios are effective instruments for a move in this direction (see Volkery and Ribeiro 2009). They should be a sine qua non of a sustainable development planning tool box. Absence of clarity about change over time measured with indicators referring to quantity or quality makes it difficult to envisage the context of decision-making as being either effective or legitimate. The same holds true for scenarios. A major task here is the integration into current thinking of intergenerational transformations and concerns for the long-term implications of decisions made now on urban planning and decision-making. Following this logic, the short-term orientation encouraged by the 'urban management' approach of the 1980s is contextualized within a broader temporal and spatial framework.

Giving content to the normative assumptions of the sustainability *Leitbild* (here: of the Helmholtz Integrative Sustainability Concept) and putting them into political and planning practice, demands solid empirical knowledge. This linking of normative reflections to empirical knowledge as a combination of top-down and bottom-up approaches is realized at the *indicator* level. Although there is no consensual view on the definition or purpose of indicators (see, for example, Köckler 2005), five core functions or objectives of (sustainability) indicators can be highlighted (Arancón Sánchez 2007; Grunwald and Kopfmüller 2006; Weiland 2006; Opschoor and Reijnders 1991):

- *Information*, i.e., the suitable description and concretization of complex subjects of research and their interdependencies, allowing them to be measured and analysed.
- *Orientation*, i.e., description, measurement and assessment of sustainability performance by supporting state and trend forecasts, comparisons at temporal and spatial scales, identification of problem and action priorities, and the analysis of existing or potentially conflicting goals. As such, indicators can act as an early warning.
- *Steering*, i.e., measurement and evaluation of political goals already realized and their effectiveness in terms of intended and unintended consequences, as a precondition for the implementation or improvement of political and societal decisions and steering processes.
- *Communication*, i.e., appropriately simplified description and communication of complex issues to initiate and support discussion, learning and awareness processes.
- *Integration* in and between science, policy and society, i.e., creation of a common orientation for individual actors or groups by using shared indicators and target values.

Sustainability indicators can be seen as part of an evolution of urban indicators over time. Urban sustainability indicators are a fusion of conventional urban indicators, such as population, population density, spatial area, land uses and services coverage, and the rising interest in environmental indicators from the 1980s (WCED 1987). These include issues such as water and air quality, and more recently, biodiversity. Rayén Quiroga (2001) points to a three-step evolution of indicators: the first generation of specific bio-physical indicators (environmental indicators); the second generation of sets of bio-physical, social and economic indicators; the third step of compound indicators established by weighting variables and generating a single aggregated measure (the ecological footprint, the genuine progress indicator, and material flow analysis are examples of this type of indicator, see Barton et al. 2007). While it is possible to generate compound indicators, these are based on second-generation data sets, making it possible to have both second and third generation indicators side by side, each fulfilling a different purpose. Third generation indicators have a strong communicational dimension, whereas second generation indicator lists are likely to be more useful for policy-makers and decision-takers.

Since the Helmholtz conceptual framework breaks down into substantial and instrumental rules, these can be translated into specific indicators for analysis, evaluation and action. For each rule there are sets of indicators, which ought to cater for adequate representation of the issues referred to in the respective rule.

This way the rules are further substantiated as well as contextualized with regard to the particular application.

Over the last 10–15 years, countless sustainability indicators and indicator lists have been developed for use at international, national or local levels. Hák et al. (2007), the European Commission (2004), Parris and Kates (2003), Heiland et al. (2003), and IISD (2002) all provide an overview of existing indicator systems and approaches. If the chief purposes mentioned above are to be fulfilled and a goal-oriented long-term strategic urban planning is aimed at, rather than short-term mainstream urban management, there are some crucial challenges for the design of suitable sustainability indicator systems:

- Select indicators based on appropriate criteria;
- Combine a mainly scientific and concept-based top-down approach with stakeholder views and a problem-based bottom-up approach in appropriate ways;
- Realize an appropriate mix of existing indicator types – single indicators, interlinkage indicators, indices, objective/subjective (i.e., public survey-based) indicators, etc. – that will fulfil the functions mentioned above in different ways;
- Select a number of indicators that allow for an appropriate description of the sustainability topics according to the concept used, manageable analysis and communicable results, and
- Determine quantitative and binding target values in a socially coordinated process in order to provide the necessary reference lines for performance evaluations and to facilitate planning security for the actors concerned.

Although various indicator lists have been proposed and used at the local level (as in Seattle, Jacksonville and Manizales), most of the systems suggested are confined to the scientific arena, while urban policy and planning measures are still rarely based systematically on such indicators or oriented towards their concrete target values.

4.5 The Helmholtz Integrative Sustainability Concept: Building on Recent Experiences

The application of a broad-based concept to a specific case will always be met with certain scepticism regarding its suitability. This is also true of the Brundtland definition or the Melbourne Principles. Nevertheless, the conceptual framework of the Helmholtz Integrative Sustainability Concept imposes neither actions nor desirable outcomes. It establishes a framework of principles and rules from which indicators for monitoring and evaluation, and for scenario design can be derived.

The Helmholtz Concept consciously avoids the typical formulation of the three sustainability pillars of environment, economy and society, and the numerous 'adjectival' variations on these themes, such as environmental quality, social equity and economic growth. In the Santiago case, this construction helps to move the sustainability agenda away from the conventional focus on bio-physical aspects of the city's development, as made clear in the GEO Santiago report (UNEP/IEUT-UC 2004), for example (which has this specific goal), or the habitual environment versus economy controversies. Instead, it is closer to the Brundtland definition and its focus on needs, capacities and justice.

This particular emphasis brings sustainable development closer to all aspects of urban development and decision-making. It is no longer an 'environmental add-on'. Instead, it becomes a paradigm for all urban and regional development formulation. Many sustainability concepts insist on a separation between resource use and environmental degradation, on the one hand, and poverty, housing and basic needs, on the other. In contrast, the Helmholtz Integrative Sustainability Concept can deal with the remnants of poverty in the city's *poblaciones* and *campamentos*, and at the same time focus on capacity-building, i.e., in terms of strengthening the physical, human and social capital resource bases through education and community initiatives, for example. The interpretations of what is important in each of the rules will clearly vary and are subject to the indicator selection process. However, the selection of rules in the Helmholtz Concept helps to structure this discussion and ensures consistency with an overarching sustainability framework and the concrete experiences of the city region itself.

The selection of indicators for the analysis was based on simultaneous processes. As a cross-cutting concept, sustainable development provided an axis for analysis of the city region, complementing governance and risk foci. However, it also had to be adapted to, and resonate with the topics (e.g., energy, water, waste) described in the following chapters, dealing with alternative issues and conscious of the overlaps. Given the focus of the conceptual framework on the ability to evaluate sustainability performance and its potential as a risk to city region development, designing indicators was a core task. The concept of sustainability has close associations with the notion of risk. Risks can be understood as obstacles or constraints to the achievement of more sustainable development. The sustainability criteria can be used to identify these risks and to design the appropriate governance and implementation options required to mitigate or remove them.

Up until now, Santiago has shown a tendency to avoid the use of sustainability indicators in urban and regional planning. There are no sustainability indicators in place as a planning process orientation or as an evaluation tool to measure the effectiveness of various instruments, either spatial (e.g., the Regional Development Strategy, Metropolitan Regulatory Plan, Regional Urban Development Plan, or Local Regulatory and Development Plans) or sectoral (e.g., health, education, housing). Emphasis in the past two decades has been on effective budget management via state-wide modernization of the public apparatus. As a rule targets are based on annual public spending rather than long-term strategic objectives, and measurements refer primarily to public agency effectiveness in managing projects and spending, rather than to the outcome of that spending over a longer time frame.

Despite the existence of a variety of planning instruments, there is little clarity about how they might generate more sustainable outcomes in terms of synergetic effects arising from plans, programmes and investments of both the public and the private sectors. A wide range of policies and instruments co-exist and are operated by individual ministries, services and local authorities. These include the Regional Development Strategy, the Regional Urban Development Plan, and the Metropolitan Regulatory Plan, as well as regulatory and development plans at the local level. Although these instruments project the region forward in many ways, such as land-use zoning, and orientation and prioritizing for regional and local development, they do not extend to determining targets to be achieved in individual spheres of activity, such as travel times, demographic and population distribution trends, resource consumption, and income generation and distribution. The most recent example is the Modification 100 of the Metropolitan Regulatory Plan approved in 2011, which involves a widespread reconfiguration of the urban periphery. The concept is to top up green space, reduce socio-spatial segregation, and increase centre-periphery connectivity. However, no measures have been taken to assess the efficacy of the instrument over time in generating these outcomes.

There is an obvious absence of an overall guiding instrument to shape sectoral and territorial policies, plans and investments, and to monitor their performance over time (not only in terms of their own goals but of more generalized, system-wide goals). This situation is an outcome of the specific, complex governance and institutional system that prevails in the region, with various actors and their respective interests and action approaches (see Fig. 4.1). Fifty-two municipal authorities, various regional ministerial secretariats (SEREMIs) (and respective services), the under-secretariat for regional development and administration (SUBDERE) and a regional government comprised of a presidential representative (*Intendente*) and a regional council (regional councillors are elected by municipal councillors) all

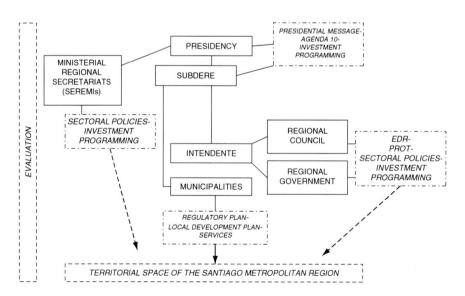

Fig. 4.1 Public institutional structure for the Santiago metropolitan region (Source: The authors)

intervene in various ways and with various formal responsibilities. The argument for a single regional or metropolitan elected authority – an *alcalde mayor* (city mayor) – has been put forward as a way to move beyond the current institutional complexity of asymmetries of power and overlapping democratic constellations (see Chuaqui and Valdivieso 2004; Heinrichs et al. 2009; Orellana 2009).

However, there are interinstitutional initiatives in the region to move forward along more sustainable development lines, e.g., the public-private municipal organization CODESUP (Corporation for the Sustainable Development of Pudahuel) or the collaboration between the Ministry of Housing and Urbanism, the Ministry of Public Works and the Ministry of Public Property known as the Inter-Ministerial Committee on City and Territory (CONECYT), or the Cities Agenda 2006–2010 of the Ministry of Housing and Urbanism. Moreover, a new, potentially vital, vehicle is the regional sustainable development strategy that is currently being formulated and to be ratified in 2011 (encompassing areas such as waste management, energy efficiency and green space).

In the two cases where sustainability indicators were drawn up, neither was utilized as a decision-making tool. The first was a product of the OTAS project (Ordenamiento Territorial Ambientalmente Sustentable/Environmentally Sustainable Spatial Planning), the second emerged within the frame of the Regional Development Strategy 2006–2010 (GORE 2006). The OTAS project was a collaboration between the regional government, the University of Chile and the German development agency GTZ. From 1998 to 2005, collaboration on territorial information development was extensive and served to create conditions more suitable to decision-making. The core concept of OTAS is embedded in the 2000–2006 Regional Development Strategy. This Strategy also embraces the principal infrastructural changes experienced in the region in the first decade of the new millennium. Of these projects, the most well-known are the urban motorway system and the construction of three major waste water treatment plants.

The outcome of the OTAS project was a set of maps and territorial data, representing a notable advance in regional data generation and development. Among these products is the development of a set of indicators (GORE 2005), which are precise and cover a broad spectrum of sustainability issues. Nevertheless, as with all sustainability indicator sets, the selection of variables is questionable and open to challenge, e.g., the inclusion of several road-building indicators as a positive contribution to city region development and the absence of socio-cultural and broad-based economic indicators, such as income distribution and employment (apart from the composition of regional GDP). Failure to use this list as a basis for the Regional Development Strategy 2006–2010 reveals the inherent weaknesses in public administration with respect to objective and obligatory criteria to assess the impact of decision-making processes and instruments.

Similar can be said of the Regional Development Strategy 2006–2010, for which a separate list of indicators was developed. Since it is based on specific axes or development topics – specifically, to become one of the top global ranking regions, to increase international competitiveness and to raise quality of life for all – it follows that it should be fed by appropriate indicators. These are generated in a list of headline indicators and a longer list of supplementary, more specific indicators. The latter introduces a broad base of measures for regional development across the relevant fields. As a rule these variables are not only measured against other major cities in the region, but also provided with target values for direction in decision-making up to 2010. In this sense, the indicators are well developed and move beyond the standard of most existing sustainability indicator sets. Nonetheless, it is one more experience of indicators not being used, despite their publication in the strategy.

This is clearly a weakness of public administration, for which there is a two-fold explanation. First of all, there is comparatively little interest on the part of decision-makers, in this case the Intendente and the regional government, in an objective assessment of their own performance. In the case of immediate issues in the public domain and the media, there is a definite preference for short-term action as distinct from long-term planning. The NIMTO (Not In My Term of Office) concept seems apt here. This is further compounded by the issue that Intendentes are political actors appointed by the president, and can be rapidly moved to other posts in public administration (or removed from them). From its tendering in 2004 to its final publication in 2007, the Regional Development Strategy and its development process outlived four successive Intendentes. Secondly, the restructuring of administrative functions with respect to planning was affected in the period 2007–2008. Regional planning was transferred from the Ministry of Planning and Coordination (MIDEPLAN) to the regional governments and their new planning departments. This involved contracting professionals and integrating new competencies and responsibilities into these departments and the corresponding tool development. Consequently, the capacity to use the available indicators and convey their importance is not yet fully developed in this new group of planners; nor were they involved in selecting the indicators, suggesting a certain hesitation to commit themselves wholly to the indicator list.

These two experiences illustrate that in technical terms there has been some progress in the thinking and creation of sustainability indicators in the region over the past 5 years. The two sets mentioned above (OTAS and the Regional Development Strategy) have been published, and were both developed in collaboration with the regional government. However, resistance persists, or at least a lack of awareness, a circumstance that inhibits their effective use in decision-making. The result is that there are still no 'objective' measures of regional development beyond the 'classic' management indicators such as investment and specific sectoral projects. Without concrete target values, it is impossible to assess distances of performance to targets or track the achievement of goals outlined in the Regional Development Strategy. There is also evidence that this planning instrument is underutilized in public decision-making as a genuine yard-stick of the development process.

To deal with the indicator-related challenges mentioned above and provide a basis to overcome deficits in current practice in the Santiago Metropolitan Region, an alternative approach based on the Helmholtz Integrative Sustainability Concept was chosen for the research project described in this book:

- In a first step the sustainability indicator sets were initially developed 'top-down', based on five core criteria: (1) appropriate concretization of the sustainability criteria laid down in the Helmholtz Concept, (2) comprehensiveness and transparency, (3) the possibility to determine target values, (4) clarity of interpreting increasing/decreasing values as more/less sustainable, (5) sufficient data availability and quality.
- In a second 'bottom-up' step, different stakeholders were involved in the process of identifying and assessing indicators in the framework of several workshops or expert interviews to include their experiences and views on thematic priorities (affecting indicator design) and on suitable target values as a necessary complement and corrective to the scientific perspectives.

Grounding the concept in the case study involved in part the adaptation of relevant topics of specific rules to the experiences in the city region itself. For example, while child malnourishment may be a suitable indicator for many cities in Latin America, it is no longer a primary concern in Santiago, having been replaced by concerns over diet and obesity in children. The original long list of possible indicators was honed according to the demands of the analytical framework, available data at the regional or metropolitan area level, and to what are regarded as priority issues for Santiago. A final list was drawn up containing general variables that provide indicators for the city region as cross-cutting data. The core indicators combined with sets of between 10 and 20 indicators for the different topical fields are systematically applied to the measures and evaluate the current sustainability performance and scenario options. The aim is to work out recommendations for action to be taken in a sustainable metropolitan development strategy. At the same time these indicators provide the basis for a comparative analysis of metropolitan areas in Latin America (see Chap. 2). The process and the indicator set applied at the general level for the Santiago Metropolitan Region are presented in detail in Chap. 14.

Based on existing political and scientific sources, target values are determined for selected indicators, making it possible to measure changes in the indicators over time and to conduct 'distance-to-target' assessments. This methodology was used to identify and assess the risks to current and future sustainable development, and is applied in the subsequent chapters on energy, waste and transportation. In this way, urban transformations can be measured, indicating how certain targets can be achieved while others struggle to be met – this was the experience, for instance, with the Millennium Development Goal process. While the fixing of target values for indicators is a highly complex normative process, given the uncertainties and political implications of such decisions, they are useful in shaping scenarios and linking them to existing urban performance and trends. The results provide the necessary basis for recommendations on suitable measures to deal with the most urgent sustainability problems in the Metropolitan Region.

4.6 From Concepts to Actions

Urban sustainability is a complex paradigm. It has been constructed in different ways by different organizations since the early 1970s. Acselrad, Lele, Naredo, Girardet, Gross and others all point to this complexity, but they also establish the value of the paradigm as an overarching conceptual framework to understand cities, to change governance systems and to drive different, albeit complementary sets of actions at multiple scales. In Chap. 14, the performance of Santiago, as established by the use of selected core indicator sets, will be described and discussed. However, the formulation and application of indicators is not a process that is de-contextualized or derived from 'problem-solving' positions. What the Helmholtz Integrative Sustainability Concept establishes with its three overriding principles and its rules is that indicators are part of a step-wise, systematic and consistent construction that links theoretical abstraction to deliberate action: the political and societal practice of sustainable development.

In essence, it is only by analysing experiences of sustainable development with case studies that the concept comes to life. This occurs through participatory processes, well-founded prioritization steps and the development and application of complementary tools, in particular indicators and scenarios. For much of the early growth in sustainability thinking, the discussion focused on governments and national contexts. However, 'glocal' processes have led increasingly to links between this national focus and other scales of engagement, with its national inventories, national indicator sets and strategies, and national sustainable development committees and action plans. City regions, given their concentration of populations, their ecological footprints, and the new balance of urban and rural settlements, provide important cases for engaging with these issues, with regard to the perspective of challenges, i.e., of risks to development processes, and the governance perspective of designing and implementing a suitable sustainability policy. More integrated responses that take complexities and uncertainties into account more appropriately are likely to lead to more nuanced understandings of how the city region as a socio-ecological system can generate improved quality of life and reduce demands on the bio-physical resource base.

References

Acselrad, H. (1999). Sustentabilidad y Ciudad. *EURE, 74*(2), 36–46.
Arancón Sánchez, S. (2007). *Grounding sustainable development in urban planning. A framework of sustainability indicators for the metropolitan region of Santiago de Chile*. Masters thesis supported by the "Risk Habitat Megacity" project. Madrid.
Barton, J., et al. (2007). *Cuan sustentable es la Región Metropolitana de Santiago? Metodologías de evaluación de la sustentabilidad*. Santiago.
Capra, F. (1982). *The turning point: Science, society and the rising culture*. London: Bantam.
Chuaqui, T., & Valdivieso, P. (2004). Una ciudad en busca de un gobierno: una propuesta para Santiago. *Revista de Ciencia Politica, 24*(1), 104–127.

CSD – United Nations Commission on Sustainable Development (2002). *Second Local Agenda 21 survey*. Background paper, No. 15. New York: International Council for Local Environmental Initiatives (ICLEI).
de la Espriella, C. (2007). Designing for equality: Conceptualising a tool for strategic territorial planning. *Habitat International, 31*(3–4), 317–332.
Ehrlich, P. (1968). *The population bomb*. New York: Sierra Club Ballantine.
European Commission (2004). EU member state experiences with sustainable development indicators. Luxembourg. http://epp.eurostat.ec.europa.eu/cache/ITY_OFFPUB/KS-AU-04-001/EN/KS-AU-04-001-EN.PDF. Accessed 2 May 2011.
GORE – Gobierno Regional de la Región Metropolitana (2005). *Ordenamiento territorial ambientalmente sustentable*. Santiago de Chile.
GORE – Gobierno Regional de la Región Metropolitana (2006). *Estrategia de desarrollo regional*. Santiago de Chile.
Gallopin, G. (2003). *Sostenibilidad y desarrollo sostenible: un enfoque sistémico*. Santiago de Chile.
Grunwald, A., & Kopfmüller, J. (2006). *Nachhaltigkeit Eine Einführung*. Campus Verlag: Frankfurt a. Main.
Hák, T., Moldan, B., & Dahl, A. (2007). *Sustainability indicators. A scientific assessment*. Washington, DC: Island Press.
Hardin, G. (1968). The tragedy of the commons. *Science, 162*, 1243–1248.
Heiland, S., Tischer, M., Döring, T., Pahl, T., & Jessel, B. (2003). *Indikatoren zur Zielkonkretisierung und Erfolgskontrolle im Rahmen der Lokalen Agenda 21*. UBA-Texte 67/03. Berlin.
Heinrichs, D., Nuissl, H., & Rodríguez Seeger, C. (2009). Dispersion urbana y nuevo desafios para la gobernanza (metropolitana) en America Latina: el caso de Santiago de Chile. *EURE, 35*(104), 29–46.
IPCC – Intergovernmental Panel on Climate Change (2007). *Fourth assessment report*. Geneva.
IISD – International Institute for Sustainable Development (2002). *Compendium of sustainable development indicator initiatives*. Winnipeg
Jiménez Herrero, L. (2000). *Desarrollo sostenible. Transición hacia la coevolución global*. Madrid: Ediciones Piramide.
Kates, R., Parris, T., & Leiserowitz, A. (2005). What is sustainable development? Goals, indicators, values, and practice. *Environment, 47*(3), 9–20.
Köckler, H. (2005). *Zukunftsfähigkeit nach Maß. Kooperative Indikatorenentwicklung als Instrument regionaler Agenda-Prozesse*. Wiesbaden: VS Verlag.
Kopfmüller, J. (2006). *Ein Konzept auf dem Prüfstand. Das integrative Nachhaltigkeitskonzept in der Forschungspraxis*. Berlin: edition sigma.
Kopfmüller, J., Brandl, V., Jörissen, J., Paetau, M., Banse, G., Coenen, R., & Grunwald, A. (2001). *Nachhaltige Entwicklung integrativ betrachtet. Konstitutive Elemente, Regeln, Indikatoren*. Berlin: edition sigma.
Lele, S. (1991). Sustainable development: A critical review. *World Development, 19*(6), 607–621.
Meadows, D., et al. (1972). *The limits to growth*. London: Signet.
Naredo, J. M. (1996). Sobre el origen, el uso, y el contenido del termino sostenible. Cuidades para una futuro mas sostenible. http://habitat.aq.upm.es/cs/p2/a004.html. Accessed 2 May 2011.
Opschoor, H., & Reijnders, L. (1991). Towards sustainable development indicators. In O. Kuik & H. Verbruggen (Eds.), *In search of indicators of sustainable development* (pp. 7–27). New York: Springer.
Orellana, A. (2009). La gobernabilidad metropolitana de Santiago: la dispar relación de poder de los municipios. *EURE, 35*(104), 101–120.
Parris, T., & Kates, R. (2003). Characterizing and measuring sustainable development. *Annual Review of Environment and Resources, 28*, 559–586.
Quiroga, R. (2001). *Indicadores de sostenibilidad ambiental y de desarrollo sostenible: estado del arte y perspectivas*. Santiago de Chile: Naciones Unidas.

Steinberg, F. (2005). Strategic urban planning in Latin America: Experiences of building and managing the future. *Habitat International, 29*(1), 69–93.
UNEP (2002). Melbourne principles for sustainable cities. http://www.iclei.org/fileadmin/user_upload/documents/ANZ/WhatWeDo/TBL/Melbourne_Principles.pdf. Accessed 2 May 2011.
UNEP/IEUT-UC (2004) *Perspectivas del medio ambiente urbano: GEO Santiago*. Santiago de Chile.
Volkery, A., & Ribeiro, T. (2009). Scenario planning in public policy: understanding use, impacts and the role of institutional context factors. *Technological Forecast & Social Change, 76*(9), 1198–1207.
von Bertalanffy, L. (1968). *General system theory*. London: George Braziller.
Wackernagel, M., & Rees, W. (1996). *Our ecological footprint. Reducing human impact on the earth*. Gabriola Island, BC: New Society Publishers.
WCED – United Nations World Commission on Environment and Development (1987). *Our common future*. Oxford: Oxford University Press.
Weiland, U. (2006). Sustainability indicators and urban development. In W. Wang, T. Krafft, & F. Kraas (Eds.), *Global change. Urbanization and health* (pp. 241–250). Beijing: China Meteorological Press.
Williams, G. (1999). Metropolitan governance and strategic planning: A review of experience in Manchester, Melbourne and Toronto. *Progress in Planning, 52*, 1–100.

Chapter 5
Megacity Governance: Concepts and Challenges

Henning Nuissl, Carolin Höhnke, Michael Lukas, Gustavo Durán, and Claudia Rodriguez Seeger

Abstract The exploration of governance issues with regard to the sustainable development of and risk mitigation in megacities is crucial, as it provides knowledge on the feasibility of urban development strategies and insights into – actual and potential – risk-creating policies and practices. This chapter therefore introduces the most salient aspects of urban governance in Latin America in general and Santiago de Chile in particular. It outlines a framework for the analysis of urban governance and presents the findings obtained by applying this framework to the case of Santiago. These findings hint at some particularly notable governance challenges – primarily the weak position of both local and regional authorities, as well as the sometimes overly strong interest coalitions between public and private partners. It is likewise evident, however, that there is no universal solution to these challenges. Chapter 15 will further explore the implications of the arguments presented in this chapter.

Key words Decentralization · Informality · Participation · Privatization · Urban policy

5.1 Introduction: The Issue of Megacity Governance

Megacities face a host of challenges in terms of sustainability deficits and related risks, making them *risk habitats*. Given their paramount national and international significance, it is of the utmost importance that the urban policies and development strategies applied there prove adequate and succeed in tackling these challenges. Sanitation infrastructure, for instance, should serve the entire urban population to

---

H. Nuissl (✉) · C. Höhnke · M. Lukas · G. Durán · C.R. Seeger
Department of Geography, Humboldt-Universität zu Berlin, Rudower Chaussee 16, 12555 Berlin, Germany
e-mail: henning.nuissl@geo.hu-berlin.de

sustain public health; land-use regulations should prevent inhabitants of the city from exposure to natural hazards; disaster management programmes should function under critical circumstances. There is no doubt, however, that solutions to urban development problems will only work if they account for the political and institutional context to which they are applied. Urban policies and development strategies must therefore consider how, by whom and along what institutional lines political decisions are made and public affairs managed. In short, they need to account for governance issues.

The examination of governance issues, however, is not only important because urban policies and development strategies might otherwise fail to match their respective context of implementation. It is also crucial because policies and strategies, i.e., the way public decisions are made and carried out, can severely aggravate or even trigger problems of urban development. In other words, as much as it may provide the solution to sustainability deficits and risks, urban governance can itself create or amplify these problems. This is particularly true of megacities because the mere scale, density, complexity and speed of change inherent in these urban systems make urban governance a particularly tricky business. Inadequate land-use regulations, for example, can spur the encroachment of settlements into hazardous locations; ill-designed energy or transport policies are a major cause of air pollution; lack of synchronization in public planning systems frequently leads to the erection of infrastructure facilities in environmentally sensitive areas vital to the recreation of the urban population.

Originating in the political science and economics disciplines, the governance concept has been developed (roughly within the last 25 years) to further the understanding of decision-making processes beyond the individual level. While in economics it is mainly applied to decision-making in organizations, particularly enterprises (corporate governance), the political science strand of the governance discourse broadly concerns the regulation of publicly relevant affairs. It is the latter notion of governance that applies to politics and decision-making processes in urban areas, including megacities. Accordingly, a vast body of urban governance literature has been produced in recent years. It is worth noting in this respect that the governance notion, although not undisputed and perceived by some as the unwelcome intrusion of a 'Northern (Yankee) concept,' has been adopted by the Hispanic academic community as 'gobernanza' (cf. Stren 2000).

This chapter expands on the assumption that any attempt to resolve major problems of mega-urban development is likely to fail unless it takes governance issues into consideration. It intends to illustrate both the importance of understanding the context of implementation for designing adequate policies and the connection between governance and risk. To emphasize the significant role of governance issues for the fate of cities and urban regions, Sect. 5.2 sketches the fundamental political and institutional matters that megacities have to deal with, in particular in Latin America. Taking the predominant governance trends in Santiago de Chile in recent decades, Sect. 5.3 highlights the different aspects of megacity governance that are potentially crucial in terms of risk creation. Section 5.4 outlines a set of analytical categories for the empirical analysis of urban governance. Section 5.5

moves on to apply these categories to four of the seven policy fields (transportation, water supply and services, land-use change and waste management), which are likewise the focus of in-depth discussions in later chapters of this volume. The intention here is to provide a detailed understanding of typical governance issues in the megacity context, taking selected policy fields as examples. Finally, Sect. 5.6 summarizes the findings of the governance analysis and draws brief conclusions on the explanatory value of such an analysis. While a comprehensive account would exceed the scope of this chapter, the following pages seek to show that awareness of governance issues is crucial to tackling the challenges of megacity development.

5.2 Matters of Megacity Governance in Latin America

When it comes to understanding how urban governance works in Latin America, probably the most important fact to be recognized is the concurrent process of profound democratization and vibrant globalization on which almost all countries in the region have embarked in the last 20 years. While this has not solved fundamental problems, e.g., the uneven distribution of political power and the exclusion of many people from the policy process, it has certainly come along with a few major political and institutional matters concerning urban governance (e.g., De Mattos et al. 2005), which deserve – and receive in the literature – particular attention. These are (1) decentralization, (2) privatization, (3) participation and (4) informality (see Stren 2000, who emphasizes the first and the third matter, Bauer 2002, who stresses the second, or De Soto et al. 1986 who are intrigued by the fourth). These governance matters concern entire political systems but are also highly significant at the regional and local level, and particularly in megacities. This chapter expands on the four matters of decentralization, privatization, participation and informality so as to highlight the interrelationships between governance and risk.

1. Latin American countries have been striving for the last 20 years to overcome their extreme degree of centralization (Finot 2002), and have made *decentralization*, the transfer of power away from the central state level, a major policy goal. Decentralization can take quite different forms, hinging on to whom authority is conveyed and to what extent. It can either consist of the mere assignment of responsibility to lower levels of administration (deconcentration) or semi-private entities (delegation) or of a fully-fledged transfer of powers to politically independent regional or local authorities (devolution) (Litvack et al. 1998). Regardless of the concrete form decentralization takes, there is wide consensus that the process of metropolization and urban expansion in megacities calls for the implementation of political and administrative structures that cover the entire urban region, i.e., for some kind of metropolitan authority (e.g., Sharpe 1995). As a rule, however, this is difficult to achieve, since a common feature of most megacities is the abundance of local authorities that rarely speak in unison. In addition, national governments often hesitate to promote decentralization and

local community self-control in large cities, precisely because these cities are so crucial to the country's entire political and economic system. Such a move would make their political representatives strong counterparts of national politicians and, generally, the national government.
2. *Privatization* has been a leading paradigm of governmental policies in most Latin American states in recent times and has shaped megacity development to a notable extent. Today, this paradigm does not remain uncontested, particularly in countries like Venezuela, Bolivia or Ecuador, now ruled by so-called 'new left' governments. Moreover, there is a strong interest on the part of the private sector, including international financial capital, to invest in infrastructure systems in big cities, due to the large number of clients. Hence a pronounced policy of infrastructure privatization makes big cities comparatively well-served nodes within increasingly fragmented infrastructure networks that characterize the landscapes of 'splintering urbanism' (Graham and Marvin 2001). A closer look at these huge agglomerations, however, reveals sharp differences in the quality and accessibility of the services provided, depending on the income and socio-economic status of the neighbourhoods. In poor neighbourhoods people usually pay more for second-rate services, since private suppliers tend to concentrate their investments on urban areas with significant purchasing power. For the same reason, technical infrastructure systems in these areas frequently suffer from grave underinvestment. Neither are there incentives for private providers to avoid externalities, such as environmental risks. Two general questions associated with privatization policies thus arise. Firstly, while the provision of infrastructure and services by private companies often leads to the expected increase in efficiency and (occasionally) quality, it also bears major risks with regard to failure of technical systems or accessibility and affordability of basic services for all (Rodríguez 1994). Secondly, privatization policies inevitably reduce the scope of the public sphere for action; i.e., decisions on where, when and how much to invest in infrastructure or development projects remains for the most part the independent choice of private investors. This points to the inherent link between privatization policies and the question of how the required amount of public control of common affairs can be maintained or even reclaimed (Pflieger 2008).
3. From a normative point of view, citizen *participation* in decision-making processes is a vital element of democracy. In most liberal democracies in the western world, the basic form of participation is the election of representatives (i.e., politicians) at various administrative levels (national, regional, local). In addition, there are more direct forms of participation. These include nominal forms of participation aimed at legitimizing decisions taken elsewhere but also wide-ranging transfers of authority to civil society in an attempt to grant citizens real decision-making power. Participation is not merely a value in itself but also desirable from an instrumental point of view in at least three respects. Firstly, it enhances the intellectual foundation of public decision-making, since non-governmental organizations and citizens can contribute particular knowledge, competences and capacities to the political process. Secondly, the involvement of these actors helps to gear public decisions to local requirements and

peculiarities. Thirdly, participation enhances both the legitimacy and the acceptance of decisions, and has the potential to prevent or arbitrate conflict. All of these aspects are crucial to any kind of precautionary politics, including the preemptive attempt to reduce human exposure to different kinds of risks (Sabatini et al. 2000). Opinions are nonetheless mixed on the success of recent efforts in Latin America to support public participation. While in Latin America there are, on the one hand, innovative institutions such as the internationally acclaimed participatory budgeting in Porto Alegre (Avritzer 2002), we see, on the other hand, how the long-standing tradition of marked social contrasts, poverty and the exclusion of the lower strata of society from the political process still interfere profoundly with the very idea of participation (Roberts and Portes 2006). These problems are most evident in the context of megacities, where polarization between rich and poor is particularly pronounced.

4. Generally speaking, *informality* means that activities take place 'beyond' legal regulation. Although not implying that these activities are necessarily illegal, it does deprive them of legal, i.e., public sector, approval. Due to its imprecision, the notion of informality is highly contested in political science and policy research. On the other hand, it is a key concept in international discourses on urban development, as it hints at the constraints of an approach towards issues of public decision-making that is narrowly focused on the public sector. Traditionally, these constraints were particularly obvious with regard to housing and infrastructure supply in unplanned and non-legalized "informal" urban settlements that largely rely on the self-organization of their (mostly poor) population. The warranty of property rights, as well as the security of tenure and infrastructure supply in these settlements constitute significant challenges for urban governance. The economy is another crucial area where informality matters. Specifically in developing countries, a vast amount of economic activity takes place in informal markets that are not subject to public supervision or control. In Latin America, the share of the informal sector in the economy is estimated at an average of almost 40% (UNCHS 2001). Against this background liberal intellectuals make a strong claim for acknowledgement of the informal sector as a powerful economic engine, creating income for the many (cf. Schneider and Enste 2002), not to mention its enabling effect on the poor to adapt to market behaviour: "The streets of Latin America have become the best business schools available" (Ghersi 1997, p. 104). This attitude, however, is frequently contested as cynical, since informalization of the labour force tends to go hand in hand with precarious working conditions and major health risks to those 'employed' in the informal sector.

Assuming that the aforementioned crucial matters of megacity governance become somewhat more tangible if traced out in a particular case, the following section looks at the evolution of governance modes and trends in Santiago de Chile in recent decades. This serves to illustrate how decentralization, privatization, participation and – to a somewhat lesser extent – informality fall into place and create a situation that is particularly demanding in terms of public-private interplay.

5.3 The Exemplary Case: Past and Present Governance Trends in Santiago de Chile

The recent evolution of governance patterns in Santiago de Chile largely converges with other megacities of the Latin American southern cone. Looking at the time span from the mid- twentieth century onwards, three phases can be identified with regard to urban development processes, the associated problems and attempts at their regulation. Not surprisingly this periodization coincides with economic and political macro-trends at the national level: (1) the phase of the developmental state and import substitution politics, (2) the phase of military dictatorship and economic liberalization, and (3) the phase of re-democratization and the deepening of the neoliberal model.

In Chile and its capital Santiago, the phase of strong state intervention and import substitution (*1930–1973*) had a serious impact on urban development and the organization of urban governance. In the early 1930s, mass migration from the countryside to Santiago led to an increase in the city's population from around 700,000 to almost 2,900,000 between 1930 and 1970. In the first 20 years at least, this growth took place almost exclusively in the form of uncontrolled urbanization on the periphery, so that in 1960 a third of the population of Santiago lived under precarious circumstances and suffered from insufficient or nonexistent provision of drinking water, electricity, sewerage, education and health care (De Ramón 2007). Consequently, urban policy and resolving the severe shortcomings of infrastructure and services became a high priority on the national agenda. Pursuing the ideal of a centralized, modern welfare state, the national government took control of a growing number of policy fields, including urban planning and road infrastructure (via foundation of the Ministry of Housing and Urbanism [MINVU] in 1965), as well as education and health services (Siavelis et al. 2002). Private business initiatives also emerged during this phase, e.g., in the transport sector (Estache and Gómez-Lobo 2005, p. 140), but were heavily regulated by state legislation.

The military coup by Pinochet and his neoliberal advisors in 1973 saw the beginning of dictatorship (*1973–1989*) and a radical change in the context of urban development and governance. The invisible hand of the market was now regarded as the 'ideal urban planner'. The chief guidelines were laid down in the National Policy for Urban Development in 1979 and in 1981, and stated that "urban land is not a scarce resource," "that markets are the best distributors of land for different uses," and that "land should be regulated through flexible measures that respond to market requirements" (MINVU 1981, cited after Sabatini 2000, p. 2). While the transportation sector was fully liberalized (Figueroa 1990, p. 26), the water and sanitation sectors initially continued to operate under a public scheme, passing through a process of profound modernization that ultimately led to their privatization in the early 1990s. Although Chile was governed in a highly centralized manner under the authoritarian rule of the Pinochet regime, the country was nevertheless divided into 13 regions, 51 provinces and 341 municipalities in 1974. This "paradoxical process of decentralization with centralized and authoritarian direction" (Siavelis et al. 2002, p. 279) was intended to enhance (rather than contain) the control of the central government over the whole country, according to the National Security Doctrine. Thus, the territorial reform was an instance of deconcentration rather than decentralization (Rodríguez Seeger 1995). In 1980, the military government redefined municipal boundaries and re-transferred the responsibility for primary health care and education to the local level. However, it did not vest local authorities with the financial resources to fulfil their new duties. Up until now, this has frequently led to municipalities acting in the interests of real estate investors, who, in turn, promised to deliver services and local infrastructure (Rodríguez Seeger and Ducci 2006).

Contemporary urban development and governance in Santiago (*1990 onwards*) are still heavily influenced by the neoliberal reforms of the 1970s (De Mattos 2004, p. 23). Following the return to democracy in 1990, the centre-left coalition (Concertación), which was to govern the country for the next 20 years, did not substantially alter the pivotal role given to the market, as the following examples show. At the peak of the regime transformation process in 1989 and 1990, the so-called second wave of privatization got under way, with decisive market-oriented reforms in the electricity and water sectors (Pflieger 2008); the franchising of highway concessions was introduced in the early 1990s (Gómez-Lobo and Hinojosa 2000); a little later, public-private partnerships were established as key components of downtown regeneration schemes and special business zones authorized (Parra and Dooner 2001); at the same time, urban planning focused more and more on large development projects (De Mattos et al. 2004) organized as public-private partnerships (Zunino 2006), or private residential developments and business parks (Borsdorf and Hidalgo 2008); in 1994, a privatization concession model was adopted in the water sector, for the most part reducing the role of the state to a regulatory function. Overall, the democratic governments of the last 20 years have endeavoured to overcome the authoritarian legacy of the 1973 coup d'état, while seeking to maintain the perceived achievements of liberalization and deregulation. Only in the urban transport sector did the government strive to reintroduce regulation to a system that had virtually been deregulated for the space of two decades.

As a result of past governance trends, Santiago now provides the opportunity for urban research and an encounter with the four governance matters introduced in the previous section. As the capital of what was traditionally a highly centralized state, Santiago sees decentralization as a 'big issue'. Moreover, the city suffers from its administrative fragmentation, as it consists of (up to) 37 independent municipalities (depending on where the line around the city region is drawn), each of which is tasked with local government and the delivery of a set of public services. The state tier above the municipal level, the Region Metropolitana, covers an area that extends far beyond the urban area, leaving the megacity itself without political representation. Furthermore, this authority functions primarily as a regional office of the national government, with its highest regional representative appointed by the president of state, and has little scope for independent decision-making. Hence, the regional state tier has almost no real power to coordinate local activity, resulting in a considerable lack of coordination – and frequently also of trust – among the various actors in the public sector. At the same time the influence of the national state level remains paramount: "Despite some notable efforts at decentralization, nationally oriented elites remain as deeply involved with Santiago as they did during military rule" (Siavelis et al. 2002, p. 293). Valenzuela (1999, p. 17) gives a succinct description of this situation, referring to Santiago de Chile as the 'victim' of a still over-centralized state. Accordingly, problems caused by the marked lack of genuine decentralization are tangible in various fields of urban policy. In the field of land-use planning, for instance, it is not uncommon to see local authorities bypassed by a coalition of real estate investors and national ministry representatives (e.g., Zunino 2006).

Since the neoliberal orientation of national policies has been maintained, Santiago is still a stronghold of privatization policies, where privatization of infrastructure and the corresponding services is promoted as a powerful remedy for the inefficiencies and flaws of earlier patterns of urban development. Today almost all public utilities are privatized (and most are privately owned) (Bauer 2002). Massive private urban developments have likewise been carried forward in the region, in particular on the city's remote fringes (Borsdorf and Hidalgo 2004; Cáceres and Sabatini 2004; Heinrichs et al. 2009). Parallel to this strong orientation towards market-based solutions, the coverage and overall standard of infrastructure provision and services and its efficiency have improved immensely in Santiago in recent years (cf. Valenzuela and Jouravlev 2007). The aforementioned privatization policy trade-offs, however, are also present in the Chilean capital.

On the whole, Santiago today is a good example of the poor public participation that characterizes numerous political systems throughout Latin America (Sabatini and Wormald 2004). The 'new system of participation' (Siavelis et al. 2002) declared by the 'Concertación' governments, which introduced a series of participatory mechanisms into the local political system, has not fulfilled the expectation of rendering democracy tangible in people's everyday lives. On the contrary, several studies have shown that these new institutions are frequently manipulated by those in power (Greaves 2004). While participation is a political buzzword in Chile today, in reality both central and local governments have done little to advance genuine citizen participation. "Although Concertación politicians fully recognize the need for participation, they have transformed it from its original militant and leftwing significance into an instrument for effective governance and efficient bureaucratic policy making" (Cleuren 2007). The prevalent mode of governance whereby national and local political elites make technical (rather than political) decisions aimed at effective service provision has been described as elitist-managerial and recently provoked the opposition of civil society actors (Rivera-Ottenberger 2007). As a direct response to the authoritarian implementation of urban modernization agendas, new citizen organizations – mostly defensive, NIMBY-like, middle-class-based initiatives – are emerging and mobilizing around vast infrastructure and private real estate projects with growing success (Ducci 2000; Poduje 2008). This illustrates how the absence of meaningful participation in the early stages of planning and decision-making bears the risk of blocking the (urban) political system and adding to political polarization.

In contrast to the three above-mentioned governance matters, the phenomenon of informality is of less significance in Santiago than in other Latin American megacities. While informal urban development was paramount in Santiago in the past, when urban migrants built what was then the mostly informal ring of working-class suburbs, today almost all poblaciones (i.e., informal settlements) are consolidated. Nonetheless, informal economic activity still carries weight in Santiago. The occasionally clandestine opposition of micro entrepreneurs and their associations in the transport sector to recent attempts to regulate public transport with the introduction of the Transantiago scheme is a compelling example; waste picker activity is another.

Overall, current governance structures and trends in Santiago seem to be unable to meet the challenges of mega-urban development fully, although it has been said that of all Latin American metropolises Santiago comes closest to the ideal of 'good metropolitan governance' (Ward 1996). Further exploration of the benefits and deficits of urban governance in the Chilean capital needs conceptual foundation; it requires conversion of the governance concept into a manageable design for empirical research. The next section will therefore introduce a heuristic framework with a more precise definition of governance, to be applied to the case of Santiago in an empirical analysis of governance issues.

5.4 Using the Governance Concept to Understand Urban Development Processes and Challenges

The attempts to conceptualize governance – beyond its general definition as decision-making on publicly relevant affairs – are numerous and in many instances remain somewhat vague. Almost all of these definitions point to the need in understanding how public affairs are settled and handled in reality to look not only at the public sector, but also at actors from the private sector and civil society (Senarclens 1998). In addition, the empirical analysis of governance issues should allow for some general principles, so as to safeguard the research orientation towards application and implementation:

- To give equal emphasis to the societal spheres of the state and its organizations, the market and civil society, including their characteristic mechanisms of coordination;
- To (at least try to) understand the interests, ideas and convictions behind the motivation of (individual and collective) actors, i.e., to comprehend diverging interests and incommensurable world views;
- To consider the at times delicate relationship between different actors from the same or different spheres (often with a long 'history');
- To take into account the complex set of laws, regulation and norms, i.e., institutions, that defines the corridor of potential activity;
- To understand how scientific results are adopted in practice, and

– To reflect on the broader cultural norms and values against which policy goals are set and success and effectiveness judged.

To typify and structure the basic elements of governance, Alain Motte (1996) proposes a simple categorical framework. He suggests identifying the (1) relevant actors who create a particular governance arrangement, studying the (2) relationships between these actors, and scrutinizing their cognitive referents – with the latter representing the socio-cultural context. In order to cover the complexity of governance processes in a city like Santiago, it seems useful to modify and expand this approach slightly by including two further categories (cf., Healey 1996, pp. 35f.): firstly, the (3) *institutional framework* category, which covers the rules of social action, including the actors' cognitive referents, as these mostly reflect common knowledge[1]; secondly, the (4) *decision-making process and outcome* category. The latter is important, since deducing the course and result of governance processes from the arrangement of actors, relationships and institutions is virtually impossible (Healey et al. 2002, p. 15).

Table 5.1 illustrates the heuristic conceptual framework proposed here for the empirical analysis of urban governance and gives an overview of elements to be considered in an application-oriented study of urban development processes and problems (Nuissl and Heinrichs 2011). It should be remarked that this attempt to 'translate' the governance concept into a conceptual device for empirical analysis is designed to ease the gathering and organizing of information and data on a specific case of urban governance (rather than to make a contribution to the theoretical discourse).

The *actors* category basically refers to the simple but crucial question of which actors take part in a particular governance process. These actors – individual or collective – come from all spheres of society and can therefore be characterized as representatives of the state, the private sector, or civil society (1). Once the relevant actors have been identified, it is essential to determine their (often diverging) interests (2), as well as their power resources, financial means, social capital and knowledge (3). While the analytical focus should not be confined to public actors alone, the competences and duties of the authorities, and their resources in terms of budget and personnel nevertheless deserve particular attention, since it is they who are generally responsible for public policies (Burki et al. 1999).

---

[1] As part of the social world, institutions, particularly informal institutions, are typically the result of the joint effort of social actors to create meaning that will give them orientation in a complex world. Dietrich Fürst applies this basic argument of the interpretative strand of social theory to the issue of governance, stating that governance "*is constituted by communication, i.e., primarily by cognitive processes. This means that the relevant actors create common assumptions as to requirements, conditions and potential courses of action*" (Fürst 2001, p. 374; translation by authors). In a similar vein, the way in which stakeholder interests are formulated or public engagement in governance processes takes place can be seen as an expression of prevailing cultural patterns (Douglas 1987).

Table 5.1 Heuristic framework for the analysis of arrangements and processes of governance

Category	Criterion	Examples/important aspect
Actors	Public authorities Private sector Civil society	Authorities from different administrative levels (with specific duties/powers): multinational enterprises, small and medium-sized businesses; citizens associations, NGOs, grassroots organizations
	Interests	Vested interests, programmatic goals of organizations, interest of private sector actors in economic benefits, interest of public sector actors in maintaining their role/authority
	Resources	Financial means, decision-making power, social capital (networks), know-how etc
Relationships	Lasting relationship patterns between actors	Interest coalitions between actors from different spheres of society (e.g., local "growth coalitions"), public-private partnerships, clientelistic networks, 'traditional' opposition between organizations or individuals, kinship
Institutional framework	Formal Institutions	Laws, directives, ordinances, constitutional mechanisms of political and/or legal control, special (temporary) policy instruments (e.g., aid programmes)
	Informal Institutions ('unwritten rules')	Norms, values, interpretative patterns, local or milieu specific knowledge
Decision-making process and its outcome	Balancing of interests and resolution of conflicts	Use (and abuse) of institutional power, realization of opportunities for participation
	Coordination between public authorities	Horizontal coordination (between municipalities or between sectoral policies), vertical coordination (between different administrative tiers), coordination of state action by a superior board or agency
	'Information and knowledge policies'	Provision of information and knowledge for ...Actors who are not directly involved (e.g., public relations) ...For the actors involved (e.g., scientific policy advice, capacity building)
	Policy effectiveness and efficiency	Costs, acceptance, knowledge absorption and long-term perspectives in the face of new policies, implementation deficits

Source: The authors, based on Nuissl and Heinrichs 2006

When it comes to understanding the behaviour of actors involved in governance processes and the outcomes of political bargaining, the *relationships* category comes into play and concerns the extent to which converging or conflicting interests between different actors are translated into cooperation or resentment. As regards 'positive' relationships, it also pertains to the way in which these relationships are organized: hierarchical/vertical (e.g., between different tiers of the administrative system), network-like/horizontal (e.g., between local and regional NGOs that belong to the same umbrella organization), or market-based but supported by contracts or oral agreements (e.g., between partners of a public-private partnership).

The category of *institutional framework* refers to the prevailing rules of conduct in a particular context that define people's expectations as to the behaviour of their fellow men. These rules influence, and sometimes determine, the decisions of the actors concerned and hence the course of action. Drawing on institutional theory, formal and informal rules of behaviour, i.e., institutions, can be distinguished. (1) Formal institutions are codified and include laws, regulations, ordinances, plans and other constitutional mechanisms of political and/or legal control. (2) Informal institutions are not the product of legislation but come into existence when members of society are aware of them and use them to judge the performance of others (as is the case with norms, values, or unwritten laws concerning, e.g., bribery). As a rule it is impossible to understand governance processes and their outcomes without considering informal institutions (Cars et al. 2002), whose importance magnifies in realms where the state is weak.

The *decision-making process and its outcome* category reflects the procedural and dynamic character of governance. It entails four aspects that seem particularly critical: (1) The extent to which different actors succeed in balancing their interests and arrive at feasible compromises is probably the most vital aspect of public decision-making. The question of whether hypothetical potential for participation (opened up by, e.g., planning laws) is taken up in reality is of particular relevance here. (2) The interaction mechanisms with which the authorities attempt to coordinate their activities, resolve conflicts, or co-operate on certain issues are another key aspect. Whether or not the different tiers of state administration are in a position to adapt their respective policies and plans to each other, for instance, depends on the feasibility of such mechanisms. (3) How information and data are dealt with is likewise crucial. Furthermore, it is germane to the course of governance whether there is a flow of information and knowledge from the 'outside' to the 'inside', i.e., whether those with political powers acknowledge the viewpoints of those affected by public policies. (4) After all, it is essential to reflect on the long-term effectiveness and efficiency of the decisions found and the solutions implemented in a particular governance process by judging them against the background of the problems to be tackled at the outset.

5.5 The Making of Megacity Governance: Insights from Different Urban Development Fields in Santiago

The heuristic framework outlined in the previous section will be applied in the following to four separate domains of urban development in Santiago. The aim is to illustrate that a sound understanding of how urban governance works in a particular megacity is indispensable to both (1) implementing solutions and strategies in distinct fields of urban development and (2) identifying risks that might be inherent in the corresponding policies.

The following brief outline of empirical findings from a range of policy fields is structured in three parts. Firstly, the policy fields under scrutiny are introduced: public transport, water supply and sewerage, land-use policies and planning, and waste management. Secondly, the heuristic framework is employed to identify some striking governance features in each policy field. And thirdly, a short conclusion will be drawn on what appear to be major leverages to improve the current practice of public decision-making. The arguments presented in this section are based on stakeholder and expert interviews, literature reviews and the evaluation of existing data. In addition, we gathered information by consulting fellow research groups within the Risk Habitat Megacity research initiative.

5.5.1 Public Transport and Decentralization

Up to the mid-1970s, a state-owned company and various private enterprises provided public transport services in Santiago. In the wake of the 1973 coup d'état the transport sector was deregulated step by step until total liberalization of the transport market was achieved. With the return to democracy in the 1990s, the national state cautiously began to re-regulate the market by introducing several requirements to be met by new public transport providers. However, private companies continued to run the bus service. The year 2000 saw the presentation of *Transantiago*, a vast scheme designed to reform Santiago's public transport system, which has since then dominated public discourse on transport issues and occasionally urban development in general. This attempt to improve the efficiency of the public transport system by increased regulation has caused huge problems in the first instance, eliciting harsh criticism both from academics and public opinion. It has been widely acknowledged that an overcentralized planning and implementation process, i.e., lack of responsibility with local authorities and the neglect of local demand and mobility patterns (Quijada et al. 2007), was (and is) a major reason for the shortcomings of *Transantiago*.

Applying the governance analysis categories introduced in the previous section to the case of Transantiago reveals that almost no actors from the regional or local levels are involved. While Transantiago is largely the product of the national Ministry of Transport (MITT) and its regional section (SEREMITT, which implements national transport strategies in the metropolitan region), there has been no delegation of public responsibilities to either local or regional authorities, or to semi-state bodies with metropolitan-wide powers. Accordingly, there has been no attempt to make systematic use of local know-how either. It is worth noting that from the outset the Transantiago scheme suffered from the reluctance of private bus operators to submit to the project, which in their view infringed on their personal economic interests. The main interest of the national transport ministry, besides swift implementation of the scheme, was the creation of an efficient market for transport services, thereby awarding the market a pivotal role. Accordingly, public-private partnerships between national authorities and private bus companies became the dominant relationship pattern in the field of public transport. The most important aspect of the institutional framework is the formal rules that vested the national government with substantial discretionary power to mould Santiago's public transport system. They also enabled the national transport ministry to draw up the Urban Public Transport Plan for Santiago (PTUS), the key element of which is the Transantiago system. Collaboration between public and private partners was mainly defined through published documents of the tender process and contracts drawn up on the basis of these documents. The most influential informal institution 'behind' Transantiago was the guiding star of (technical) rationality, i.e., the firm conviction that it was possible to design complex systems on the basis of scientific knowledge, i.e., transport models (cf. Silva 2008). The decision-making process that led to Transantiago was highly efficient in that state and ministry officials actually succeeded in developing this complex scheme. Nonetheless, the huge problems that surfaced in the process of implementation were primarily addressed by a 'muddling through' approach, behind which no consistent strategy was apparent. Transantiago is moreover characterized by its scant attention to social aspects and its lack of adequate information for public transport users.

On the whole, Santiago's transport sector epitomizes most infrastructure projects in megacities, with their inclination to be technocratic and far too big. In the case of Transantiago, the planning and implementation process was typified by a strong focus on efficiency considerations and the absence of a regulatory body in charge of metropolitan issues. This scenario carries the risk that disappointed public transport users will change over to private transport, a move that in turn leads to a decline in the use of public transport. The more centralized the infrastructure policy, the greater the subsequent risks. A decentralized infrastructure policy would require a general appreciation of participatory practices and the local embeddedness of policies, including the involvement of regional and local actors.

5.5.2 Water Supply, Sewerage and Privatization

The privatization of water and sanitation services in Santiago was completed in 1994, notably after the return to democracy. At the same time, the 'classic' challenge of urban infrastructure supply – to connect the entire population to water supply networks and sanitation systems – has largely been solved (although the city's poorest population still sometimes lacks on-site access to drinking water). At first sight this seems an overwhelming achievement of privatization policies. There are, however, some troublesome issues associated with the water sector (Valenzuela and Jouravlev 2007). In particular there has been a sustained increase in water fees. Drinking water is now about four times as expensive as it was at the beginning of the privatization process and is today an expensive commodity for the poorer sections of the population. Accordingly, water consumption rates in some neighbourhoods have dropped considerably. This 'consumption breach' (Rodriguez 1994) has become particularly visible in recent years. For 2007, the seven major concessionaires for water supply in the Metropolitan Area of Santiago reported huge differences in water consumption figures. While the average water consumption in high-income municipalities and new housing developments amounted to 120 m^3 per customer/month, the respective figure in some of the poorer municipalities was around 20 m^3 per customer/month. Hence, judged according to the equity criterion for efficiency and service quality, the model of privatized water supply and management seems to have failed. With regard to waste water treatment, there is still considerable scope for improvement. Although there has been substantial investment in sewage plants recently, much of the waste water is discharged into the receiving waters without pre-treatment.

One of the chief actors to push privatization of public services forward was the former military government. In 1977, the National Service of Water Works (SENDOS) was introduced as a first step towards a far-reaching withdrawal of the state from the water sector. Santiago followed suit with the creation of the Metropolitan Company of Waste Water Works (EMOS) which, although it continued to operate under a public scheme, gradually incorporated private sector business practices. Other – municipal and private – water service companies emerged in Greater Santiago, albeit their share in the water market was negligible. Shortly before the end of the dictatorship in 1989, the Superintendency of Sanitary Services (SISS) was introduced as the next major step towards liberalization of the water sector, ushering in the complete separation of regulatory and operative functions on the part of state entities. With the exception of a few municipal companies, service providers in the field of water provision and sanitation subsequently became private companies, with Grupo Aguas, the EMOS successor, holding almost 90% of the region's water and waste water markets. The relationship between the actors that have emerged in the field of water services and sanitation within the last 20 years is characterized in the main by smooth collaboration, as all of them are the product of the same privatization policy. In 2008, the state partly redefined its role and is now striving for tighter regulation of the water sector, especially in the context of waste water disposal and treatment. As yet, however, this policy rethinking has neither led to lower water prices nor to a fairer distribution of this precious good. With respect to the formal institutional framework that regulates water supply and sanitation, the decisions and regulations that have got under way since the 1980s were all clearly dedicated to the goal of complete privatization of the sector. Although privatization in other infrastructure sectors, e.g., electricity, was to proceed even faster, the water sector served almost as a model, characterized as it was by early efforts to organize services (then still public providers) according to market principles. In addition, privatization policies in the water sector were rarely challenged by informal institutions. In recent decades, decision-making processes related to Santiago's water sector focused on several important issues, e.g., deregulation, the attraction of foreign capital, and the design of the tariff system. The reorganization of the public water sector was accompanied by investments from multilateral banks (in particular, the World Bank), which conditioned their contributions to the process of liberalization. Efforts to attract investments in the water sector were intensified in the 1990s, when EMOS realized that its financial base was too weak to cover the monumental cost of improving the water system in environmental terms. At that time, less than 20% of sewage was treated (Valenzuela and Jouravlev 2007, p. 12).

All in all, the water sector in Santiago is an excellent example of the successful step by step enactment of privatization policies. It seems remarkable that the privatization of water services has led to an almost 100% share of people in Santiago with access to clean drinking water, contributing significantly to the mitigation of health risks to the disadvantaged population. Moreover, privatization has attracted direct investment to the sector and increased the quality of irrigation water substantially. On the other hand, the water sector clearly shows the pitfalls of this strategy, as the figures for water use and water distribution point to a sharp increase of inequity – which in the medium run carries the risk of denying the poor access to basic water services.

5.5.3 Land-Use Policy, Planning and Participation

Land-use policy and planning in Chile have undergone several profound conceptual revisions, from the state interventionist modes in the 1960s and 1970s, via the radical liberalization under the military regime in the 1980s, to the reappearance of the urban growth boundary in the Regional Land-Use Plan for Santiago (*PRMS*) under democratic rule in 1994. The most recent major shift occurred in 1997, when urban containment policies were once again abandoned, this time to create a polycentric city region of 'world class' through private investments in peri-urban megaprojects. Consequently, several multi-billion dollar projects in the form of mammoth business parks and vast residential estates were built on the urban fringes or are in the final planning stages (Borsdorf et al. 2007). These projects, some of which are designed to host more than 100,000 inhabitants, have profound impacts on their locations, i.e., mostly poor municipalities. Due to the mercantile planning style and the lack of any meaningful participation mechanisms for affected citizens, however, local issues are seldom taken into account. Hence these new megaprojects are an instructive example of current land-use planning patterns in Santiago and the risk they harbour of socio-spatial and political marginalization.

A glimpse at the actors involved in urban fringe developments of megaproject dimensions reveals the dominance of the private sector. Private land developers and specialized planning consultants initiate, plan and carry out the individual projects, entering into negotiations with the public sector over the necessary construction permissions. Permission must be obtained from the municipality concerned and the regional government, which subsequently modifies the PRMS. In practice, however, regional and local authorities find themselves confronted with powerful interest coalitions between members of the private sector and the respective national ministries, especially the Ministry of Housing and Urbanism (MINVU) and the Ministry of Public Works (MOP). Some of the most influential land developers and consultants are in fact ex-politicians with close relationships to public decision-makers at central state level. It was an informal coalition of this kind that pushed ahead the comprehensive modification of the institutional framework for land-use planning necessary for the realization of megaprojects on the urban fringes of Santiago. Prior to this institutional reform, the PRMS had circumscribed urban development to the confines of the urban area. New land-use categories were introduced to the PRMS, i.e., 'Conditioned Development Zones' (ZODUC) in 1997 and 'Conditioned Development Projects' (PDUC) in 2003. These planning instruments make urbanization possible in the entire metropolitan region, provided it meets specific criteria (defined for ZODUCs or PDUCs) regarding size, social mix and the mitigation of negative externalities. The basic principle of these new planning instruments has been called 'planning by conditioning'; it entails a shift from public planning to public-private negotiations of urban development activity. The actual negotiations and decision-making processes both on new planning instruments and the subsequent approval of individual projects were far from transparent. Almost no information was available to the wider public or even in some cases to the local authorities in question. As a rule, however, the affected communities do not reject the megaprojects per se but condemn the style of decision-making associated with them. Similarly, on the regional scale, citizens organizations demand a voice in the planning of the metropolitan region. They perceive the new megaprojects as elitist consumer spaces, triggering urban sprawl and socio-spatial fragmentation rather than providing self-sufficient and sustainable new towns.

To sum up, sub- and ex-urban megaprojects realize, on the one hand, the ideals of new public management and 'the lean state', where the private sector takes organizational and financial responsibility for urban development. On the other hand, the largely private character of this type of 'city building' virtually strips the development process of any form of public voice and thus profoundly contradicts governmental agendas of deepening participation and decentralization. This top-down, privatized and exclusionary style of planning and decision-making bears the risk that projects miss out on the consideration of local social and environmental conditions and the human scale of urban development. This again gives rise to severe problems of legitimacy and increases the potential for social conflict. All in all, it seems necessary to make planning processes more participatory, more transparent and, above all, more accountable.

5.5.4 Waste Management and Informality

Only around 13% of the total amount of waste in Santiago currently undergoes recycling treatment, far less than is satisfactory in terms of sustainability. More than 90% of the waste that is reused is recovered by waste pickers. Against this background the question arises as to whether informal activities that boost the recycling rate could be increased by integrating the informal waste sector more efficiently into the official urban waste management system.

Looking at the actors who are important for the recycling of waste we first of all can identify the waste pickers who earn their daily living by searching refuse (cartoneros) and litter (cachureros) for valuable materials. At the same time, there are the municipal or private enterprises in the formal sector that specialize in recycling materials. In between these two actor groups are middlemen who buy what the waste pickers have collected and sell it in large lots to the enterprises. This circumstance determines the relationships between the actors: waste pickers are primarily dependent on middlemen and their payments, and usually have no direct access to the users of the goods they deliver. Moreover, the waste pickers themselves rarely coordinate their individual activities and only in a couple of municipalities are they organized in waste-picker associations. Hence the middlemen have a monopoly on organizing the interface between the informal and the formal sector, making them the major beneficiaries of the current system. The institutional framework further impedes the exchange of goods at this interface. Since no official target for the recycling rate has been defined, there is little public acknowledgement of the waste pickers' contribution to the sustainability of the urban system. In addition, working conditions for waste pickers carry sizeable individual risks in terms of health and social security. Likewise, waste pickers are typically excluded from decision-making processes dealing with municipal waste management. Instead, policies in the field of waste management are for the most part committed to the 'high-tech' waste treatment model that provides feasible solutions for the industrialized world but leaves no scope for inclusion of the informal sector.

To conclude, current waste management governance in Santiago creates a situation in which the potential of waste pickers to recover valuable waste is underutilized. The waste pickers themselves are exposed to serious risks. Bringing about change would require improving the working and living conditions of those active in the informal waste sector, increasing their potential to recover material and facilitating the exchange of raw material across the edge between the informal and formal waste sectors. A range of instruments could be employed to achieve these ends. These include the definition of minimum standards for the recycling of raw material; the promotion of the informal waste sector's contribution to sustainable urbanism; the provision of devices for labour protection (gloves, glasses) or training courses for waste pickers; on the part of the informal waste sector, the pooling of individuals in cooperative associations to coordinate collection activities and enforce joint interests. Santiago's informal waste sector is thus an excellent example of the ambiguity of informal phenomena, typically combining serious risks with sizeable opportunities, not least in terms of urban sustainability.

5.6 Conclusion

This chapter set out with the argument that governance can be both a remedy and a source of risk. The analysis of governance issues can therefore provide a significant contribution to the refinement of urban policies committed to sustainable development and risk mitigation. The discussion of the methodology of governance analysis and current governance issues in Santiago has shown that decision-making on publicly relevant urban affairs is a complex process. In particular, it became clear that 'optimal' sustainable management of a megacity is an idealistic rather than a realistic task. This is evident, for instance, in current governance patterns in the fields of water supply and services or public transport, which mirror the frictions and shortcomings related simultaneously to decentralization, participation and privatization. Likewise, the examples discussed have illustrated how a particular governance strategy, while rendering urban development more sustainable, may also bear grave risks.

Regarding the empirical study of governance issues, their complex structure clearly calls for an analytical approach that can disentangle the interlinked factors shaping an individual governance problem. Such an approach needs, first of all, to allow for a variety of actors and should not focus on public sector activities alone (lest it become an urban management rather than an urban governance approach).

Having gained some insights into how urban governance works in Santiago de Chile the question arises as to whether and, if so, how these insights can be used to refine urban development strategies. Generally, the respective findings obtained from the fields of transportation, land-use management, water services and waste management hint at some major downsides in the structure of public decision-making in Santiago, all of which should be addressed by an urban development strategy. These are in particular the weak position of the municipalities and the (regional) metropolitan government, on the one hand, and the sometimes overly strong interest coalitions between public and private partners that largely impede an effective form of public participation, on the other (see Chap. 15 for a detailed discussion of these issues). It has also become clear in this chapter that the analysis of governance issues rarely results in a recipe for working the urban system, let alone a universal recommendation on how to design urban governance. Recalling the examples of transportation and land-use policies: it is almost impossible to define a specific set of policy instruments that will safeguard the provision of public transport for all, and even less likely, its use by all; nor is it possible to define the 'optimal' land-use policy, since land use is always a political issue. The analysis of governance issues, nevertheless, provides insights into how the 'decision-making part' of the system can be modified; it gives an indication of how to mould and reshape institutional framework conditions and the principles of decision-making in the public realm. Finally it should be acknowledged that tackling the challenges of megacity development is not merely a question of appropriate governance, but calls for in-depth knowledge of the material, social and environmental problems on the ground, the subject of the following chapter in this book.

References

Avritzer, L. (2002). *Democracy and the public space in Latin America*. New Jersey: Princeton University Press.
Bauer, C. J. (2002). *Contra la corriente: Privatización, mercados de agua y el Estado en Chile. Ecología y medio ambiente*. Santiago: LOM Ediciones.

Borsdorf, A., & Hidalgo, R. (2004). Formas tempranas de exclusión residencial y el modelo de la ciudad cerrada en América Latina. El caso de Santiago. *Revista de Geografia Norte Grande, 32*, 21–37.

Borsdorf, A., & Hidalgo, R. (2008). New dimensions of social exclusion in Latin America: From gated communities to gated cities. The case of Santiago de Chile. *Land Use Policy, 25*(2), 153–160.

Borsdorf, A., Hidalgo, R., & Sanchez, R. (2007). A new model of urban development in Latin America: The gated communities and fenced cities in the Metropolitan areas of Santiago de Chile and Valparaíso. *Cities, 24*(5), 365–378.

Burki, S. J., Perry, G., & Dillinger, W. (1999). *Decentralizing the state: Beyond the centre. World Bank Latin American and Caribbean studies*. Washington, DC: World Bank.

Cáceres, G., & Sabatini, F. (Eds.). (2004). *Barrios cerrados en Santiago de Chile: Entre la exclusión y la integración social*. Santiago de Chile: Pontificia Universidad Católica de Chile.

Cars, G., Healey, P., Madanipour, A., & De Magelhaes, C. (Eds.). (2002). *Urban governance, institutional capacity and social milieux*. Aldershot: Ashgate.

Cleuren, H. (2007). Local democracy and participation in post-authoritarian Chile. *European Review of Latin American and Caribbean Studies, 83*, 3–18.

De Mattos, C. (2004). Santiago de Chile: Metamorphosis bajo un nuevo impulso de modernización capitalista. In C. De Mattos, M. E. Ducci, A. Rodríguez, & G. Yánez (Eds.), *Santiago en la globalización: ¿una nueva ciudad?* (pp. 17–46). Santiago: SUR-EURE libros.

De Mattos, C., Ducci, M. E., Rodríguez, A., & Yáñez, G. (Eds.). (2004). *Santiago en la Globalización: ¿Una nueva Ciudad?* Santiago: SUR-EURE libros.

De Mattos, C., Figueroa, O., Orellana, A., & Yánez, G. (Eds.). (2005). *Gobernanza, Competitividad y Redes: La Gestión en las Ciudades del Siglo XXI*. Santiago: EURE libros.

De Ramón, A. (2007). *Santiago de Chile. Historia de una sociedad urbana*. Santiago: Catalonia.

De Soto, H., Ghersi, E., & Ghibellini, M. (1986). *El otro sendero: La revolución informal*. Lima: Editorial El Barranco.

Douglas, M. (1987). *How institutions think*. London: Routledge & Kegan Paul.

Ducci, E. (2000). *Governance, urban environment, and the growing role of civil society*. Washington, DC: Woodrow Wilson International Center for Scholars.

Estache, A., & Gómez-Lobo, A. (2005). Limits to competition in urban bus services in developing countries. *Transport Reviews, 25*(2), 139–158.

Figueroa, O. (1990). La desregulación del transporte colectivo en Santiago: Balance de diez años. *Revista Eure, 16*(49), 23–32.

Finot, I. (2002). Descentralización y participación en América Latina: Una mirada desde la economía. *Revista de la CEPAL, 78*, 139–149.

Fürst, D. (2001). Regional governance – ein neues Paradigma der Regionalwissenschaften? *Raumforschung und Raumordnung, 59*(5–6), 370–379 (Regional governance – a new paradigm for regional science?).

Ghersi, E. (1997). The informal economy in Latin America. *Cato Journal, 17*(1), 99–108.

Gómez-Lobo, A., & Hinojosa, S. (2000). *Broad roads in a thin country. Infrastructure concessions in Chile* (World Bank Policy Research Working Paper No. 2279).

Graham, S., & Marvin, S. (2001). *Splintering urbanism. Networked infrastructures, technological mobilities and the urban condition*. London/New York: Routledge.

Greaves, E. (2004). Municipality and community in Chile: Building imagined civic communities and its impact on the political. *Politics & Society, 32*(2), 203–230.

Healey, P. (1996). An institutionalist approach to spatial planning. In P. Healey, A. Khakee, A. Motte, & B. Needham (Eds.), *Making strategic spatial plans: Innovation in Europe* (pp. 21–36). London: UCL Press.

Healey, P., Cars, G., Madanipour, A., & De Magelhaes, C. (2002). Transforming governance, insitutionalist analysis and institutional capacity. In G. Cars, P. Healey, A. Madanipour, & C. De Magelhaes (Eds.), *Urban governance, institutional capacity and social milieux* (pp. 6–28). Aldershot: Ashgate.

Heinrichs, D., Nuissl, H., & Rodriguez Seeger, C. (2009). Urban sprawl and new challenges for (metropolitan) governance in Latin America – The case of Santiago de Chile. *Revista EURE, 35*(104), 29–46.

Litvack, J., Ahmad, J., & Bird, R. (1998). *Rethinking decentralisation in developing countries*. Washington, DC: The World Bank.

Motte, A. (1996). The institutional relations of plan-making. In P. Healey, A. Khakee, A. Motte, & B. Needham (Eds.), *Making strategic spatial plans: Innovation in Europe* (pp. 231–254). London: UCL Press.

Nuissl, H., & Heinrichs, D. (2006). Zwischen Paradigma und heißer Luft: Der Begriff der Governance als Anregung für die räumliche Planung. In U. Altrock, S. Güntner, S. Huning, T. Kuder & H. Nuissl (Eds.), *Sparsamer Staat, schwacher Staat?* Berlin: Verlag Uwe Altrock (Planungsrundschau 13), 51–72

Nuissl, H., & Heinrichs, D. (2011). Fresh wind or hot air – does the governance discourse have something to offer to spatial planning? *Journal of Planning Education and Research, 31*(1), 47–59.

Parra, C., & Dooner, C. (2001). *Nuevas experiencias de concertación público-privada: las corporaciones para el desarrollo local* (Serie Medio Ambiente y Desarrollo, Vol. 42). Santiago de Chile: CEPAL.

Pflieger, G. (2008). Historia de la universalización del acceso al agua y alcantarillado en Santiago de Chile (1970–1995). *Revista EURE, 34*(103), 131–152.

Poduje, I. (2008). *Participación ciudadana en proyectos de infraestructura y planes reguladores* (Serie Temas de la Agenda Pública, Vol. 22). Santiago de Chile: Pontificia Universidad Católica de Chile.

Quijada, R., Tirachini, A., Henríquez, R., & Hurtubia, R. (2007). Investigación al Transantiago: Sistematización de Declaraciones hechas ante la Comisión Investigadora, Resumen de Contenidos de los Principales Informes Técnicos, Información de Documentos Públicos Adicionales y Comentarios Críticos. http://www.ciperchile.cl/wp-content/uploads/Reporte_Transantiago.pdf. Accessed 14 Sept 2010.

Rivera-Ottenberger, A. (2007). Decentralization and local democracy in Chile: Two active communities and two models of local governance. In V. Beard, F. Miraftab, & C. Silver (Eds.), *Planning and decentralization: Contested space for public action in the global south* (pp. 119–134). New York: Routledge.

Roberts, B., & Portes, A. (2006). Coping with the free market city: Collective action in six Latin American cites at the end of the twentieth century. *Latin American Research Review, 41*(2), 57–83.

Rodríguez, A. (1994). Cuatro historias de servicios urbanos en América Latina y una explicación. In A. Puncel Chronet (Ed.), *Las ciudades de América Latina: Problemas y oportunidades* (pp. 25–34). Col lecció oberta. València: Universidad de València.

Rodríguez Seeger, C. (1995). *Raumentwicklung und Dezentralisierungspolitik in Chile (1964–1994)*. Kieler Arbeitspapiere zur Landeskunde und Raumordnung, Kiel: Christian-Albrechts-Universität.

Rodríguez Seeger, C., & Ducci, M. E. (2006). La descentralización en Chile: El peso de las tradiciones centralistas y autoritarias. Historias de la descentralización: Transformación del régimen político y cambio en el modelo de desarrollo América Latina, Europa y EUA (pp. 571–614). Bogotá: Edición Universidad Nacional de Colombia

Sabatini, F. (2000). Reforma de los mercados de suelo en Santiago, Chile: Efectos sobre los precios de la tierra y la segregación residencial. *Revista EURE, 26*(77), 49–80.

Sabatini, F., & Wormald, G. (2004). La Guerra de la Basura de Santiago: Desde el Derecho a la Vivienda a Derecho a la Ciudad. *Revista EURE, 30*(91), 67–86.

Sabatini, F., Sepúlveda, C., & Blanco, H. (Eds.). (2000). *Participación Ciudadana para Enfrentar los Conflictos Ambientales*. Santiago: Publicaciones Cipma.

Schneider, F., & Enste, D. (2002). *Hiding in the shadows: The growth of the underground economy*. Washington, DC: International Monetary Fund (Economic Issues No. 30).

Senarclens, P. (1998). Governance and the crisis in the international mechanisms of regulation. *International Science Journal, 55*(155), 91–104.
Sharpe, L. J. (Ed.). (1995). *The government of world cities: The future of the metro model.* Chichester: Wiley.
Siavelis, P. M., Valenzuela Van Treek, E., & Martelli, G. (2002). Santiago: Municipal decentralization in a centralized political system. In D. Myers & H. Dietz (Eds.), *Capital city politics in Latin America: Democratization and empowerment.* London: Lynne Rienner Publishers.
Silva, P. (2008). *In the name of reason: Technocrats and politics in Chile.* Pennsylvania: Pennsylvania State University Press.
Stren, R. (2000). New approaches to urban governance in Latin America. Paper presented CIID Conference, Montevideo-Uruguay, 6–7 April 2000. http://www.idrc.ca/en/ev-22827-201-1-DO_TOPIC.html. Accessed 14 Sept 2010.
UNCHS – United Nations Centre for Human Settlements. (2001). *Compendium of human settlements statistics.* New York: United Nations.
Valenzuela, E. (1999). *Alegato historico regionalista. Colección de estudios sociales.* Santiago: Ediciones SUR.
Valenzuela, S., & Jouravlev, A. (2007). *Servicios urbanos de agua potable y alcantarillado en Chile: Factores determinantes del desempeño. Serie Recursos naturales e infraestructura.* Santiago de Chile: Cepal.
Ward, P. (1996). Contemporary issues in the government and administration of Latin American mega-cities. In A. Gilbert (Ed.), *The mega-city in Latin-America* (pp. 53–72). Tokyo/New York/Paris: United Nations University Press.
Zunino, H. (2006). Power relations in urban decision-making: Neo-liberalism',Techno-politicians' and authoritarian redevelopment in Santiago, Chile. *Urban Studies, 43*(10), 1825–1846.

Part III
Exploring Policy Fields

Chapter 6
Earthquake Risks: Hazard Assessment of the City of Santiago de Chile

Marco Pilz, Stefano Parolai, Joachim Zschau, Adriana Perez, and Jaime Campos

Abstract Santiago de Chile is located at the top of a deep sedimentary basin close to the active tectonic San Ramón Fault. In the case of rupture, this fault at the eastern edge of the city can generate earthquakes with a magnitude of up to 7.1. As seen in past earthquakes, the soil characteristics within the Santiago basin change the level of seismic hazard, since they heavily modify the level of ground shaking over short distances. This chapter takes a closer look at the relationship between these parameters and their influence on local site conditions. The methodology it presents to provide a rough estimate of the seismic hazard combines a high resolution map of the fundamental resonance frequency of the soil and a 3D shear wave velocity model for the northern part of the city. By comparing the results with mapped intensities of recent events, the chapter estimates the areas of the city that are more endangered and recommends a more thorough investigation for these parts of the basin. Although the findings cannot be generalized for all possible earthquakes affecting the city, it concludes with some practical recommendations such as retrofitting the existing building stock in areas particularly under threat of seismic hazard.

Keywords Earthquakes • Fundamental resonance frequency • Microzonation • Soil parameters • Velocity model

6.1 Introduction: Prevalence of Earthquakes in Chile

Chile is one of the most seismically active areas in the world, as demonstrated recently by the Maule earthquake on 27 February 2010. The entire country spans along the boundary between two tectonic plates, the Nazca plate in the eastern

---

M. Pilz (✉) • S. Parolai • J. Zschau • A. Perez • J. Campos
Department 2 Physics of the Earth, Helmholtz Centre Potsdam, GFZ German Research Centre for Geosciences, Telegrafenberg, 14473 Potsdam, Germany
e-mail: pilz@gfz-potsdam.de

Pacific Ocean off the west coast of the continent and the South American tectonic plate that embraces both the continent itself as well as a sizeable region of the Atlantic Ocean. Along the country, both plates converge at a rate of several tens of millimetres a year; this leads to a subduction of the western Nacza plate below the eastern South American plate. As a result of the steady movement of the plates, the Chilean coast has suffered numerous thrust earthquakes along the plate boundary. In the last 100 years alone, 16 events gave rise to significant human and economic loss, one of which was the strongest earthquake ever measured, i.e., the 1960 Valdivia earthquake that was rated with a magnitude of 9.5. The boundary in the northern part of Chile ruptured in 2007, and in 2010 in the southern part, generating earthquakes of high magnitudes. The spatial and time distribution of these earthquakes has been explored in detail (e.g., Kelleher 1972; Barrientos 1981; Nishenko 1985; Ramírez 1988), so the long-term seismic hazard has already been assessed. Return periods for magnitude $M \sim 8$ events are in the order of 80 to 100 years for any given region in Chile (Barrientos et al. 2004). This, therefore, sees the city of Santiago, the country's capital with more than six million inhabitants, also confronted with a high degree of seismic hazard. Combined with dense population and a high concentration of industrial facilities, this translates to a high seismic risk for the entire country.

Moreover, the city of Santiago de Chile is located at the top of a narrow basin between the Andes and the coastal mountains. The basin originated from the depression caused by tectonic movements in the Tertiary, a geologic period 65 million to 1.8 million years ago; it is filled with thick layers of soft sediments that have a strong effect on seismic ground motion over short spatial scales. Given the recent growth of the city and the concomitant spread of the urban population into areas of unfavourable soft soils, earthquakes may cause extensive human and economic damage in the future, since the geometrical and mechanical features of alluvial deposits impact heavily on seismic wave propagation and amplification. A notable example thereof in recent years is the comparatively modest (magnitude 6.6) earthquake that struck Central Mexico on 19 September 1985. It caused only slight damage in the epicentral area but triggered the collapse of more than 400 buildings and damaged many more in Mexico City, which is situated at the top of a dried-out lake 240 km away from the epicentre. Santiago de Chile also experienced local variations in ground motion (albeit on a smaller scale) during the 1985 Valparaiso event, showing clear differences in ground motion intensity within the city (Çelebi 1987; Bravo 1992).

Unfortunately, it is not possible at the moment to 'predict' earthquakes. Scientists are only in a position to estimate the likelihood, or probability, of earthquake occurrence, but unable to forecast that an earthquake will or will not occur at a given time. Hence, particular attention should be paid to mitigating the destructive impact of the earthquake, i.e., to raise awareness of the geological characteristics of the region that might influence the propagation of seismic waves and seismic energy (e.g., underlying rocks, soil texture).

This chapter focuses on a microzonation study carried out in the north of the city. To this end, the geological conditions in the Santiago basin will first be described.

It will be shown that microtremor measurements can determine the fundamental resonance frequency of the soil and, moreover, that local shear wave velocity profiles can be derived. A combination of these parameters allows for a rough estimate of areas in the city specifically endangered and prone to site amplification, a more thorough investigation of which is highly recommended.

6.2 Methodological Frame

6.2.1 Conceptual Frame for Risk

Seismic risk is defined as the interaction between seismic hazard and exposure, i. e., vulnerability. As highlighted in Chap. 3, the definition of seismic risk is therefore broader and more subjective than that of seismic hazard alone. Although seismic risk can generally be defined as the probability of the occurrence of adverse consequences to society (Reiter 1990; McGuire 2004), it has different meanings for different stakeholders (Smith 2005). Engineers, for example, are interested in the probability that a specific level of ground motion at a site of interest could be exceeded in a given period, a definition that is analogous to flood and wind risk, whereas insurance companies are more interested in the probability that a specific level of loss in a region or at a specific site could be exceeded in a given period.

In general, seismic risk is quantified by three parameters: probability of occurrence, level of hazard or consequence to society, and vulnerability (Wang 2006, 2007). However, earthquakes do not affect all members of society equally. The possibility of exposed objects or systems, such as buildings and infrastructure, to be affected by seismic hazard can be taken as a manifestation of vulnerability and considered an internal risk factor. Risk as a convolution of hazard and vulnerability implies the potential loss of an exposed object or system. The term convolution refers here to the contemporaneity and mutual conditioning of the two phenomena. In other words, vulnerability only arises if there is a threat and in the absence of vulnerability and exposure, there can be no threat. Thus hazard and vulnerability can be understood as mutually conditioning situations, neither of which can exist on its own (Cardona 2004). Beyond that, it is difficult to predict the degree of damage an earthquake will cause to elements at risk, e.g., infrastructural objects such as lifelines or buildings, which, in turn, cause loss of lives, since the resilience of such objects depends, for example, on the interaction between the structural elements of the object and the direction, frequency and duration of the ground motion. According to how elements at risk are defined, risk can be measured in terms of expected economic loss, human lives lost or physical damage to property where the appropriate measure of damage is available. Since for large-scale applications an exact determination thereof will include large uncertainties, it is of utmost importance first of all to quantify the seismic hazard itself by considering seismic source parameters (i.e., strength of the seismic source, depth at which the seismic fault might rupture, orientation of the fault) and local geological parameters (i.e., fundamental resonance frequency of the soil, shear wave velocity), all of which impact heavily on the spatial distribution of site specific ground motion in the case of earthquakes.

6.2.2 Methodology

In this regard, measuring seismic noise is a suitable procedure. Seismic noise, also referred to as microtremor, includes all random and unmeant signals of ground motion induced by wind, ocean waves, and anthropogenic activity across different frequency bands. By accurate processing the signals recorded at a specific site, it is possible to derive its fundamental resonance frequency and, with the use of further geological and geophysical data sets, its local 1D shear wave velocity profile. A 3D shear wave velocity model of the area under study was derived after interpolation between the individual profiles. Thus, we were further able to test whether the average shear wave velocity in the uppermost 30 m of the soil – now a standardized parameter for mapping seismic site conditions and used in several building codes worldwide – could serve as a reliable first-order proxy for site response estimation for Santiago de Chile.

In general, two main parameters influence ground motion: stratigraphic effects produced by the contrast of seismic wave velocities between the soft sedimentary soil layers and the hard bedrock, on the one hand, (Bard and Bouchon 1985; Sánchez-Sesma and Luzon 1995; Bielak et al. 1999; Chávez-García et al. 1999) and amplification effects caused by the specific shape of the surface and the bedrock topography, leading to a focusing and/or scattering of seismic waves around crests and hills, on the other (Bouchon 1973; Paolucci 2002; Semblat et al. 2002). Hence, the seismic response at a particular site in the city is affected by the local subjacent geology as well as by the shape of the 3D basin. It is important to note that the shape of the basin might produce a very different seismic response from that of a 1D layer. This is due, for example, to the reflection and trapping of seismic waves (and consequently seismic energy) in the basin. Interference of these trapped waves can lead to resonances, the shapes and frequencies of which are well correlated with the geometrical and mechanical characteristics of the structure. In this regard, the fundamental resonance frequency of the soil, the shear wave velocity structure of unconsolidated sediments down to the bedrock, the impedance (i.e., velocity) contrast between these sediments and the bedrock, and the shape of the sediment-bedrock interface, can be seen as the principal controlling indicators of such an imprint. An exact identification of these parameters is clearly a crucial step in seismic hazard assessment.

Up to now, only rough data on the surface geology and shape of the sediment-bedrock interface has been available for Santiago de Chile. The bottom of the basin, whose shape is known only indirectly from gravimetric measurements (Araneda

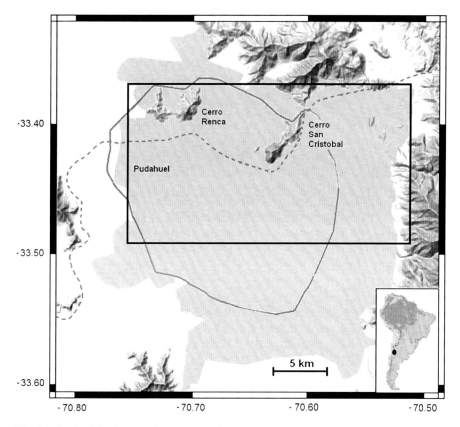

Fig. 6.1 Basin of Santiago de Chile. (Areas of high housing density are toned. The *dark grey line marks* the highway ring. The *dashed line* represents the Mapocho River. Locations mentioned in the text are indicated. The area of investigation is marked by the *black rectangle*). Source: The authors

et al. 2000), corresponds to an uneven surface exposed by the local outcropping of several hill chains buried within the basin (e.g., Cerro Renca, Cerro San Cristóbal, see Fig. 6.1). The mapping of the basement topography suggests a large paleo-drainage network associated with small troughs, the deepest of which is located in the western part of the city, with a depth of more than 550 m.

The sediments at the top of the basin are composed primarily of gravel, sand, and clay (Fig. 6.2); some deposits are believed to result from volcanic mud flows. Most of the north-western part of the city is made up of soft clayey material, while the eastern part of the basin contains soil of high density and low deformability, originating from deposits from the River Mapocho (and to a lesser extent, the River Maipo), and grading to coarser deposits closer to the apex of the two rivers. Between these stretches, in the centre of the valley, a transition zone forms out displaying large horizontal and vertical stratigraphic variations. Moreover, in the western part of the investigated area (Pudahuel district), a layer of ash (pumice) is known to sit at the top

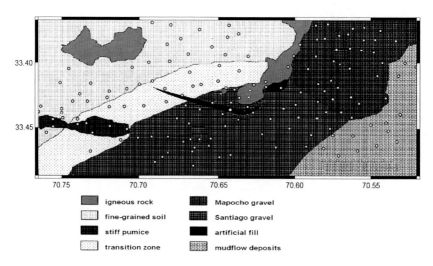

Fig. 6.2 Surface geology of the city of Santiago de Chile. Source: The authors

of the sedimentary column, probably resulting from a major eruption of the Maipo volcano. However, the spatial extension of each formation and, therefore, further detailed shear wave velocity values at depth are unknown.

There are several ways to fill this knowledge gap and obtain detailed velocity models of the subsurface, e.g., active in-situ measurements or boreholes. Particularly in urban areas, however, it may not be feasible to apply these techniques, since they entail explosive sources or high drilling costs. For this reason, non-invasive and passive seismic techniques that are cost-effective have become an attractive option in recent years for seismic site effect studies, providing reliable information on the subsurface with good lateral coverage in cities.

Over the last 20 years recordings of seismic noise have been made using a single seismic instrument only. The most popular of these single station methods is the Nakamura technique (Nakamura 1989, also referred to as H/V technique). Numerous experiments carried out in the last two decades have demonstrated that in the case of a high impedance contrast between the surface and deep rock materials, the noise spectra show an obvious peak that is well correlated with the fundamental resonance frequency of a site (e.g.; Lermo and Chavez-Garcia 1994; Field and Jacob 1995; Horike et al. 2001; Bard 2004). In this context, we made use of the technique and recorded seismic noise at 146 sites in the city of Santiago (measurement sites are indicated in Fig. 6.2). By placing the sensor directly on the ground, the soil vibration (horizontal and vertical components) is recorded for at least 30 min. The signals are then systematically analysed (for details about the processing, see Pilz et al. 2009), with the fraction of the horizontal and vertical components finally forming the H/V spectral ratio.

Tokimatsu and Miyadera (1992) recently found that the variation of H/V spectral ratios with frequency in fact corresponds to a change in the shear wave velocity profile at the site. Based on a theory of seismic surface waves proposed by Harkrider (1964), Arai and Tokimatsu (2000) found that the theoretical H/V spectrum computed for the shear wave velocity profile of a site closely matched the observed microtremor H/V spectrum and, vice versa, that the shear wave velocity structure of a site can be estimated by inversion of the microtremor H/V spectrum. Several studies have exploited the availability of ambient noise measurements to derive a vertical shear wave velocity profile (e.g., Fäh et al. 2001, 2003, 2006; Scherbaum et al. 2003; Arai and Tokimatsu 2004; Parolai et al. 2006).

For each of the sites where fundamental resonance frequency of the soil had been determined, we systematically inverted the H/V spectral ratios to derive local 1D shear wave velocity profiles. To constrain the results, the total thickness of the sedimentary cover has to be known in advance. For the Santiago basin, this thickness was determined by gravimetric measurements (Araneda et al. 2000). Although the velocity resolution for deeper parts of the calculated model remains lower, it is still possible to retrieve an average shear wave velocity that would serve engineering seismology purposes.

Interpolation between the individual 1D shear wave velocity profiles led to a 3D velocity model. The model contains a detailed description of the sedimentary basin shape as defined by the contact between lower shear wave velocity sediments and the higher shear wave velocity bedrock. Moreover, orography, i.e., the shape of the surface, is incorporated into the model, based on digital elevation model data available with a resolution of 1 arcsec (approximately 30 m) for all of Santiago (Pilz et al. 2010).

Since soil parameters can vary over relatively short distances, a factor that influences ground motion substantially, recognition of the importance of ground motion amplification has led to the development of systematic approaches for the mapping of seismic site conditions (e.g., Park and Elrick 1998; Wills et al. 2000; Holzer et al. 2005) and the small-scale quantifying of both amplitude and frequency-dependent site amplifications (e.g., Borcherdt 1994). Nowadays, the standardized approach to mapping seismic site conditions is to measure and map v_s^{30}, the average shear wave velocity in the uppermost 30 m of soil. The Uniform Building Code (International Conference of Building Officials 1997), for example, uses v_s^{30} to group sites into several broad categories, each of which is assigned the characteristic factors that modify the response spectrum differently. In addition, many US building codes (Building Seismic Safety Council 2004) and the Eurocode 8 (CEN 2003) now rely on v_s^{30} for seismic site characterization. In this context, we also focus on this parameter and provide a spatially comprehensive study for the central part of the Santiago province bordered by the municipalities of Renca in the northwest, Vitacura in the northeast, Pudahuel in the southwest, and La Reina in the southeast, since available information allows us to assume that in this part of the basin geotechnical conditions change on short scale; however, the findings could be extended to the entire city.

6.3 Hazard Analysis Findings

6.3.1 Fundamental Resonance Frequency

In the case of earthquake, seismic waves will cause soil resonances, whose shapes and frequencies strongly depend on the characteristic parameters of the soil. As mentioned above, the peak in the H/V spectra is well correlated with the fundamental resonance frequency of a site. Analysis of the fundamental resonance frequencies calculated for all sites allowed for drawing a map of the fundamental frequencies of the investigated area (Fig. 6.3). The frequency tends to vary only slightly in space and displays no significant divergence from neighbouring sites. Figure 6.3 clearly indicates that almost the entire area is characterized by frequencies around or below 1 Hz. Only in sites close to the city centre and around the San Cristóbal hill the fundamental frequency shows higher fundamental frequency values. In this part of the basin, the thickness of the sedimentary cover is in decline, with bedrock outcropping. Note that there is also a rapid decrease in the sediment thickness towards the northwest of the area under review. Since no noise measurements have been carried out in the immediate vicinity of the Renca hill, evidence of an increase in the fundamental frequency could not be detected. Only two adjacent sites show a trend towards higher frequencies.

On the other hand, frequencies below 0.5 Hz can be found in the west and towards the south-southeast of the area. The sedimentary cover thickness reaches more than 500 m in these parts of the basin, supporting the general trend of declining fundamental frequency and increasing sedimentary cover thickness.

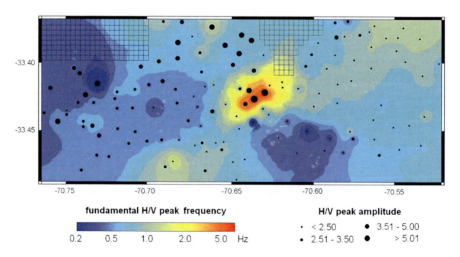

Fig. 6.3 Map of the fundamental resonance frequency in the investigated area (*circles* indicate microtremor measurement sites that have been considered for interpolation. Diameter of the spots corresponds to peak amplitude. Since no measurements were carried out in the hatched area, results are due to interpolation, only). Source: The authors

Not only the soil, but all objects (including infrastructure and buildings) have a 'natural period', i.e., the time it takes to swing back and forth. As seismic waves move through the ground, the ground will move at its natural period. In particular, if the period of the ground coincides with that of a building on the ground this can become a problem. When a building and the ground sway or vibrate at the same rate they are said to resonate, and damage is likely to occur. This is because their vibrations are amplified as they resonate, placing greater stress on the building structure.

Although the phenomenon of resonance can cause tremendous damage, its effect can be reduced. In designing seismically safe buildings, architects and engineers must be concerned with 'tuning' a building in such a way that the tendency of resonance to amplify its vibration is reduced or eliminated. In the absence of appropriate design, the building is exposed to greater risk of earthquake damage, particularly if it has already been subjected to earthquakes in the past. Cumulative damage occasioned by previous earthquakes was stressed by Edwards (1951) in his analysis of structural damage ensuing from the 1949 Puget Sound earthquake. However, not only the fundamental frequencies but also higher harmonics, which are often invisible in microtremor H/V spectral ratios but not necessarily damped out, may also overlap. Hence, this map can only serve as a lower bound in terms of frequency and amplitude for site amplification.

A key factor affecting the period is height. Although taller buildings are predominantly characterized by low frequencies, no blanket conclusions should be drawn. Simple theoretical height–period relationships that determine a structure's resonance frequency tend to overestimate experimental data in terms of building periods by taking the structural system into account but not its relevant specifications such as foundation flexibility or infill stiffness. Determining endangered structures in advance on a large scale could therefore be a difficult challenge.

Furthermore, it is obvious from Fig. 6.3 that the northern and northwestern parts of the study area display larger H/V peak amplitudes, whereas flat H/V curves and peaks of low amplitude feature in the eastern and southern parts. When the H/V peak is clear, a large impedance contrast between sedimentary cover and bedrock can be expected at this site, and hence higher ground motion amplification. On the other hand, flat H/V curves are located at the surface of dense sediment locations, i.e., mainly gravel, and point to the presence of a low impedance contrast between stiff sediments and underlying bedrock, suggesting smaller amplification compared to the northwestern parts. Therefore, to gain a more detailed description of site conditions we aim at characterizing the shear wave velocity of the sediments.

6.3.2 Shear Wave Velocity of Sediments in the Santiago Basin

Cross sections of the shear wave velocity model of the Santiago basin are shown in Fig. 6.4. As expected, all profiles show a trend towards increasing shear wave

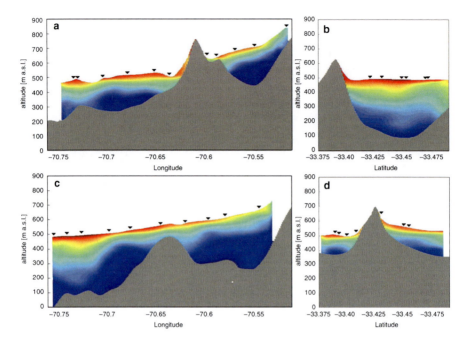

Fig. 6.4 Cross sections of the interpolated 3D shear wave velocity model within the area of investigation (*black triangles* mark measurement sites within a distance of 500 m to the cross section. (**a**) Lat 33.405 S, (**b**) Lon 70.730 W, (**c**) Lat 33.440 S, (**d**) Lon 70.635 W. For all profiles, topography is exaggerated about 16 times. The shape of the bedrock, shown in the *grey pattern*, is derived from gravimetric data. Orography is based on high resolution digital elevation data). Source: The authors

velocities with depth as a result of compaction and increasing pressure, but spatial differences in the velocity gradient over short scales, i.e., within a few kilometres, are clearly visible, even at first glance. To the east of Cerro San Cristóbal, the shear wave velocity is generally higher than to the west (Fig. 6.4a). Flat H/V curves and H/V spectral ratios of low amplitude are located predominantly in the eastern parts of the basin over dense sediments, i.e. gravel. This area of the basin is characterized by high seismic velocities of sediments, leading to small velocity contrasts with the bedrock. Small amplification is therefore expected to occur here. In contrast, superficial fine-grained material found in the western parts of the investigated area and characterized by low shear wave velocities can be clearly identified (Fig. 6.4b). To the south-west of the area (Pudahuel district, see Fig. 6.1), there is a thick layer of pyroclastic flow deposits, outcropping at the top of the filling and showing a shear wave velocity well below that of the surrounding materials (Fig. 6.4c). As already mentioned, H/V spectral rations here are characterized by higher amplitudes (Fig. 6.3). Hence, higher amplification is expected to occur in this part of the basin.

In general, our model can reliably point out local characteristics shown in other data sets, but takes advantage of the fact that all individual shear wave velocity profiles reach the bedrock, albeit the resolution for deeper parts of the model

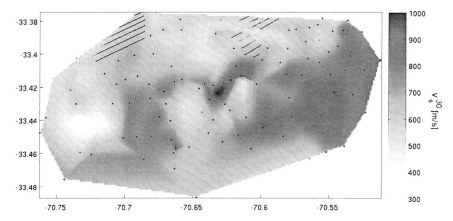

Fig. 6.5 v_s^{30} for the area limited by a linear connection between the outermost measurement sites (*Black dots* show the location of the ambient noise measurement sites used for mapping v_s^{30}. No measurements were carried out in the *dashed area*. Results are due to interpolation only and therefore not plausible, as outcropping bedrock is found there). Source: The authors

remains low. By interpolation between the individual measurement sites, v_s^{30} was calculated for the entire area under investigation (Fig. 6.5). Areas characterized by different shear wave velocities in the uppermost 30 m can be identified clearly. An ellipsoidal area of low v_s^{30} values in the western part represents the layer of pumacit on top of the sedimentary filling (see Fig. 6.2). In the central part, the outcropping igneous bedrock of the Cerro San Cristóbal is apparent. The v_s^{30} values found there contrast with those in the surroundings and depict the shape of the hill quite well. In the northern part of the Cerro San Cristóbal and close to the Cerro Renca, no ambient noise measurements have been carried out. Consequently, since there are no reliable v_s^{30} interpolation values for these parts of the investigated area, they are masked out in Fig. 6.5.

6.3.3 Correlation of Soil Parameters and Macroseismic Intensity of Past Earthquakes

It is beyond question that the parameter of the shear wave subsoil velocity profile should have a key role in the site effect assessment, since this parameter is tied to soil rigidity. A comparison of our v_s^{30} map with the damage distribution of the Mw = 7.8 (epicentre 120 km west of Santiago) earthquake of 3 March 1985 (e.g., Astroza and Monge 1991) allows for an agreement by trend, since most of the damage to humans, objects of nature, and man-made structures (i.e., higher intensities) was concentrated in the northwestern part of the city with a small-sized building stock of low quality (Astroza et al. 1993). At first glance, this cannot be explained by a correspondence between the fundamental resonance frequencies of the soil and the building stock. Theoretical height-period relationships suggest that small-sized structures in that area might have a natural frequency of at least 5 Hz, whereas the fundamental frequency of the soil is well below 1 Hz (Fig. 6.3). It should be kept in mind, however, that not only the fundamental resonance frequency but also higher modes (higher harmonics) could correspond to the fundamental resonance frequency or even the higher harmonics of these buildings (i.e., partials whose frequencies are whole number multiples of the fundamental frequency).

Furthermore, large H/V peak values were observed in areas with higher intensities (western part of the city) and, on the contrary, small H/V peak values in those with lower intensities (eastern part). Measurements carried out on Cerro San Cristóbal showing high fundamental frequencies and high H/V peak amplitudes do not follow this trend.

In general, the lowest v_s^{30} values were found on top of the pumacit layer, where the intensity reached a maximum (Fig. 6.5). However, only a few H/V ratios showed high amplitudes. Although most of the small-sized houses in that area and the surrounding municipalities seem to be old and of low quality (Astroza et al. 1993), the properties of the soil are completely different to those in the rest of the city, a factor that affected the existing building stock. Several adobe houses were completely destroyed, and many adobe and simple masonry buildings suffered significant damage. This result is not surprising, since damage is contingent on a combination of ground motion intensity and building vulnerability.

The highest H/V peak amplitudes were found for the central northern and northwestern parts of the investigated area (Fig. 6.3). Following the 27 February 2010 earthquake, damage in these parts of the basin was above average, even to newer structures most likely erected according to recent building codes. Although v_s^{30} values found for the area were not too low, high H/V peak amplitudes suggest that ground motion would have been relatively strong.

Turning to the significance of the model results for risk management, there is a very general tendency for ground motion to be amplified where shear wave velocity is low. There is evidence, however, that seismic amplification and local site conditions are too complex to be related to a single parameter measured in the first 30 m of subsoil alone. Further characterization could be added by seismic noise measurements. There is a consensus that, although the analysis of seismic noise measurements can provide the fundamental frequency and a lower-bound estimate of the amplification for soft soil sites, it fails to provide higher harmonics. A leading advantage is that this technique is inexpensive and can be applied almost everywhere. As simple and affordable techniques for deriving local site response and amplification are in general of broad interest, a combination of these parameters, including sediment thickness and basin structure (e.g., to identify the multi-dimensional effects of the scattering or focusing of seismic waves), might allow for a more detailed estimation of seismic site conditions.

6.4 Discussion and Conclusions

The research investigation carried out has shown that it is possible to derive a reliable model of the shear wave velocity structure for the basin of Santiago de Chile by means of low-cost measurements of ambient seismic noise. Our findings are consistent with other available data sets. A combined use of shear wave velocity and fundamental frequency of the soil can be used to subdivide the seismic prone study area in the northern part of the Santiago basin into zones with similar characteristics for microzonation purposes.

In this regard, measurements of seismic noise were carried out in the city of Santiago, 125 of which allowed for further analysis, since a confident estimate of the fundamental resonance frequency at these sites was possible. Amplification peaks occur predominantly at low frequencies below 1–2 Hz, posing a threat to very tall buildings, although slight amplification also affects frequencies from 2 to 13 Hz. Following inversion of the H/V spectra, local shear wave velocity profiles were derived for each measurement site and formed the basis for a 3D shear wave velocity model for the entire investigated area. Quite different shear wave velocity-depth gradients were found within the Santiago basin.

As simple and affordable techniques to derive local site response and amplification are of particular interest, we tested whether a simple relationship existed between the site amplification and our v_s^{30} map on a local scale. A higher resolution of the v_s^{30} map can be accomplished by adding more measurement sites. At first glance, a very rough correlation between v_s^{30} and local geological conditions is visible. Taking into account the distribution of intensities observed for the 1985 Valparaiso event allows for a better identification of a correlation between v_s^{30} and the intensity distribution, albeit the correlation still remains rough. Although these results cannot be generalized for every earthquake that might affect ground motion in the Santiago basin, it indicates that site effects clearly modify ground motion in the city. Moreover, in addition to near-surface site conditions, seismic waves in basins are also known to be heavily influenced by the thickness of the sedimentary cover and the shape of the sediment-bedrock interface.

Since it is not possible to modify the seismic hazard itself in order to reduce risk, the sole option is to modify the vulnerability conditions of exposed elements in hazard-prone areas and to adopt structures (building stock, infrastructure) suitable to these seismic site conditions. Identifying areas that are likely to suffer stronger ground motion in the case of earthquake could theoretically prevent risk if spatial expansion was steered in such a way that hazard-prone areas and those that increase hazard in urban areas remain free of settlements. As this is unlikely to happen in reality, the results can still be used to enforce the application of building codes. A seismic code involves a vast number of aspects that should be carefully considered in advance. In particular, concerning the inclusion of site specific characteristics in building code provisions, attention should be given to the adequate definition of soil classes. As an initial, albeit very rough, approximation, it would not be overconservative to use a single amplification factor for all points at the surface of a particular soil deposit. It is worth noting, however, that close to the edge of a basin or valley, differential motion rather than absolute amplification may be the most decisive effect in the case of strong ground motion resulting from a pronounced surface and bedrock topography (e.g., Moczo and Bard 1993).

A realistic implementation of the findings and the most effective way to reduce human casualties might therefore be achieved by a retrofit of the existing building stock. A comprehensive retrofit campaign is a formidable task and would entail the earthquake performance screening of a large portion of the building stock, with prioritization based on exhibited risk.

References

Arai, H., & Tokimatsu, K. (2000). Effects of Rayleigh and Love waves on microtremor H/V spectra. *Proceedings 12th World Conference on Earthquake Engineering*, paper 2232.
Arai, H., & Tokimatsu, K. (2004). S-wave velocity profiling by inversion of microtremor H/V spectrum. *Bulletin of the Seismological Society of America, 94*, 53–63.
Araneda, M., Avendano, F., & Merlo, C. (2000). Gravity model of the basin in Santiago, Stage III. *Proceedings of the 9th Chilenian Geological Congress*, 2, pp. 404–408, Sociedad Geólogica de Chile, Santiago.
Astroza, M., & Monge, J. (1991). Seismic microzones in the city of Santiago. Relation damage-geological unit. *Proceedings of the 4th International Conference on Seismic Zonation*, Stanford, pp. 595–599.
Astroza, M., Moroni, M., & Kupfer, M. (1993). Calificación sísmica de edificios de albañilería de ladrillo confinada con elementos de hormigón armado, *Memorias de las XXVI Jornadas Sudamericanas de Ingeniería Estructural*, 1, Montevideo, Asociacion Sudamerican de Ingenieros Estructurales.
Bard, P. Y. (2004). The SESAME project. An overview and main results. *Proceedings of the 13th World Conference on Earthquake Engineering*, Vancouver, Paper 2207.
Bard, P. Y., & Bouchon, M. (1985). The two-dimensional resonance of sediment filled valleys. *Bulletin of the Seismological Society of America, 75*, 519–541.
Barrientos, S. (1981). *Regionalización sísmica de Chile*. MSc thesis, Facultad de Ciencias Físicas y Matemáticas, Universidad de Chile, Santiago.
Barrientos, S., Vera, E., Alvarado, P., & Monfret, T. (2004). Crustal seismicity in central Chile. *Journal of South American Earth Sciences, 16*, 759–768.
Bielak, J., Xu, J., & Ghattas, O. (1999). Earthquake ground motion and structural response in alluvial valleys. *Journal of Geotechnical and Geoenvironmental Engineering, 125*, 413–423.
Borcherdt, R. D. (1994). Estimates of site-dependent response spectra for design (methodology and justification). *Earthquake Spectra, 10*, 617–654.
Bouchon, M. (1973). Effects of topography on surface motion. *Bulletin of the Seismological Society of America, 63*, 615–622.
Bravo, R. D. (1992). Estudio geofisico de los suelos de fundación para un zonificacion sísmic del area urbana de Santiago Norte. PhD thesis, Universidad de Chile, Santiago.
Building Seismic Safety Council (BSSC) (2004). *NEHRP recommended provisions for seismic regulations for new buildings and other structures.* 2003 edition (FEMA 450), Building Seismic Safety Council, National Institute of Building Sciences, Washington, DC.
Cardona, O. D. (2004). The need for rethinking the concepts of vulnerability and risk from a holistic perspective: A necessary review and criticism for effective risk management. In G.

Bankoff, G. Frerks, & D. Hilhorst (Eds.), *Mapping vulnerability: Disasters, development and people*. London: Earthscan.

Çelebi, M. (1987). Topographical and geological amplifications determined from strong-motion and aftershock records of the 3 March 1985 Chile earthquake. *Bulletin of the Seismological Society of America, 77*, 1147–1167.

CEN. (2003). *Eurocode (EC) 8: Design of structures for earthquake resistance – Part 1 general rules, seismic actions and rules for buildings, EN 1998-1*. Brussels: CEN.

Chávez-García, F. J., Stephenson, W. R., & Rodríguez, M. (1999). Lateral propagation effects observed at Parkway, New Zealand: A case history to compare 1D vs. 2D effects. *Bulletin of the Seismological Society of America, 89*, 718–732.

Edwards, H. H. (1951). Lessons in structural safety learned from the 1949 Northwest earthquake. *Western Construction, 26*, 70–74.

Fäh, D., Kind, F., & Giardini, D. (2001). A theoretical investigation of average H/V ratios. *Geophysical Journal International, 145*, 535–549.

Fäh, D., Kind, F., & Giardini, D. (2003). Inversion of local S-wave velocity structures from average H/V ratios, and their use for the estimation of site-effects. *Journal of Seismology, 7*, 449–467.

Fäh, D., Steimen, S., Oprsal, I., Ripperger, J., Wössner, J., Schatzmann, R., Kästli, P., Spottke, I., & Huggenberger, P. (2006). The earthquake of 250 A. D. in Augusta Raurica, a real event with 3D site effect? *Journal of Seismology, 10*, 459–477.

Field, E. H., & Jacob, K. H. (1995). A comparison and test of various site-response estimation techniques, including three that are not reference-site dependent. *Bulletin of the Seismological Society of America, 85*, 1127–1143.

Harkrider, D. G. (1964). Surface waves in multilayered elastic media, part 1. *Bulletin of the Seismological Society of America, 54*, 627–679.

Holzer, T. L., Padovani, A. C., Bennett, M. J., Noce, T. E., & Tinsely, J. C. (2005). Mapping v_s^{30} site classes. *Earthquake Spectra, 21*, 353–370.

Horike, M., Zhao, B., & Kawase, H. (2001). Comparison of site response characteristics inferred from microtremor and earthquake shear waves. *Bulletin of the Seismological Society of America, 91*, 1526–1536.

International Conference on Building Officials (1997). *Uniform building code*. International Conference on Building Officials, Whittier.

Kelleher, J. A. (1972). Rupture zones of large South American earthquakes and some predictions. *Journal of Geophysical Research, 77*, 2087–2103.

Lermo, J., & Chavez-Garcia, F. J. (1994). Are microtremors useful in site response evaluation? *Bulletin of the Seismological Society of America, 84*, 1350–1364.

McGuire, R. K. (2004). Seismic hazard and risk analysis. Earthquake Engineering Research Institute, MNO-10.

Moczo, P., & Bard, P.-Y. (1993). Wave diffraction, amplification and differential motion near strong lateral discontinuities. *Bulletin of the Seismological Society of America, 83*, 85–106.

Nakamura, Y. (1989). A method for dynamic characteristics estimation of subsurface using microtremor on the ground surface. *Quarterly Reports of the Railway Technical Research Institute, 30*, 25–33.

Nishenko, S. (1985). Seismic potential for large and great interplate earthquakes along the Chilean and Southern Peruvian margins of South America: A quantitative reappraisal. *Journal of Geophysical Research, 90*, 3589–3615.

Paolucci, R. (2002). Amplification of earthquake ground motion by steep topographic irregularities. *Earthquake Engineering and Structural Dynamics, 31*, 1831–1853.

Park, S., & Elrick, S. (1998). Predictions of shear wave velocities in southern California using surface geology. *Bulletin of the Seismological Society of America, 88*, 677–685.

Parolai, S., Richwalski, S., Milkereit, C., & Fäh, D. (2006). S-wave velocity profiles for earthquake engineering purposes for the Cologne area (Germany). *Bulletin of Earthquake Engineering, 4*, 65–94.

Pilz, M., Parolai, S., Leyton, F., Campos, J., & Zschau, J. (2009). A comparison of site response techniques using earthquake data and ambient seismic noise analysis in the large urban area of Santiago de Chile. *Geophysical Journal International, 178*, 713–728.

Pilz, M., Parolai, S., Picozzi, M., Wang, R., Leyton, F., Campos, J., & Zschau, J. (2010). Shear wave velocity model of the Santiago de Chile basin derived from ambient noise measurements: A comparison of proxies for seismic site conditions and amplification. *Geophysical Journal International, 182*, 355–367.

Ramírez, D. (1988). Estimación de algunos parámetros focales de grandes terremotos históricos chilenos. MSc thesis. Geofísica, Universidad de Chile, Santiago.

Reiter, L. (1990). *Earthquake hazard analysis*. New York: Columbia University Press.

Sánchez-Sesma, F. J., & Luzon, F. (1995). Seismic response of three-dimensional alluvial valleys for incident P, S, and Rayleigh waves. *Bulletin of the Seismological Society of America, 85*, 269–284.

Scherbaum, F., Hinzen, K. G., & Ohrnberger, M. (2003). Determination of shallow shear wave velocity profiles in the Cologne Germany area using ambient vibrations. *Geophysical Journal International, 152*, 597–612.

Semblat, J. F., Duval, A. M., & Dangla, P. (2002). Seismic site effects in a deep alluvial basin: Numerical analysis by the boundary element method. *Computers and Geotechnics, 29*, 573–585.

Smith, W. (2005). The challenge of earthquake risk assessment. *Seismological Research Letters, 76*, 415–416.

Tokimatsu, K., & Miyadera, Y. (1992). Characteristics of Rayleigh waves in microtremors and their relation to underground structures. *Journal of Structural Engineering, 439*, 81–87.

Wang, Z. (2006). Understanding seismic hazard and risk assessments: An example in the New Madrid Seismic Zone of the central United States. *Proceedings of the 8th National conference on earthquake engineering*, San Francisco, Paper 416.

Wang, Z. (2007) Seismic hazard and risk assessment in the intraplate environment: The new Madrid seismic zone of the central United States. In S. Stein & S. Mazzotti (Eds.), *Continental intraplate earthquakes: Science, hazard, and policy issues, Geological Society of America, Special Paper 425*, pp. 363–373.

Wills, C. J., Petersen, M. D., Bryant, W. A., Reichle, M. S., Saucedo, G. J., Tan, S. S., Taylor, G. C., & Treiman, J. A. (2000). A site-conditions map for California based on geology and shear wave velocity. *Bulletin of the Seismological Society of America, 90*, 187–208.

Chapter 7
Land-Use Change, Risk and Land-Use Management

Ellen Banzhaf, Annegret Kindler, Annemarie Müller, Karin Metz, Sonia Reyes-Paecke, and Ulrike Weiland

Abstract This chapter focuses on flood risk analysis and risk prevention in Santiago de Chile. It presents a conceptual framework for flood risk analysis in urban areas and demonstrates the utility of a mixed set of methods, including remote sensing and GIS techniques, to improve the methodological basis for flood risk assessment and risk prevention. Population growth and land-use changes are analysed as key elements of urban development and indicators of flood risk production. A conceptual framework comprising the core elements of exposure, elements at risk and vulnerability serves as a tool for risk analysis and risk assessment, and is applied to the municipalities of La Reina and Peñalolén. The chapter reviews existing institutional responses to land-use and risk management and, based on expert interviews, detects their deficits. As a conclusion, recommendations to improve flood risk prevention in Santiago de Chile are made. The absence of a systemic view of flood risk resulting from complex ecological and social processes is the chief weakness of current risk prevention in Santiago de Chile.

Keywords Demographic change • Flood risk • Geoinformatics • Land-use change • Recommended actions • Risk-related indicators

7.1 Introduction

The rapid population growth and ongoing expansion of the urban area in Santiago puts enormous pressure on the environment: agricultural land is now gradually being transformed into built-up areas, while the share of green space dwindles.

E. Banzhaf (✉) • A. Kindler • A. Müller • K. Metz • S. Reyes-Paecke • U. Weiland
Department Urban and Environmental Sociology, Helmholtz Centre for Environmental Research – UFZ, Permoserstraße 15, 04318 Leipzig, Germany
e-mail: ellen.banzhaf@ufz.de

Peri-urban construction, e.g., in the Andean piedmont, and the clearing of avalanche forests in mountain regions produces long-term environmental impacts such as heightened imperviousness of land surfaces and a decline in rainwater retention areas (Ebert et al. 2009; Niehoff et al. 2002). The resultant excess of surface run-off, especially during storms, hits a higher concentration of built-up areas, affecting not only a greater number of people in the process but also economic values. These complex processes of flood risk production in urban regions call for appropriate land-use management.

The overall objective of this chapter is to analyse and assess flood risk and to discuss measures to improve flood management and particularly flood risk prevention by means of land-use management. To meet this objective, the chapter explores a set of related questions:

- What population and land-use/land-cover changes in the Metropolitan Area of Santiago (MAS) influence flood risk?
- How can flood risk be analysed and assessed, and how has it evolved over time?
- How adequate are current risk and land-use management instruments in Santiago and how can land-use management in particular be improved with respect to flood risk?

The chapter is organized as follows: Sect. 7.2 describes the conceptual risk framework developed for this research and gives an overview of the methods applied. Demographic and land-use change as major drivers of flood risk production are presented in Sect. 7.3. The analysis of flood risk as a threat arising from urban expansion in two selected municipalities is the focus of Sect. 7.4. The analysis pays particular attention to flood risk generation induced by processes of attraction, ecosystem disturbance, and finally exposure of people and values. Against this background, Sect. 7.5 investigates how appropriate land-use management contributes to risk prevention. The concluding Sect. 7.6 provides recommendations on land-use management with regard to flood risk prevention.

7.2 Conceptual Frame of Risk

While the mechanisms typically associated with risk generation in megacities are lined out in Chap. 3, this chapter focuses on flood risk generation induced by processes of attraction, ecosystem disturbance and exposure of people and values.

The conceptual risk frame is derived from hazard risk research against a social science background (Wisner et al. 2005; UNDP 2004). Some adaptations of the theoretical framework, such as the incorporation of elements at risk as a separate component, were made to facilitate application of the concept in a densely populated urban area and to use it as a base for decision-making in land-use and risk management. Risks resulting from natural hazards occur only in areas that are prone to the respective hazard and additionally fail to accommodate people and values with the necessary capacity to withstand it. In short, risk is a complex phenomenon dependent on the spatial and temporal co-occurrence of several

7 Land-Use Change, Risk and Land-Use Management

components. In order to describe and analyse the process of risk generation in a complex urban setting, risk is conceptualized as a function of hazard, the elements at risk and their vulnerability:

$$\text{Risk} = f(\text{hazard, elements at risk, vulnerability})$$

Figure 7.1 indicates the underlying concept of flood risk assessment. Hazardous events only involve risk when damage occurs. If people and goods are not exposed,

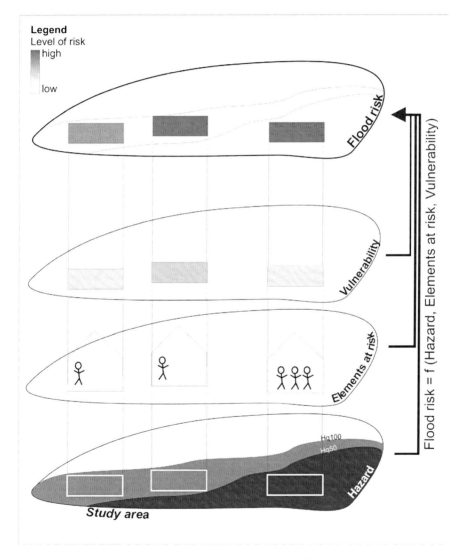

Fig. 7.1 The components of risk exemplified for the case of floods: hazard, vulnerability, elements at risk (Source: Ebert 2008)

i.e., if they are not vulnerable, or if the capacities to cope are adequate, there is no risk. In contrast, the risk factor intensifies in areas where hazardous events are likely to occur or where people and values are exposed to these events.

Hazard as a prime component of risk is expressed as the probability of a potentially damaging event of a certain magnitude occurring within a certain period of time (ISDR 2004). Hazards heavily depend on site-specific and – in the case of floods – seasonal climatic conditions. The probability of floods occurring in Santiago peaks in the winter months between May and August when the likelihood of precipitation is highest, since they are triggered by a temporary run-off surplus.

The number and spatial distribution of *elements at risk* refers to people and values located in hazard-prone areas, initially independent of their level of vulnerability. Considering elements at risk as a separate component has the advantage that decision-makers can set their own priorities in terms of risk reduction measures. In other words they can distinguish between areas with a critical infrastructure, for example, or a high population density.

Vulnerability refers to the physical and social fragility of the elements at risk and their lack of capacity to cope with the negative impact of a hazardous event (Cardona 2001). The concept of vulnerability is difficult to capture or measure, as it comprises not only physical or socio-demographic characteristics, but also non-tangible factors such as lack of knowledge about the hazard in question. A common understanding and definition of vulnerability has not yet been found (Birkmann and Wisner 2006). The concept of vulnerability describes the characteristics but not the number of people or volume of infrastructure exposed to a hazard (Wisner et al. 2005). Similar to the analysis of elements at risk, the analysis of the spatial distribution of vulnerability patterns allows decision-makers to focus measures for risk reduction by systematically minimizing deficits that cause vulnerability.

The concept of *risk* allows for exploration of its root cause and its magnitude. It makes possible an investigation of the change in hazard scales as an external factor, on the one hand, and the system-intrinsic reasons why disasters happen (vulnerability), on the other. The analysis and reduction of flood hazards, the physical vulnerability with respect to the above-mentioned hazards and the minimization of elements at risk is in the foreground of this study, while social vulnerability, which is closely related to physical vulnerability, cannot be assessed for reasons of data non-availability.

Quantitative risk assessment is a sophisticated task, as it requires the quantitative assessment of all three components of risk (hazard, elements at risk, vulnerability). While hazard can be measured as a probability and the number of people counted, the quantification of vulnerability is a methodological challenge, as discussed in Sect. 7.4.

7.3 Methodology: Indicators for Urban Growth and Flood Risk Production

Methods from different fields of knowledge are applied in order to treat the research questions depicted in Sect. 7.1 in an adequate manner. Analyses, assessments and the derivation of recommendations are structured according to a modified DPSIR[1] model, developed to determine the interrelation between environmental impacts, the state of the environment and the response to negative impacts (see NERI 1995).

7.3.1 Indicators for Urban Growth

In order to investigate population change as a driving force for urban growth, use was made of urban and regional statistics taken from census data. Census data is collected once every 10 years (recent censuses 1992 and 2002) by the respective national statistical authorities (INE). The census is the only statistical instrument that allows for data comparisons in terms of geographic scales, population projections, analyses of socio-economic compositions/structures, and characteristics of residential areas and households. A further data source is the National Socioeconomic Characterization Survey CASEN of the Ministry of Planning of the Government of Chile (MIDEPLAN) from 2006.

Using census data and Geographic Information Systems (GIS), the number of inhabitants per hectare and municipality were determined. Investigations of population dynamics made on two spatial levels comprise the whole of the MAS and each of the 34 municipalities individually. This includes the 32 municipalities of the Province Santiago, Puente Alto municipality in the Province Cordillera, and San Bernardo municipality in the Province Maipo (see Fig. 7.2). A detailed analysis of population dynamics was made for the time period 1992–2002–2006 to establish the spatial distribution of the population in the different municipalities and the changes over time, and to identify municipalities with a growing, declining or stagnating population in the MAS. As these processes occur simultaneously and the population is exposed to flood risks in general, people possess different levels of vulnerability to such risks.

In this study, urban remote sensing took into account satellite images and aerial photographs. Satellite images were used to track land-use and land-cover (LULC) changes, as well as to analyse floodplains with respect to vulnerability for 1993 (Landsat/Spot fused data), 2002, 2005, and 2009 (ASTER data).

The time series were chosen for the following reasons: firstly, two time steps (1993 and 2002) coincided with the census data (1992 and 2002) and characterized patterns of demographic change that had physical footprints in the urban area and

---

[1] D: Driving forces; P: Pressure; S: State; I: Impact; R: Response.

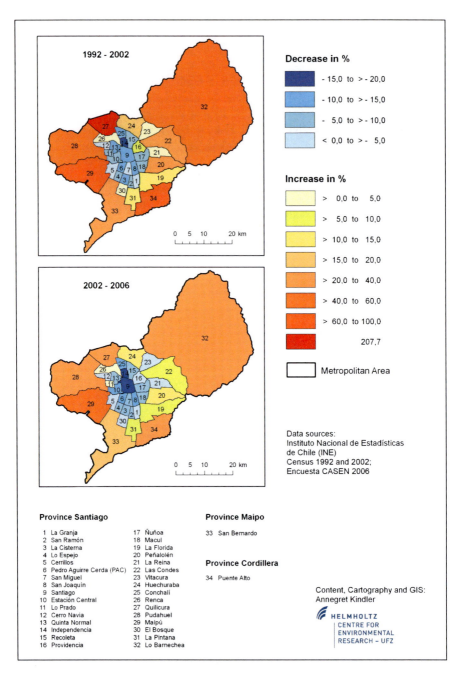

Fig. 7.2 Change in population by municipality 1992–2002 and 2002–2006 in the Metropolitan Area of Santiago de Chile (Source: see figure)

in ongoing land-use dynamics. Secondly, more recent years (2005 and 2009) provided further information on urban land-use development as well as up-to-date spatial information for this decade. This made it possible to mark the velocity of land-use change and urban growth over a period of approximately 20 years. The shorter time steps in the last decade (2002–2005–2009) support a more differentiated understanding of urban growth rates, particularly building activities in risk-prone areas. Moreover, these land-use classifications show relevant changes in land-use dynamics with respect to the amount of lost pervious soils over rather short intervals, i.e., fertile land, above all agricultural fields, irreversibly lost (see Fig. 7.4). Aerial photographs from 2006 were used to monitor vegetation.

LULC classifications over time are a useful tool to examine how urban forms modify the landscape. They help to detect and evaluate the distribution of impervious surfaces, a key parameter for risk analysis and the prevention of, e.g., flooding. Remotely sensed data is used to detect and evaluate the physical structure and composition of urban areas, such as residential, commercial or mixed neighbourhoods, green space or agricultural areas and woodland. How land is allocated is another factor that both determines and influences the risk of flooding in megacities. The key question with regard to seeping water rates and surface run-offs is whether land is used for green space and parks, agricultural purposes or housing. The main LULC driver consists of urban expansion in the form of built-up areas.

7.3.2 Elaboration of Indicators for Flood Risk Analysis

In order to analyse the driving forces behind flood risk, the state of the environment and current urban development processes in the MAS, indicators were elaborated as depicted in the previous subsections and listed in Table 7.1 (cf. Weiland 2006). The degree of imperviousness and the vegetation monitoring focus on the loss of urban green space to construction were used as a further indicator of flood risk production. Detection and allocation of change from one land-use type to another support findings on the degree of imperviousness and growth rate of built-up areas. In addition, the number of new settlements and infrastructure developments in areas with a high flood hazard level is calculated for each municipality.

The indicators are defined and calculated as follows:

1. The number of inhabitants per hectare indicates population density. Analysis of this indicator over time produces information on the inner-urban development pattern in the MAS. It also shows the need for urban infrastructure such as flood mitigation measures. Neighbourhoods with high population densities in high flood risk areas call for special consideration in risk management and adequate land-use planning processes.
2. The built-up area per municipality refers to the total number of buildings, road networks and the remaining urban built infrastructure. The amount of elements at risk can be estimated by means of the derived density.

Table 7.1 Selected risk-related indicators

Aspect	Indicator	Temporal scales	Data source
Population density	Inhabitants/municipality [inh./ha]	1992, 2002, 2006	Census data CASEN
Building density	Built-up area/municipality [ha]	1993, 2002, 2005, 2009	Remote sensing data
Imperviousness	Degree of imperviousness/municipality [ha]	1993, 2002, 2005, 2009	Remote sensing data
Green space cover	Number and size of green spaces/municipality [ha]	2006	Remote sensing data
Elements at risk	Proportion of new settlements and infrastructure/municipality in areas facing a high flood hazard level [%]	1993, 2002, 2009	Remote sensing & GIS data

Source: Weiland 2006

3. The degree of imperviousness per municipality in contrast distinguishes between various levels of imperviousness. The four derived levels comprise LULC categories with (1) a minimal degree of or no imperviousness (woodland, urban green space, barren/open space, agricultural areas), (2) a low degree (sparse vegetation, dispersed built-up areas), (3) an intermediate degree (medium density of built-up areas), and (4) high imperviousness (densely built-up areas, streets, rocky areas). This indicator points to surface infiltration capacities.
4. The number and size of green spaces, both private and public, is measured by remote sensing as both should be considered in flood risk prevention. The greater the number and area of green spaces, the more effective the storm water infiltration. Thus an increase in the value of this indicator points to improvements in flood risk prevention in municipalities with a low distribution of green space or located in risk-prone areas.
5. The proportion of new settlements and infrastructure developments with high flood hazard levels is theoretically restricted by law. Nevertheless, exceptions are possible and captured by the indicator that measures the relative amount of new settlements and infrastructure developments facing a high flood hazard level (one or more events every two years) per municipality.

7.4 Analysis of Demographic and Land-Use Dynamics as Major Driving Forces Behind Flood Risk

It is a central assumption that land use and especially land-use change are key factors in the transformation of natural hazards into risks. Land-use dynamics occur as a result of population and economic growth, and the subsequent increase in transportation infrastructure. Furthermore, population growth and social differentiation increase housing demands to satisfy both basic and higher accommodation requirements. Identifying the principal features of land-use change is an important task and findings should in particular provide information on time, space and quantity. These changes in the demographic dynamics of MAS will be analysed as the major drivers of risks and potential mitigation measures.

As a result of urban growth, the spatial expansion of urban areas into peri-urban and rural localities is now a common phenomenon. Pressured by demographic development, new settlement construction is pushed towards the periphery and beyond. The applied risk concept (see Sect. 7.2) substantiates that population is the most critical element at risk, since affluence and population growth lead to planned and unplanned settlements. These areas of urban expansion are prone to natural hazards such as floods, intensifying risk regardless of the level of vulnerability. The greater the number of people living in hazard-prone areas, the higher the risk that they will be affected by a hazardous event. With respect to flood risk management (see Sect. 7.5.2) and the development of measures (prevention, mitigation and preparedness), population dynamics, including spatial distribution throughout the city and especially in areas prone to natural hazards, must be investigated.

7.4.1 Demographic Change

The MAS comprises 34 of the 52 municipalities that make up the Metropolitan Region of Santiago (MRS). It covers an area of 2,274 km^2 and amounts to 15% of the MRS area. In 2002, the MAS was home to approximately 5.4 million inhabitants. This corresponds to 89% of the population of the MRS and 36% of the total population of Chile. The 50 year period from 1952 to 2002 saw an almost linear population growth in the MAS from 1.41 million inhabitants in 1952, to 3.92 million in 1982, reaching 5.39 million inhabitants in 2002 (CEPAL 2005: 89).

Between 1992 and 2002, the MAS population rose by around 650,000 inhabitants from 4.7 to 5.4 million (INE 1992; INE 2002) (see Fig. 7.2). This overall population increase of 13.7%, which corresponds to an average annual growth rate of 1.38% in the MAS, was accompanied by contrasting processes of growth and decline at the individual municipality level. Whereas 18 of the 34 municipalities showed a decline in population figures, the remaining 16 municipalities experienced an increase. Although population has been in decline in the centrally located municipalities with smaller areas and higher population densities, it has risen in municipalities located on the periphery, which have larger areas and lower population densities. Those with the highest growth rates are *Quilicura* (municipality No. 27) in the northern part, *Puente Alto* (No. 34) in the southeastern part and *Maipú* (No. 29) in the western part of the MAS. *Independencia* (No. 14), *San Joaquin* (No. 8), *Santiago* (No. 9) and *Conchalí* (No. 25) in the central part of the MAS are the municipalities with the highest population losses between 1992 and 2002 (see Fig. 7.2).

Between 2002 and 2006, the population in the MAS increased from 5.4 to 5.8 million (INE 2002; MIDEPLAN 2006), corresponding to a growth rate of 7.3% and an average annual growth rate of 1.8%. The latter exceeds that of the 1992–2002 period and points to a more rapid annual population growth. The spatial pattern of growth and decline for this 4-year-period, on the other hand, is comparable to that of the 1992–2002 period.

Population changes in the MAS are primarily defined by intra-urban migration and immigration processes from municipalities outside the MAS, i.e., from the MRS, from other regions in the country or from abroad (see Chap. 8), and, additionally, by the positive natural population balance stemming from a higher birth than death rate.

7.4.2 Land-Use Change

Demographic change coupled with economic and technological developments leads to land-use change and urban growth. Land-use classifications and changes were calculated for four time steps (1993, 2002, 2005 and 2009) and indicate urban growth in built-up urban areas, as well as in the agricultural environment and other open space around Santiago. Figure 7.3 shows the amount of land transformed into built-up areas for each of the years under investigation. It further demonstrates the direction of urban growth. To quantify the increase in built-up areas since 1993, the figure displays settlement expansion from approximately 500–575 km^2, the equivalent of a 15% increase between 1993 and 2009.

The built-up area in 1993 is highlighted in white expansion over an almost 10 year period is represented in light grey. Further built-up activities for this period are marked in grey, while dark grey denotes the most recent changes. The background to the map shows the central longitudinal valley where Santiago is situated, and the Andean mountains in the east. No land-use changes were calculated for the municipality of *San Bernardo* (No. 33) due to lack of data, so that land-use information can only be given for 33 of the 34 municipalities.

In the northwestern, western and southwestern areas, urban expansion extends far beyond the suburbs (light grey) to the peri-urban areas, transforming agricultural land into urbanized forms. Hence the urbanized form of the *Quilicura* (No. 27), *Pudahuel* (No. 28) and *Maipú* (No. 29) municipalities has undergone significant change as a result of added housing developments and a decline in the open space and agricultural land. Figure 7.3 also shows the patchiness of urban expansion and the loss of urban density in the outskirts of the city. Urban growth to the east of the city has thrust its way into the Andean foothills, striking examples of which are *Peñalolén* (No. 20) (see Sect. 7.4) and *Lo Barnechea* (No. 32). More recent urban development occurred in the municipality of *Las Condes* (No. 22), where housing developments have soared in the last five years, again to the higher altitudes of the Andean foothills. Urban expansion in the

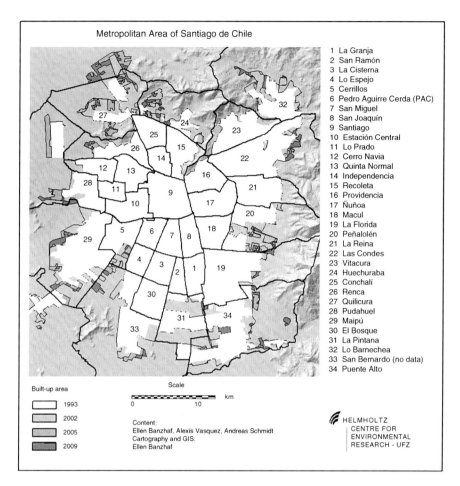

Fig. 7.3 Changes in built-up areas showing dynamics of urban growth (Source: The authors)

municipality of *Puente Alto*, in contrast, has forayed eastwards into the valley south of the Andean foothills.

The Metropolitan Land-Use Plan (PRMS) promulgated in 1994 marked a turning point, allowing as it did urbanization on former agricultural land. It defined a new urban boundary that now drew vast surrounding areas into the category of 'developable land', a legal condition that provided opportunities for real estate investment and consequently accelerated urban dynamics (Reyes-Paecke 2004).

The extent of built-up area per hectare and municipality is evidence of the rapid increase in urbanization across the time frame 1993–2009 (see Fig. 7.4). As a recent development, urban growth at a steady but lower pace was monitored for this decade.

Figure 7.4 shows the proportion of change within the urbanized area and the development of agricultural land in the 33 analysed municipalities. Built-up areas

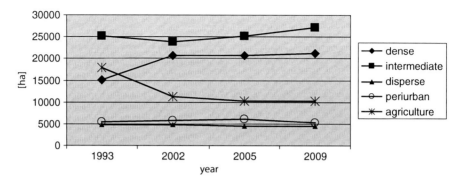

Fig. 7.4 Dynamics of land-use change with reference to urban built-up densities and loss of agricultural land (Source: The authors)

are subdivided into urban areas with specific density categories ('dense', 'intermediate', 'disperse') and peri-urban settlements. The category definition is based on the spectral properties for each of the satellite images. Thus a 'dense' area is built-up to between 70% and 100%, 'intermediate' to between 40% and 69%, and 'disperse' to between 10% and 39%. The category 'peri-urban' refers to the distance from the urban built-up form. The graph illustrates the sharp increase in dense urban settlements during the 1990s, and the ongoing growth of intermediate and dense settlements. So, dense and intermediate dense settlements are the dominant types of urbanization in Santiago. As part of the urbanization trend, areas that were peri-urban during the 1990s have gradually become part of the city and therefore no longer qualify as peri-urban. Instead they now belong to the disperse or intermediate dense settlement categories. The most striking feature is the loss of agricultural land in the area under investigation, which has been transformed for a large part into urbanized areas due to relentless urban expansion.

Exposure to risk-related hazards is extremely difficult to minimize in the historically built-up areas of the centre. In contrast, high-risk exposure in suburban areas could be minimized by appropriate land-use management or even the restriction of further settlement activities. To gain an understanding of urban expansion processes extending to the urban periphery, a quantitative analysis of the suburban municipalities was undertaken (Table 7.2). Whereas the overall population density of each municipality is low, it is significantly higher in the built-up areas as a result of a natural constraint on further expansion.

7.4.3 Diminishing Proportion of Green Space

Green space has a widely recognized role in flood risk prevention in urban areas (Arendt 1996; Sorensen et al. 1998; Fernández 2004), especially in semi-arid environments like Santiago de Chile, where annual precipitation is concentrated

Table 7.2 Development of population density in the peripheral municipalities in total and in their built-up areas

Municipalities	Population density in the municipality (A) [inh./ha] versus population density in the built-up area (B) [inh./ha of built-up area]					
	1992/1993		2002		2005/2006	
	A	B	A	B	A	B
Huechuraba (No. 24)	13.8	108.5	16.5	87.9	18.4	96.9
La Florida (No. 19)	46.5	106.3	51.6	111.8	55.6	121.2
La Pintana (No. 31)	55.4	97.3	62.1	96.2	65.5	101.9
La Reina (No. 21)	39.5	62.8	41.4	62.7	41.1	65.5
Las Condes (No. 22)	20.9	65.2	25.1	72.7	27.2	80.2
Lo Barnechea (No. 32)	0.5	45.2	0.7	45.9	0.9	57.3
Maipú (No. 29)	19.3	58.5	35.2	85.2	50.8	119.6
Peñalolén (No. 20)	33.2	108.3	39.9	106.8	44.0	116.1
Pudahuel (No. 28)	7.0	47.2	9.9	62.3	12.1	59.2
Puente Alto (No. 34)	28.9	71.4	55.9	115.6	71.4	147.6
Quilicura (No. 27)	7.2	25.9	22.0	54.7	30.7	67.1
Vitacura (No. 23)	28.0	59.5	28.8	54.2	28.4	55.8
Total	9.9	69.8	14.1	83.7	17.0	97.3

Source: The authors

over a few days. From a hydrological point of view, green space has several environmental functions, e.g., it contributes to an increase in storm-water infiltration, and helps to recharge ground water and reduce run-off (McPherson 1992; McPherson et al. 1997; Sorensen et al. 1998; Fernández 2004). Green space also contributes to the reduction of surface water pollution before the water reaches the receiving water courses. Consequently these environmental services render green space a key component of strategies for sustainable urban development.

The MAS is characterized by a rather low overall presence of green space and its unequal distribution throughout the city. The total surface of green space comprises 3,841 ha (38.41 km^2) (Figueroa 2009; Figueroa Aldunce and Reyes 2010) or 6.68% of the built-up area. Most of the vegetation cover is found outside the city, in the mountains and protected areas, but this too has declined in recent decades from 39 km^2 in 1993 to 25 km^2 in 2009.

Green space in the MAS is clearly limited in its potential to meet the above-mentioned environmental functions, as a result of its respective size and broad distribution throughout the city. There are 11,607 individual green spaces, 58% of which are less than 1,000 m^2 in size, 93% less than 5,000 m^2 and a mere 3% over 10,000 m^2 (see Table 7.3) (Figueroa 2009; Figueroa and Reyes 2010). There are no regulations or land-use plans in place to define the location, minimum size, vegetation cover or minimum standard of perviousness of green space. Consequently green space is dependent on individual urban development design, with no connection between the individual spaces. Nor are geographical factors such as creeks, streams, wetlands or hillsides taken into account.

The absence of regulations aimed at ensuring the existence of green space and protected wilderness areas in the MAS has led to greater imperviousness, which in

Table 7.3 Number and surface area of green spaces in the Metropolitan Area of Santiago by size rank

Size ranks of green space (GS) [m^2]	Metropolitan Area of Santiago			
	Number	Total amount GS [%]	Surface [ha]	GS [% of total surface]
<501	3,813	32.85	1,263.70	32.90
501–1,000	2,912	25.09	964.10	25.10
1,001–5,000	4,073	35.09	1,267.50	33.00
>5,000	809	6.97	345.70	9.00
Total	11,607	100.00	3,841.00	100.00

Source: The authors, based on Figueroa 2009; Figueroa and Reyes 2010

turn generates flooding even in small storms, as has occurred in Santiago in recent decades (Fernández 2004).

7.5 Flood Risk Analysis

7.5.1 Prevalence of Floods in Santiago de Chile

Floods in the area of Santiago occur regularly as a result of winter precipitation. River floods are a natural phenomenon, but in the case of Santiago recent urban expansion (especially in the last 20 years) has intensified. This results in severe negative impacts on the ecosystem (see Sect. 7.2).

The average annual rainfall in Santiago de Chile varies between 332.3 mm in the city centre (station Terraza Oficinas Centrales DGA, 560 m) and 442.0 mm in the areas of the Andean piedmont (station Antupiren, 920 m) (annual average rainfall 1979–2008). Although these figures seem low, flood risk is on the increase, not least because of the enormous spatial expansion taking place in the MAS and leading to the loss of valuable retention areas and a growth in the number of impervious surfaces. Apart from diminishing retention areas, urban expansion further aggravates the risk of flooding, as more people and values reside in hazard-prone areas: firstly, because they choose to settle down there; secondly, because their new settlements accelerate flood hazards in other parts of the city, e.g., low-lying areas, having reduced infiltration capacities in the catchment area.

Building dams and private mitigation measures protect those who have the funds to pay for them but at the same time add to the amount of run-off transported to the low-lying parts of the city. Flood levels rarely exceed 20 cm, but regularly interrupt urban functions and harm vulnerable households in one way or another. Front yards, outer walls, floors and furniture are affected most, moisture is retained in the walls (physical damage) and people in gated communities are trapped, unable to go to work (immaterial and economic damage). Awareness of the problem and the respective financial means to compensate this loss of capacity to infiltrate with flood mitigation measures such as dams and a functioning storm water evacuation system are inadequate in the study area.

Analysing the issue of floods and suggesting preventive measures with respect to land use requires an understanding of how floods are generated and where hazard-prone areas and vulnerable elements at risk are located. The interrelation of factors and processes that generate flood risk in Santiago de Chile are shown in Fig. 7.5.

Clarification of flood risk generation is based on the risk concept explained in Sect. 7.2.1. Apart from the literature research, household surveys and expert interviews were conducted to gain knowledge on the location-specific variables that define hazard and vulnerability in the study area.

Hazard-related variables include precipitation intensity and probability, the amount of run-off generated in waterways or on the streets, and the location-specific high water threshold that results in floods when exceeded. The **elements at risk** are described by the variables: number of people and number of buildings and infrastructure. **Vulnerability** is the most diverse component of risk. It is determined by variables such as building materials, building usage, position of building in relation to street level (location on or below street level increases exposure), previous experience of people with floods, knowledge of hazard and potential protection measures, as well as age, gender and employment status (see Fig. 7.5). How different processes interact and how they influence flood risk can be demonstrated qualitatively by linking variables with indicators.

A closer look at specific municipalities in Santiago, e.g., La Reina and Peñalolén, permits a deeper understanding of flood events ensuing from land-use change. In recent decades both municipalities have experienced urban expansion and the loss of green space and land for agriculture. Likewise located in the Andean piedmont, they are affected almost bi-annually by flooding of the water courses coming from the mountains.

A specific analysis of LULC changes in the municipalities of *La Reina* and *Peñalolén* shows that one risk-related indicator, i.e., the extent of built-up area per hectare and municipality, increased by 7.47% in *La Reina* (No. 21) and 13.46% in *Peñalolén* (No. 20) between 1993 and 2009 (see Fig. 7.3).

As shown in Fig. 7.6 and Table 7.2, the peripheral municipality of *Peñalolén* was mostly covered by vegetation in 1993 (agriculture and green space) and had therefore high infiltration capacities. The newly evolving housing areas and commercial sites of intermediate density have lowered vegetation coverage considerably and led to higher degrees of impervious surface. Urban expansion was less strong in *La Reina* and primarily spawned a loss of urban green space (densification of existing built-up structures). Figure 7.6 furthermore shows the location of the flood hazard zones from 1986 used in the PRMS (blue hatched polygons) and the hazard-prone areas derived from a flood hazard study of waterways coming from the Andean mountains, which was carried out in 2008 (A&C Consultores 2008). Although the geometries of the two data sets do not allow for direct comparison, it is obvious that flood hazard has undergone a radical change in the last 20 years. In *La Reina*, in particular, where construction activities occurred in former green space, almost all creeks originating in the Andean mountains (blue lines in Fig. 7.6)

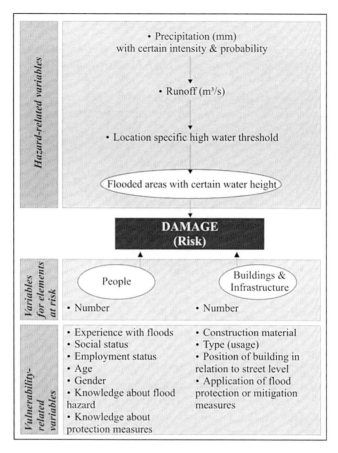

Fig. 7.5 Variables and measures relevant to flood risk generation in Santiago de Chile (Source: The authors)

imply flood hazard. The analysis of the LULC changes in *Peñalolén* shows new construction in areas declared as hazardous in the PRMS. It points to the negative trend in the proportion of new settlements and infrastructure developments with risk-related indicators per municipality in areas with a high flood hazard level and to the number of inhabitants living in flood-prone areas per municipality.

7.5.2 The Impact of Urban Expansion on Flood Risk

Besides the expansion of urban areas to the peripheral regions, built-up areas have now been constructed in locations prone to flooding. This, of course, not only affects people living in *hazard*-prone areas but also any adjacent buildings, since upper areas that previously functioned as water retention areas can no longer

7 Land-Use Change, Risk and Land-Use Management

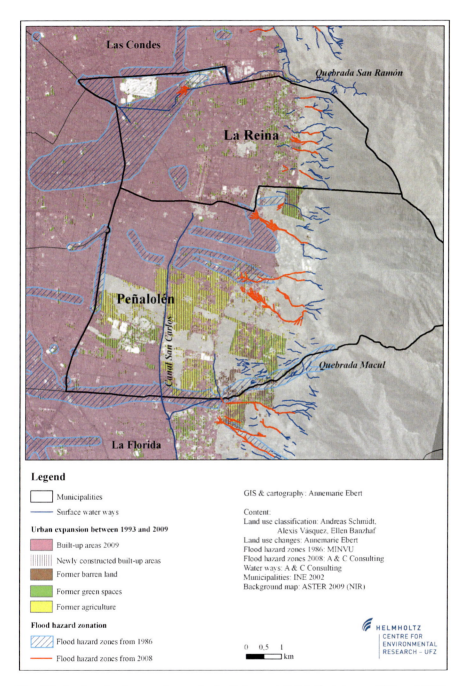

Fig. 7.6 Land-use/land-cover changes in La Reina and Peñalolén between 1993 and 2009, changes in flood-prone areas between 1986 and 2008 (Source: see figure)

provide protection by slowing down the speed and amount of surface run-off. Flood hazard levels depend on the respective land-use pattern but also on relief conditions and the availability of technical measures. Interviews conducted during field work and personal observations indicate that neither municipality has a storm-water canal system or funds for technical installations that would alleviate the situation. Flood risk also depends on the vulnerability of the people and infrastructure concerned. The analysis of census data, remote sensing and GIS data showed that *elements at risk* are located in flood-prone areas. The variables most relevant to flood vulnerability are outlined in the following.

As revealed in the interviews, the parameters referring to the location of a building (position of building in relation to street level and distance from waterway), construction materials and the availability of private protection or mitigation measures are key factors for the assessment of vulnerability. While age group, gender and building type/usage categories were rated least important (42.5%, 30% and 40%, respectively), the household's socio-economic level was rated important to 65% on average. This last value and the interpretation of comments from expert interviews indicate that floods affect not only low-income groups, but also those with middle and higher incomes, a point supported by household survey findings (Reiter 2009). The ultimate difference between these socio-economic groups is the type of damage they suffer.

Qualitative results were obtained with respect to the second underpinning research question of how flood risk should be analysed and assessed, and how it has evolved over time. They show that a key problem is the lack of retention areas to decrease the amount of surface run-off generated by rainfall events. Since a positive change in the hazard-related variables and those relevant to the elements at risk cannot be expected for several – primarily financial – reasons, risk minimization could focus on vulnerability reduction as a further risk component. The findings of the analysis reveal that the investment in better quality construction material and physical mitigation measures would improve the situation (e.g., elevation of buildings located in high-risk flood-prone areas). From the perspective of building and urban infrastructure, the availability of green space is a vulnerability indicator that functions as a buffer zone. Investment in the creation and maintenance of green space as a land-use planning measure would be useful from several perspectives. As it was not possible to assess the knowledge and experience of floods with geodata, field surveys have shown that experience of floods reduces vulnerability, as people are more prepared. Focused campaigns in potentially affected areas could also help to reduce flood risk.

It can be stated that merely reducing the number of inhabitants, buildings and urban infrastructures in hazard-prone areas would lead to an immediate and considerable lessening of the level of risk. Appropriate land-use management would go a long way to achieving this.

7.6 The Role of Land-Use Management in Risk Prevention

Considerations in the previous sections revealed that urban flood risk is to a large extent influenced by population growth and land-use change, both of which lead to urban expansion, increased surface sealing and a loss of urban green space over time. Against this background, a decisive question is whether growing risks can be contained by policy intervention, i.e., land-use management.

In order to analyse the existing flood risk management system in Santiago, actors with competences in land-use and risk management and their respective instruments were investigated. The analysis is based on expert interviews that underwent a content analysis. The experts in question belong to political institutions or are key members of the scientific community and civil society. The components of the land-use management systems are classified as: institutions, instruments and processes. Institutions and instruments are grouped according to the disaster risk management cycle.[2] The following phases of the cycle are relevant:

- Prevention: long-term measures aimed at the complete avoidance of risk (e.g., exclusion of settlement activities in areas prone to flood risk, reservation of retention areas). Land-use planning and land-use management can play an important role here.
- Mitigation: structural measures aimed at the utmost reduction of flood risk (e.g., drainage system, dikes).

7.6.1 Institutions for Land-Use Management and Risk Prevention

The land-use planning and management in Chile is mainly determined by the urban planning sector. Although a legal framework for urban planning is in place, overall spatial planning has no such legal grounds. Urban planning and management is a shared responsibility, whereby the regional and Metropolitan planning levels are accountable to the Regional Office of the Ministry for Housing and Urbanism, while the municipal (communal) planning level is in charge of the respective municipality.

The so-called General Law of Urbanism and Construction (*Ley General de Urbanismo y Construcciones*, LGUC) is the principal legal framework for the area of urban planning and development in Chile. It defines responsibilities and instruments related to the urban areas. The LGUC assigns responsibility for the regulation of urban development at national level to the Ministry for Housing and Urbanism (MINVU), while at the regional level it is dealt with by the SEREMI

---

[2] Because regions prone to flood risk often have to cope with damaging events on an almost regular basis, disaster risk management is structured in the form of a cycle, with 'before' and 'after' stages of a damaging event (FIG Working Group 8.4, 2006).

MINVU (Regional Secretary of the Ministry for Housing), the representation of the MINVU in each region. Regional planning is assigned to the regional government (GORE).

Responsibility for land-use planning and management at local level is assigned to the municipality through two divisions: the Local Planning Division (*Secretaría Comunal de Planificación*, SECPLA) and the Municipal Works Division (*Dirección de Obras Municipales*, DOM). The former is responsible for the social and urban planning that defines the Local Land-Use Plan (*Plan Regulador Comunal*) and the Local Development Plan (*Plan de Desarrollo Comunal*). The latter is responsible for the allocation of building and urbanization permits and the monitoring of construction standard compliance and land-use and environmental regulations. Referring to the Disaster Risk Management Cycle, regional and local planning have the potential to contribute to risk prevention.

As a transversal topic, effective flood-risk prevention calls for the contribution of a multitude of political actors. In addition to the above-mentioned field of urban and regional planning, the Division for Hydraulic Construction in Santiago de Chile (*Dirección the Obras Hidraulicas*, DOH) at the Ministry for Public Works (*Ministerio de Obras Públicas*, MOP) is responsible for the planning and implementing of structural defence measures, thereby contributing to risk mitigation. The Service of Housing and Urbanism (*Servicio de Vivienda y Urbanismo*, SERVIU) is responsible for drainage systems in social housing projects and can also contribute to risk prevention.

7.6.2 Flood Risk Management in Santiago de Chile

The following recommendations related to risk prevention and risk mitigation are derived from the previous considerations, in which the above-mentioned expert interviews played a major role.

7.6.2.1 Prevention: Instruments in the Sector of Urban and Regional Planning

The sector of urban and regional planning is in a position to provide instruments for flood risk prevention due to its regulatory function at an early stage of urban development.

The principal instrument in terms of land-use planning is the *Plan Regulador Metropolitano de Santiago* (PRMS). It regulates the Metropolitan Region's physical development. The PRMS is crucial to risk prevention because it controls land-use patterns, including the designation of risk-prone areas with restricted land use. The PRMS contains two types of areas of restriction or exclusion relevant to risk prevention: (1) areas exposed to natural hazards or hazards induced by human activity, and (2) areas preserved for their ecological value. Although these areas are exempt from construction on paper, restrictions could be relaxed if technical measures for risk reduction were introduced and vegetation loss compensated for elsewhere in the region. The Regional Development Strategy (*Estrategia de Desarrollo Regional*, EDR) of the Regional Government (GORE) is an indicative instrument containing guidelines for the fostering of socio-economic development, while at the same time protecting the ecological functions of the landscape in the Metropolitan Region. The Regional Land-Use Plan (*Plan Regional de Ordenamiento Territorial*, PROT), an indicative instrument to guide regional development and balance out diverse interests that find their physical expression in different land uses, is currently being elaborated by the GORE.

Urban and regional planning can contribute to flood risk prevention as follows:
Urban Planning:

- Declaration of risk-prone areas and definition of possible risks in specific areas, and
- Decision on types of land use allowed in the different areas of the Metropolitan Region.

Regional Planning:

- Determination of guidelines for regional development, and
- Balancing of sectoral interests, e.g., economic development vs. environmental protection.

7.6.2.2 Mitigation: Instruments in the Sector of Drainage-Related Public Construction

The institution legally authorized to design and implement flood protection measures in the MRS is the Ministerio de Obras Publicas (MOP). Through its Division for Hydraulic Construction (DOH), the MOP is in charge of elaborating the overall planning for the implementation of a drainage system in Santiago de Chile. Likewise, the DOH is in charge of approving technical surveys dealing with drainage infrastructure and flood risk prevention measures in new urbanization projects.

In 2001 the DOH/MOP published the master plan for storm-water drainage and evacuation (*Plan Maestro de Drenaje y Evacuación de Aguas Lluvias*), which contains the framework planning of such a drainage system. The instrument contains structural measures designed to reduce the possibility of flood events occurring in the urban area of Santiago. According to the PRMS, the time horizon for the Plan Maestro is the year 2030. The measures are therefore designed to serve the existing built-up area as well as urban expansion in the near future. However, flood mitigation has not been pursued consistently for two reasons: (1) transformation of flood-prone areas into urban areas was achieved primarily by implementation of structural defence measures, and (2) funding for the drainage system has not yet been secured. It is currently unclear whether the measures can be carried out as planned and, if so, whether they will be sufficient to reduce the flood risk effectively once the built-up area begins to expand.

7.6.2.3 Deficits in Flood Risk Prevention and Mitigation in Santiago de Chile

Despite an array of instruments to facilitate flood management, flooding events continue to occur with frequency. The core problems identified in the expert interviews are:

- There is no systematic understanding of risk in the institutions involved in land-use planning and risk prevention.
- The institutions are focused on hazard monitoring and control, but less attention is given to aspects of vulnerability.
- Vulnerability is chiefly seen as a physical condition. Hence, public agencies concentrate on structural measures such as building codes in hazard-prone areas or flood control works.
- Generally there is a strong reliance on technical solutions (e.g., collectors for rain water), while long-term planning and adequate land-use management are hardly considered in risk management. The search for technical solutions still clearly takes priority over preventive thinking.
- Risk prevention, understood as long-term risk avoidance, is likewise poorly integrated into land-use planning and the actions of the institutions responsible.

There are still no stable links between land-use management and risk prevention. Neither are aspects such as how building densities, urbanization at the Andean foothills and the sealing of soils or occupation of river basins impact on the occurrence and intensity of flood events or indeed their change over time currently being given thorough consideration. The complex relation between urbanization, future urban development and risk has not been integrated into current land-use plans to a relevant degree. One reason is lack of knowledge about risks, their generation, their intensity and their spatial distribution. Another issue is the lack of coordination and communication between actors and institutions in the sectors concerned.

7.6.3 Recommended Actions for Improved Flood Risk Prevention in Santiago de Chile

The following indications are adapted to the requirements and challenges of effective flood risk prevention in Santiago de Chile.

7.6.3.1 Prevention: Sector of Urban/Regional Planning

The urban and regional planning sector can make a valuable contribution to the prevention of flood risk by regulating types of land use and restricting the accumulation of values in risk-prone areas. The challenge today is to avoid the continuation of undesirable developments in the current and future development of Santiago,

7 Land-Use Change, Risk and Land-Use Management

since the possibility of intervening in fully consolidated areas is limited and would require dramatic action, e.g., the reallocation of people living in high-risk zones. Measures can only be effective if they are implemented consistently over a long period of time. Actions to foster the adequate implementation of flood risk prevention in the context of urban development include the following:

1. Update and coordination of flood risk analysis
   Detailed and updated information on risk-prone areas, including information on the types of hazards affecting certain areas and the degree of vulnerability involved, is indispensable if urban development planning is to be consistent. Since the assignation of risk-prone areas as included in the PRMS is based at least partially on risk analysis dating back to the late 1980s, an update of the underlying risk analysis is a matter of urgency. Analysis dealing with population and urban growth in the MAS, as well as the flood risk analysis carried out in this research should be applied and extended in order to elaborate on recent analyses of risk-prone areas for the MRS as a whole. Determining whether risk-prone areas have shifted or the probability of flood events has changed in certain areas calls for a risk analysis update that would allow for regulation adjustments referring to land-use types and zoning. The risk analysis should be carried out in close cooperation with the DOH/MOP to create a common planning basis for the two institutions, facilitating intersectoral cooperation in the process.
2. Retention areas in new urbanization projects
   The regulations pertaining to the implementation of new urbanization projects in the outskirts of Santiago count as a central topic of flood risk prevention. The legal obligation to implement structural drainage systems should be supplemented by the obligation to integrate retention areas into the territory taken over by the projects; green space can play a significant role here. Retention areas effectively reduce the peak of surface water flow during intense precipitation periods, but can rarely be implemented in fully built-up areas, although their implementation and maintenance would entail far less investment than structural measures. Additionally, the integration of flood risk prevention and green infrastructure into planning, e.g., by introducing vegetation cover (or perviousness) standards for urbanization projects, would increase infiltration capacity and flood control. Such regulations must be elaborated and implemented in close cooperation with the DOH, since it is they who are responsible for the design and realization of an overall drainage system in Santiago.
3. Watershed management strategy for the MRS
   In order to overcome sectoral approaches with unintended side effects, planning decisions and actions affecting the hydrological system of the MRS need to be coordinated. Ideally, they should follow a common vision of integrated watershed management. To encourage this on the political agenda, a watershed management strategy that includes the creation of a specialized working group to integrate representatives of water-related sectors should be implemented. Based on a recompilation of information on the functions, determinants and interdependencies of the watershed, this group gives consultancy to the official

political institutions. The planning decisions and actions of each sector with impacts on the watershed can be coordinated and harmonized via these consultancies, with the objective of introducing sustainable management to the MRS watershed. A watershed management strategy of this kind would foster a decentralized organization of politics, guided by the natural structure of the region rather than by political administrative boundaries.

4. Raising awareness among the responsible institutions and decision-makers

   A fundamental change must be brought about in the ability of the official planning institutions, i.e., the MINVU, SEREMI MINVU and the relevant decision-makers in the municipalities, to integrate the issue of flood risk prevention in their decisions and planning efforts. It is crucial that the complex interdependencies between physical urban development and the occurrence of flood events are analysed in detail and taken into account by the decision-makers. The relation between urban development and flood events is particularly dramatic when it comes to the settlement of slope areas in the surroundings of Santiago. The increase in surface water flow due to soil sealing and degeneration is alarming. The vast majority of valleys descending from the mountain ranges do not continue in a well-defined watercourse on entering the plain in the Basin of Santiago. During intense precipitation periods this leads to a transformation of streets and sidewalks into drainage paths for the surplus water masses. It is therefore essential that:

   – Slope areas are treated with utmost prudence in the planning process;
   – Extension of the built-up area to hillsides and slopes does not depend on technical feasibility but on a critical evaluation of benefit and harm, bearing in mind that changes in the natural drainage system can cause unforeseeable and undesired impacts, not only in the proximity of the slopes but in the entire region, and
   – Educational measures such as workshops on the topic of physical urban development by the SEREMI MINVU that elaborate on the driving forces behind flood risk and the link to urbanization processes are carried out with the planners and decision-makers concerned. It would also be reasonable to include municipality employees of the local urban planning and development sector in these workshops.

Regarding flood risk prevention, the key motivation for planning decisions should be to prevent aggravation of the existing situation rather than to create new problems, and to mitigate those arising from the misguided urban development planning of the past.

7.6.3.2 Mitigation: Sector of Drainage-Related Public Construction

For decades large parts of the urban area in Santiago did not dispose of a functioning drainage system. This deficit is currently being tackled by measures indicated in the *Plan Maestro de Drenaje y Evacuación de Aguas Lluvias*. To ensure the implementation of effective mitigation measures, several challenges still need to be met.

1. Coordination between MINVU and MOP
   Since responsibilities for constructing the primary and secondary drainage system are divided between the MOP and the MINVU, coordination is crucial. However, it must be intensified, above all with regard to the elaboration and utilization of risk maps. Up until now both institutions have had their own risk mapping, each designed to serve the specific sectoral needs. The elaboration and provision of common risk maps would facilitate coordination between the institutions with regard to determining priority areas for the implementation of the drainage system, as well as a common understanding of the risk-prone areas that need to be treated with great care in the planning process. Common flood risk maps would be one step towards intersectoral management of flood risks instead of each institution treating this topic within its narrow sectoral boundaries. Closer cooperation between the two ministries would also help to ensure that deficiencies due to ignorance of the drainage topic and misguided planning do not occur again in the future development of Santiago. Great attention needs to be given to the regulations on the responsibilities for the construction and maintenance of drainage infrastructure in future urban development, especially in new urbanization projects. These regulations should further consider the possibility of implementing best management practices, e.g., retention areas, since they are a low-cost but effective method of stemming flood intensity.
2. Facilitating information on risk areas to (potential) home owners
   Today, potential home buyers can approach the DOH and obtain information on the flood risk in the area where they plan to purchase property. The population, however, rarely takes advantage of this service. The information is particularly relevant to those planning to acquire new property, allowing them to make an informed decision about investing in a potential risk area. It could also serve established home owners considering an investment in mitigation measures in their own homes. Placing brochures, for example, in estate agent offices and sales offices for new urbanization projects could be an effective method of disseminating information. A regulation making it obligatory for developers of new urbanization projects – in cooperation with the DOH – to inform potential home buyers of the original flood risk is another possibility. Furthermore, reference can be made to the structural measures now in place and the presence of a residual risk in the project area. Facilitating access to such information would raise awareness of risk-related topics in the population and contribute to overall risk awareness in society.
3. Clarifying maintenance responsibilities
   Apart from the general deficiencies of the drainage infrastructure in the built-up area of Santiago, ambiguous regulations referring to responsibilities for the maintenance of existing infrastructure constitute a further problem. Responsibilities

are currently distributed throughout the different sectors and institutions. As a result and not least because of a shortage of finance and staff, this frequently leads to a situation where no institution is prepared to assume responsibility for maintenance measures, while the infrastructure itself continues to deteriorate. It would require a major effort to combine these aspects and responsibilities in one institution, e.g., to keep streets clean of litter that might eventually congest the drains. Far more pressing, however, are strategies to improve the current situation. Analysing existing regulations and organizing workshops for the relevant institutions in order to brainstorm over how to tackle the current situation and secure finance for long-term maintenance of the drainage infrastructure would help most. This is of particular interest in view of new urbanization projects that are obliged to design drainage measures but are not responsible for their long-term maintenance. Long-term maintenance responsibilities must be clarified if the overall drainage system is to function smoothly.

7.7 Conclusions

This chapter analysed population growth and land-use change as the principal driving forces behind urban expansion and the indicators for flood risk production. Despite the overall high population growth rate in the MAS, two strikingly different processes can be observed. Population figures are in decline in the centrally located municipalities and on the increase in the peripheral municipalities, especially in the peri-urban areas in the Andean foothills, in the southern and northern agricultural areas, and in the foothills of the Coastal Cordillera in the west of Santiago. This corresponds to extensive urban expansion of the built-up area in the MAS, which transformed agricultural land into urban areas. After a phase of dynamic urban expansion during the 1990s, the built-up area continued to expand, albeit at a much slower pace.

A conceptual framework comprising the core elements of exposure, elements at risk and vulnerability allows for the analysis and assessment of flood risk. It is underpinned with further risk-related indicators such as increased imperviousness, a decline in green space and a growing proportion of new settlements and infrastructure in areas facing high flood levels. Factors augmenting the flood risk are urban expansion to areas prone to flood risk that coincides with the uneven distribution of annual precipitation, with sudden heavy rainfalls in winter. The risk assessment framework was applied to the municipalities of La Reina and Peñalolén. It shows that the expansion of the built-up area to previously open space and to areas declared as hazardous intensifies flood risk.

With regard to flood risk management, some severe errors have been made in the urban development of Santiago, e.g., misguided urban growth, ignorance of the natural and artificial drainage systems and a general lack of awareness on the part of political actors of the complexity of flood hazards. Existing land-use and risk management systems are not sophisticated enough to carry out effective flood risk prevention. This has led to a large number of hazardous flood events, causing population losses and grave damage to the urban infrastructure. Since Santiago will continue to expand, the need for a broader approach cannot be overlooked.

Due to the implementation of structural measures, the frequency of hazardous floods has been reduced. However, effective flood risk management is still a long way off. The absence of a systemic view of flood risk arising from complex ecological and social processes is the primary weakness in current risk prevention in Santiago de Chile.

The topic of flood hazards and their interdependencies with physical urban development must be incorporated more strongly into planning decisions on the future development of Santiago, so as to avert introducing new problems, while those arising from urban development in the past have not yet been solved.

Existing challenges will only be met if structural mitigation measures are introduced and approaches used that are based on long-term planning and the integration of renewable natural resources and best management practices. Faced with climate change and its expected impact on precipitation, temperature and stream flow, the ambition of regional authorities should be to endorse sustainable land-use planning. Instruments that permit the integration of environmental aspects into the future development of the MRS are required. The *Evaluación Ambiental Estratégica* can certainly make a valuable contribution to improving the current situation.

A fundamental adjustment of regional development must include aspects of risk prevention, which has hitherto been integrated into political decision-making to a negligible degree only. The consideration of long-term flood risk prevention is vital to the sustainable development of the region.

Focusing on risks or hazard mitigation as a technical problem is not enough when it comes to metropolitan agglomerations such as Santiago de Chile. An effective reduction of the impacts of flooding and other disasters demands a more holistic approach. Integrating risk management, watershed management and land-use planning would produce synergetic effects. In other words, the growing degree of impervious soil, river basin alteration, piedmont urbanization and climate change should be taken into consideration simultaneously. The findings of this study are a contribution towards initiating the necessary change in the Metropolitan Area of Santiago.

References

AC Ingenieros Consultores Ltda. (2008). *Diagnóstico de Cauces Naturales*, Sector Pie Andino. Región Metropolitana. Santiago de Chile.
Arendt, R. G. (1996). *Conservation design for subdivisions. A practical guide to creating open space networks*. Washington, DC: Island Press.
Birkmann, J., & Wisner, B. (2006). *Measuring the un-measurable. The challenge of vulnerability*. Bonn: United Nations University, UNU-EHS.

Cardona, O. (2001). *Estimación Holística del Riesgo Sísmico utilizando Sistemas Dinámicos Complejos*. UPC: Universidad Politécnica de Cataluña.

CEPAL – Comisión Económica para América Latina y el Caribe (2005). Buletín demográfico No. 75 Enero 2005. Àmerica Latina: Urbanización y evolución de la población urbana 1950–2000. Santiago de Chile.

Ebert, A. (2008). The understanding of risk-related terms in the Field of Application (FoA) Land-Use Management (LUM) in the project Risk Habitat Megacity (RHM). unpublished working paper.

Ebert, A., Banzhaf, E., & McPhee, J. (2009). The influence of urban expansion on the flood hazard in Santiago de Chile. A modelling approach using remote sensing data. Joint Urban Remote Sensing Event, 20–22 May 2009, Shanghai.

Fernández, B. (2004). Drenaje de aguas lluvias urbanas en zonas semiáridas. *ARQ, 57*, 64–67.

FIG – International Federation of Surveyors, Working Group 8 .4 (2006). The contribution of the surveying profession to disaster risk management. FIG Publication No. 38, Copenhagen. http://www.fig.net/pub/figpub/pub38/figpub38.htm. Accessed 5 May 2010.

Figueroa, I. (2009). *Conectividad y accesibilidad de los espacios abiertos urbanos de Santiago de Chile (AMS, 2006)*. M.Sc. thesis: Urban Studies Institute, Catholic University of Chile.

Figueroa Aldunce, I. M., & Reyes, S. (2010). Distribución, superficie y accesibilidad de las áreas verdes en Santiago de Chile. Distribution, extent and accessibility of green spaces in Santiago de Chile. *EURE, 36*(109), 89–110.

INE – National Statistics Institute of Chile (1992). *Census of population 1992*. Microdatos. Censos de Población. Redatam. Censo de Población y Vivienda 1992. http://www.ine.cl/. Accessed 2 May 2011.

INE – National Statistics Institute of Chile (2002). *Census of population 2002*. Microdatos. Censos de Población. Censo 2002. http://www.ine.cl/. Accessed 2 May 2011.

ISDR. (2004). *Living with risk: A global review of disaster reduction initiative*. Geneva: United Nations.

McPherson, E. G. (1992). Accounting benefits and costs of urban greenspace. *Landscape and Urban Planning, 22*(1), 41–51.

McPherson, E. G., Nowak, D., Heisler, G., Grimmond, S., Souch, C., Grant, R., & Rowntree, R. (1997). Quantifying urban forest structure, function, and value: The Chicago urban forest climate project. *Urban Ecosystems, 1*, 49–61.

MIDEPLAN – Ministero de Planificación y Cooperación (2006). Encuesta CASEN 2006. Santiago de Chile http://www.mideplan.gob.cl/. Accessed 2 May 2011.

NERI – The National Environmental Research Institute (1995). *Recommendations on integrated environmental assessment*, EEA/061/95. Copenhagen.

Niehoff, D., Fritsch, U., & Bronstert, A. (2002). Land-use impacts on storm-runoff generation: Scenarios of land-use change and simulation of hydrological response in a meso-scale catchment in SW-Germany. *Journal of Hydrology, 267*, 80–93.

Reiter, J. (2009). Vulnerabilität gegenüber Überflutungen in ausgewählten Stadtteilen von Santiago de Chile. Unpublished master thesis, Humboldt-Universität zu Berlin.

Reyes-Paecke, S. (2004). Santiago; la difícil sustentabilidad de una ciudad neoliberal. In C. De Mattos, M. E. Ducci, A. Rodríguez, & G. Yáñez Warner (Eds.), *Santiago en la Globalización: ¿Una nueva ciudad?* (pp. 189–218). Santiago Chile: Ediciones SUR-EURE Libros.

Sorensen, M., Barzetti, V., Keipi, K., & Williams, J. (1998). Manejo de áreas verdes urbanas. Technical document ENV-109. BID – Inter-American Development Bank, Washington, DC. http://idbdocs.iadb.org/wsdocs/getdocument.aspx?docnum=1441394. Accessed 19 Oct 2010.

UNDP (2004). *A global report. Reducing disaster risk. Challenge for development*. Technical Report, 05/06, New York.

Weiland, U. (2006). Sustainability indicators and urban development. In W. Wuyi, T. Krafft, & F. Kraas (Eds.), *Global change, urbanization and health* (pp. 241–250). Beijing: China Meteorological Press.

Wisner, B., Blaikie, P., & Cannon, T. (2005). *At risk. Natural hazards, people's vulnerability, and disasters*. London/New York: Routledge.

Chapter 8
Socio-spatial Differentiation: Drivers, Risks and Opportunities

Sigrun Kabisch, Dirk Heinrichs, Kerstin Krellenberg, Juliane Welz, Jorge Rodriguez Vignoli, Francisco Sabatini, and Alejandra Rasse

Abstract The unmistakeable pattern that has long divided Santiago de Chile into the 'rich' northeastern municipalities and the 'poor' rest of the city has recently begun to change. Little is known about the mechanisms that drive these processes of socio-spatial differentiation and their associated opportunities and risks. This chapter explores three trends in socio-spatial change for 39 municipalities of the Greater Metropolitan Area of Santiago: demographic variables such as population trends and intra-metropolitan migration streams, housing and land market trends with respect to construction volume, building permits and land prices, and finally state housing policy with particular reference to social housing programmes. The analysis shows that in combination, these trends have supported the formation of two extreme types of socio-spatial conditions in various locations throughout the city. The chapter stresses the prominent role of state housing programmes. While the contemporary debate on these aspects focuses to a large extent on the agglomeration of low-income groups in social housing schemes on the periphery, the results suggest that the housing policy should give attention to some of the more central locations, where the move towards gentrification could cause the displacement of low-income groups in the future.

Keywords Drivers of socio-spatial differentiation • Housing land market • Intra-urban migration • Risks • Santiago de Chile • Social inclusion • State housing policy

S. Kabisch (✉) • D. Heinrichs • K. Krellenberg • J. Welz • J. Rodriguez Vignoli • F. Sabatini • A. Rasse
Department Urban and Environmental Sociology, Helmholtz Centre for Environmental Research – UFZ, Permoserstraße 15, 04318 Leipzig, Germany
e-mail: sigrun.kabisch@ufz.de

8.1 Introduction

Socio-spatial differentiation is one of the key processes behind the development of cities and determines the welfare of their residents. The term describes the dynamics involved in the distribution of individuals, households and groups, differentiated by socio-economic status, ethnicity or demographic characteristics (age, gender) across a contingent (urban) area and over time (Rodriguez 2001; Sabatini et al. 2001; Rodriguez and Arriagada 2004). This process is closely linked to segregation as the locally fixed pattern of spatial structures and social inequality at a given time (Kabisch et al. 1997).

In Santiago de Chile, processes of socio-spatial differentiation have a long tradition (De Ramón 1978). The city is acknowledged as a highly segregated place with a discernible pattern of socio-economic groups, dividing the city into the 'rich' northeastern municipalities and the rest of the city described as 'poor'. As a result of public policies, this 'traditional pattern' of socio-spatial difference broadened considerably in the 1970s and 1980s of the last century under the Pinochet military dictatorship. Recent decades, however, have begun to witness change (Sabatini et al. 2001; Caceres and Sabatini 2004). The 'new' trend manifested itself in the emergence of government housing programmes for extremely poor households and housing for high and middle-income groups provided by the private sector (Tokman 2006). While initial research results (Caceres and Sabatini 2004) show evidence of a gradual shift in the socio-economic composition of neighbourhoods and communities, little is known about the mechanisms that drive these changes or the resultant social and spatial consequences and risks.

The aim of the present chapter is to address this gap. Following the hypothesis that cities are locations of systematic risk production (see Chaps. 1 and 3 of this volume), the chapter explores conceptually and empirically the linkages between drivers of socio-spatial change and the risks engendered by lack of and opportunities for distinct social inclusion. The examination is guided by a set of related questions:

- What is the link between migration patterns of different socio-economic groups, the land market, government housing policy and social housing programmes?
- What socio-spatial 'hotspots' and differentiation types emerge as a result of combined trends in migration, social housing and the land market?
- What positive effects in terms of opportunities for inclusion or negative consequences such as risk of social exclusion are associated with these 'hotspots'?

The chapter is organized as follows: Sect. 8.2 outlines the 'risk perspective' adopted and connects it to concepts of inclusion. It also provides guidance on methodological considerations and, contrary to previous studies on the topic, highlights that the investigation should focus on the Greater Metropolitan Area of

Santiago (GMAS). This agglomeration of 39 municipalities includes locations recently incorporated into the urban area. To permit a spatial disaggregation of socio-spatial differentiation trends and drivers, the study divides GMAS into five clusters. These are characterized by a set of criteria consisting of the geographical location of the municipalities, as well as similar socio-economic and demographic features. This will be illustrated by examples at the municipal level.

The subsequent Sects. 8.3, 8.4 and 8.5 describe the selected drivers of socio-spatial differentiation. Section 8.3 examines a set of demographic variables, such as population trends and intra-metropolitan migration streams in the GMAS for the period 1982–2006, as a means of tracing the spatial distribution of different socio-economic status groups. Section 8.4 summarizes the housing land market with respect to construction volume, building permits and land prices. Section 8.5 provides an account of the state housing policy with particular reference to the social housing programmes that, as the section will illuminate, play a significant role in providing households from lower socio-economic groups with affordable housing.

Section 8.6 provides a synthesis of all three drivers and identifies socio-spatial 'hotspots', risks and opportunities. It highlights for five different clusters that the combination of the drivers with four types of socio-spatial conditions is crucial.

Linking these types to the conceptual ideas and the risk perspective developed in Sect. 8.2 and 8.7 concludes with a summary of the risks and opportunities in current socio-spatial differentiation trends in Santiago de Chile.

8.2 Understanding and Analysing 'Risks' Via Social Inclusion

The understanding and conceptualization of 'risk' in this section follows the perspective that cities are locations of systematic risk production as outlined in detail in Chap. 3 of this volume. Two constitutive arguments of the risk perspective are of importance in the context of this study on socio-spatial differentiation in Santiago de Chile.

The first argument understands megacities as places where the extreme concentration of people and values produces a vast number of desired, positive consequences (opportunities) and undesired, negative consequences (risks). As the terms 'desired' and 'undesired' indicate, the consequences tie into the expectations of individuals, social groups and society at large. This section translates positive consequences as opportunities for social inclusion and negative consequences as the risk of not achieving social inclusion.

The concept of social inclusion describes the participation of individuals and social groups in functional systems such as the economy, politics, welfare and social networks (Kleve 2005). To be included in a functional system, individuals need access to certain means of communication, e.g., money, knowledge, political influence and citizen rights. In addition, inclusion involves integration into social systems. "To be included is to be accepted and to be able to participate fully within our families, our communities and our society" (Guilford 2000, p. 1). Social inclusion is not fixed but describes various intensities or levels of accumulated advantage stemming from 'mechanisms' that allow individuals or groups to be part of functional systems.

The second argument in the systematic perspective asserts that risk is a product of the more 'intrinsic' processes and functions that take place in the city, operating to meet citizen expectations, rather than the result of 'external' events such as floods or earthquakes. Paradoxically, the production of opportunities and risks occurs simultaneously, whereby risks occasionally materialize as 'unintended' side effects. In other words, the necessity to include people in basic functional spheres, e.g., availability of employment, shelter, water supplies, may simultaneously produce negative consequences. For example, cities offer vast employment opportunities (corresponding to the positive expectation that attracts immigrants) but often under precarious conditions, e.g., informality, lack of protection, hazardous working environments. The production of opportunities and risks occurs through co-existence of specific drivers.

The spectrum of opportunities and risks is by no means confined to a certain group or class of people or situation (Kuhn 2000). Rather, it is distributed unevenly across different groups in the urban population. In addition to socio-economic status, the distribution of risks and opportunities varies across locations. Spatial mismatch of structural elements is said to produce disadvantaged areas, where people's access to opportunities for participation in civil society is curtailed (Kain 1992; Ihlanfeldt and Sjoquist 1998). Also, risks and opportunities are not static but undergo dynamic change over time.

8.3 Methodological Approach

The analysis of the relationship between socio-spatial differentiation, opportunities for social inclusion and the risk of social exclusion provided in this chapter proceeds in four steps (see Fig. 8.1).

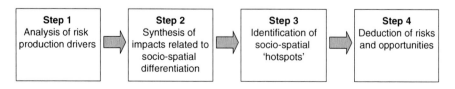

Fig. 8.1 Sequence of analytical steps (Source: The authors)

The first step analyses and examines a set of related 'drivers' that underlies socio-spatial differentiation processes at the interface between society, the market and the state.

The second step combines the impacts of migration patterns, land prices and changing social housing unit patterns. The results of synthesis demonstrate and explain the different levels of social inclusion and the associated consequences.

The third step identifies four types of socio-spatial 'hotspots' arising from 'cumulative' trends. In particular

1. Large-scale agglomeration of lower-income households;
2. (Re-)concentration of higher socio-economic status groups;
3. A 'new' spatial mix of households from different socio-economic groups in close proximity, and
4. Co-existence of urban renewal and marginalization.

The fourth step covers the final determination of risks and opportunities in relation to the level of social inclusion.

The analysis draws on a wide range of mostly official statistical data sources:

- The section describing demographic development and migration patterns uses the National Population Census (INE) and survey data in the case of Encuesta CASEN, 2006 (carried out by MIDEPLAN).
- The section on land markets draws on a study conducted by Arriagada (2009). This study contains data from the National Population Census (INE) and the Cámara Chilena de la Construcción (CCHC), from the Boletín del Mercado de Suelo for 34 'urban' municipalities, and the Real Estate Supplement (Suplemento de Propriedades) of the El Mercurio newspaper for the remaining five municipalities.
- The section on housing policy and social housing programmes comprises results from a study by PROURBANA (Brain et al. 2009b) based on the interpretation of data from the National Housing Ministry (MINVU), in particular the 'Observatorio Habitacional.

The time frame for data processed by statistical analysis is oriented towards the availability of data from 1982 to 2006. The analysis concentrates on the period from 1992 until 2006, while migration patterns are analysed for the periods 1987–1992, 1997–2002 and 2002–2006.

Apart from the temporal dimension, the spatial dimension must be underlined. Various scales are addressed. The overall regional scale covers the Greater Metropolitan Area of Santiago (GMAS) consisting of 39 urban and urbanizing municipalities. In order to carry out spatially comparable research aimed at generalized and transferable results, we divided the urban area into five clusters. A set of criteria consisting of the geographical location and similar socio-economic and demographic features was used for this specification. The five clusters are: Centre, Peri-Centre, Eastern Peri-Centre, Periphery and Extra-Periphery (see Fig. 8.2). In addition to the cluster level, several case municipalities were chosen to deepen the analysis.

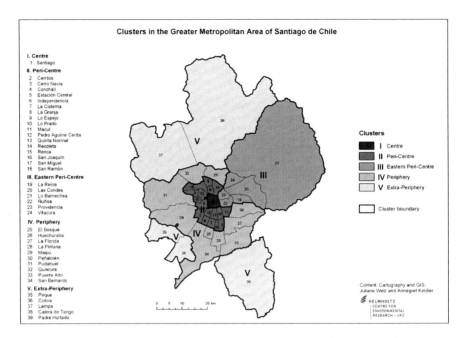

Fig. 8.2 The Greater Metropolitan Area (GMAS) of Santiago and the five municipal clusters Centre, Peri-Centre, Eastern Peri-Centre, Periphery, Extra-Periphery (Source: The authors)

8.4 Drivers of Socio-spatial Differentiation: Demographic Development and Migration

Population exposure, e.g., the condition of dwellings or spatial location, is described as a prime mechanism of risk production. Following a review of the literature, however, it is surprising that research on processes of socio-spatial differentiation frequently ignores issues of demographic development, such as population growth or shrinkage and the residential mobility of specific groups. Only a handful of studies have paid attention to ongoing residential mobility in neighbourhoods that registered change in socio-spatial structures (Massey et al. 1994; Friedrichs and Nonnenmacher 2008; Rodriguez 2008; Bolt and van Kempen 2010). According to Bolt and van Kempen (2010), residential mobility can be seen as a basic influencing factor of social inclusion and therefore as a key indicator for risk production, particularly among minority groups. This section considers three perspectives for the analysis of residential mobility:

1. The general population development of Greater Metropolitan Area of Santiago (GMAS);
2. Intra-urban migration patterns in GMAS, differentiated by five spatial clusters, and

3. The effects of intra-urban migration of various educational groups on cluster and municipal levels.

In terms of demographic drivers, the total population in GMAS increased substantially between 1982 and 2006, with the majority of the population (>90%) concentrated in the Metropolitan Region (MR) in 2006 (see Table 8.1). Nevertheless, population growth rates have oscillated slightly since 1982. This can be explained largely by the importance of net migration up to 1992, in the aftermath of which a host of influencing factors have to be considered (e.g., ageing, decline in fertility rate). In general, intra-urban migration flows increased up to the last census period in 2002.

Table 8.1 Population, intra-urban migration and net migration for GMAS 1982–2006

	Greater Metropolitan Area of Santiago (39 communes)			
	1982	1992	2002	2006
Total population GMAS	3,978,754	4,857,676	5,599,761	6,036,574
Share of population with MR (in%)	91.0	92.4	92.4	92.1
Growth rate		1982–1992	1992–2002	
Annual growth rate (%)		2.0	1.4	
Migration	1977–1982	1987–1992	1997–2002	2002–2006
Intra-urban migration (GMAS)	483,834	790,480[a]	818,224	486,024
No migrants (GMAS)	2,821,341	3,112,456[a]	3,911,953	4,908,837
Net migration of GMAS	119,378	50,751[a]	−27,241	−20,169
Net migration/10.000 inhabitants	300	104	−49	−33

Source: The authors, based on INE 1982, 1992, 2002, and MIDEPLAN 2006
[a]Several municipalities were amalgamated for the 1992 census (e.g., Las Condes, Vitacura and Lo Barnechea) to reduce the effect of migration resulting from alterations to official municipal boundaries

With regard to specific migration patterns as drivers of socio-spatial differentiation processes, intra-urban migration flows differentiated by municipal clusters were analysed. On the whole developments point to the unmistakeable direction of intra-urban migration flows from the Centre to the Periphery (see Fig. 8.3). The Centre and Peri-Centre suffer from massive ongoing outmigration. While the population of the Centre decreased in the early 1980s and 1990s, the Peri-Centre saw a continuous reduction in the negative values of intra-urban migration at the beginning of the twentieth century. Similar tendencies were observed for the Eastern Peri-Centre – a high-income area –, which registered population losses due to intra-urban migration, especially in the early 1980s and 1990s. Since 2002, this cluster has reconfirmed its attractiveness and produced more positive values for intra-urban migration. The peripheral cluster of GMAS received the highest positive values of net migration, not least as a result of intra-urban residential mobility. Finally, the role of the Extra-Periphery as a receiver of intra-urban migrants is still underexploited. Its relatively low positive net migration values, however, are the exclusive result of intra-urban migration since the 1990s.

Overall, the results of the analysis indicate quite different development paths in terms of space and time. Three migration tendencies stand out as significant:

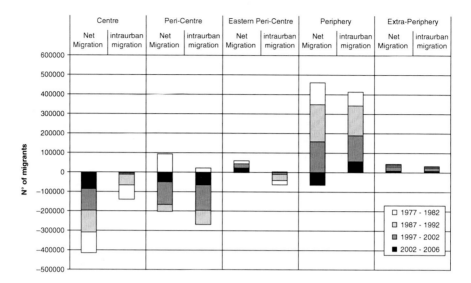

Fig. 8.3 Migration patterns differentiated by municipal cluster (Source: The authors, based on INE 1982, 1992, 2002 and MIDEPLAN 2006)

1. The continued increase of households in the Periphery;
2. The continued out-migration from the Centre to the Periphery and, more recently, to the Extra-Periphery, and
3. Reassertion of the attractiveness of the Eastern Peri-Centre.

An important risk production mechanism is the direction of intra-urban migration flows under the aspect of educational attainment. In the census data this aspect constitutes the proxy indicator of socio-economic strata. The intra-urban migration patterns of the group that concluded their educational career (population >25 years old) was analysed.

On the whole, none of the municipalities reduced the average education span between 1992 and 2006 (see Table 8.2). Nonetheless, the average number of years in education varies significantly between the municipalities (from 6.5 in Lampa to 13.2 in Vitacura). The highest increases for the period 1992–2006 correspond to the centre, peripheral and extra-peripheral clusters. Whereas the extra-peripheral cluster of the GMAS still harbours a notable proportion of rural inhabitants, the central and peripheral clusters are characterized by an almost exclusively urban population. For the peripheral and extra-peripheral cluster, the statistical factor is influenced both by the initially low level of education in 1992 (on average 8.4% in the Periphery and 7.1% in the Extra-Periphery), which facilitated a positive increase, and the positive net migration in the period 1997–2002 in almost all cases. On the other hand, educational levels have not increased significantly in all peripheral municipalities (e.g., El Bosque, Maipú and San Bernardo indicate only a medium increase). The peri-central cluster registered medium increases, since several of the municipalities began with low levels of education (an average of 8.5%), all of

Table 8.2 Greater Metropolitan Area of Santiago (39 municipalities and five clusters): average years of education, increase in years of education in percentage and average years of education of immigrants and emigrants aged 25 or over

Municipalities and clusters	Average years of education			Increase in %			Average years of education of immigrants and emigrants					
	1992	2002	2006	1992–2002	2002–2006	1992–2006	1992 Immigr.	2002 Emigr.	2002 Immigr.	2006 Emigr.	2006 Immigr.	Emigr.
Centre	10.2	11.8	13.3	16.2	12.1	30.3	10.1	10.2	11.6	11.4	13.6	12.8
Santiago	10.2	11.8	13.3	16.2	12.1	30.3	10.1	10.2	11.6	11.4	13.6	12.8
Peri-Centre	8.5	9.6	10.1	11.9	5.2	17.7	8.4	8.4	9.5	9.6	11.3	11.2
Cerrillos	8.8	9.6	10.1	9.7	5.0	15.2	8.6	8.5	9.5	9.7	11.0	12.1
Cerro Navia	7.2	8.4	8.6	15.7	2.7	18.8	7.1	7.2	8.3	8.4	10.4	9.0
Conchalí	8.4	9.3	9.7	11.4	3.7	15.5	8.2	8.4	9.2	9.3	11.5	9.9
Estación Central	9.0	10.1	10.3	12.1	2.2	14.6	8.8	8.9	9.9	10.0	13.0	11.2
Independencia	9.6	10.5	11.2	9.3	7.3	17.2	9.5	9.4	10.4	10.6	11.8	11.2
La Cisterna	9.5	10.7	11.5	12.0	7.6	20.5	9.4	9.2	10.6	10.7	13.0	13.0
La Granja	8.0	9.0	9.0	11.8	0.5	12.4	7.9	7.8	8.9	9.0	9.9	11.8
Lo Espejo	7.5	8.5	9.4	12.5	11.0	24.8	7.4	7.3	8.4	8.5	10.6	9.3
Lo Prado	8.4	9.4	9.8	11.6	4.1	16.2	8.2	8.2	9.3	9.5	10.8	10.2
Macul	9.9	10.8	10.7	9.3	−0.5	8.7	9.7	9.6	10.7	10.8	12.0	13.7
Pedro Aguirre Cerda	8.3	9.2	9.4	11.2	1.3	12.6	8.2	8.1	9.2	9.3	8.5	11.0
Quinta Normal	8.6	9.7	10.1	13.1	4.3	18.0	8.5	8.6	9.7	9.8	10.8	10.7
Recoleta	8.5	9.4	9.9	10.9	5.4	16.9	8.4	8.3	9.3	9.5	10.6	11.5
Renca	7.7	8.7	9.4	12.9	8.2	22.1	7.6	7.8	8.7	8.8	12.2	10.2
San Joaquín	8.6	9.5	10.4	10.9	9.7	21.7	8.4	8.5	9.5	9.6	11.4	11.7
San Miguel	9.8	11.3	12.3	14.9	9.3	25.5	9.7	9.6	11.2	11.2	13.6	12.5
San Ramón	7.5	8.6	9.1	13.6	6.4	20.9	7.4	7.5	8.5	8.7	10.3	11.1
Eastern Peri-Centre	12.0	13.0	13.5	8.1	3.8	12.2	12.0	11.8	13.0	12.9	14.1	14.0
La Reina	11.9	12.7	12.5	7.0	−1.3	5.6	11.8	11.6	12.7	12.7	13.9	14.2
Las Condes	12.7	13.5	14.5	6.6	7.7	14.9	12.7	12.4	13.5	13.4	15.1	13.3
Lo Barnechea	10.1	11.6	10.9	14.5	−5.9	7.7	10.2	9.3	11.6	11.3	12.1	13.5

(continued)

Table 8.2 (continued)

Municipalities and clusters	Average years of education			Increase in %			Average years of education of immigrants and emigrants						
	1992	2002	2006	1992–2002	2002–2006	1992–2006	1992 Immigr.	Emigr.	2002 Immigr.	Emigr.	2006 Immigr.	Emigr.
Nunoa	11.7	12.7	13.7	9.1	7.4	17.2	11.6	11.4	12.6	12.6	14.5	13.8
Providencia	12.7	13.9	14.1	9.5	1.3	10.9	12.7	12.6	13.8	13.7	14.6	14.0
Vitacura	13.2	13.6	15.3	3.4	12.1	15.9	13.4	13.2	13.8	13.8	14.5	14.9
Periphery	8.4	9.7	10.0	15.5	3.6	19.6	8.2	8.1	9.6	9.5	11.2	11.7
El Bosque	8.3	9.2	9.3	11.1	1.3	12.6	8.1	8.0	9.1	9.3	11.2	11.8
Huechuraba	7.3	9.2	9.4	26.9	2.0	29.4	7.1	7.1	9.2	8.7	12.4	13.3
La Florida	9.6	10.7	11.2	11.6	4.2	16.3	9.5	9.4	10.7	10.7	12.7	12.0
La Pintana	7.2	8.0	8.3	11.3	3.9	15.7	7.1	7.0	8.0	8.1	8.6	10.4
Maipú	9.8	10.9	11.2	11.1	2.9	14.3	9.7	9.5	10.8	10.9	11.2	12.0
Penalolén	7.9	9.6	9.5	21.1	−0.7	20.2	7.8	7.6	9.6	9.3	13.0	11.5
Pudahuel	7.8	9.3	10.0	19.3	7.6	28.4	7.6	7.5	9.3	9.2	10.3	10.4
Quilicura	8.4	10.2	11.1	21.0	9.2	32.1	8.3	7.9	10.2	9.7	11.6	13.1
Puente Alto	9.0	10.0	10.5	11.1	4.4	16.0	8.9	8.4	10.0	10.0	10.6	10.7
San Bernardo	8.3	9.4	9.4	13.2	0.5	13.7	8.2	8.4	9.3	9.3	10.2	12.2
Extra- Periphery	7.1	8.9	9.0	25.3	0.7	26.1	7.0	6.7	8.8	8.6	11.5	11.8
Pirque	7.8	9.6	9.0	23.0	−6.6	14.9	7.6	7.0	9.5	9.2	11.7	15.0
Colina	7.0	8.4	9.4	21.2	11.4	34.9	6.8	6.7	8.4	8.2	12.4	11.8
Lampa	6.5	8.2	8.1	25.9	−1.1	24.5	6.4	6.3	8.2	7.9	9.5	11.0
Calera de Tango	7.1	9.5	9.4	32.5	−0.8	31.4	7.0	6.8	9.4	9.0	13.7	11.7
Padre Hurtado	–	8.8	9.0	–	1.8	–	–	–	8.7	8.6	9.9	9.6

Source: The authors, based on calculations by processing census data INE 1992, 2002 (with Redatam) and Encuesta CASEN 2006

8 Socio-spatial Differentiation: Drivers, Risks and Opportunities

which have experienced a population decline in recent decades. The Eastern Peri-Centre indicated comparatively moderate increases due to high levels of education at the outset (on average 12%).

Regarding the influence of migration on the average number of years spent in education in these areas and on the spatial proximity of the different socio-economic groups, two observations call for immediate attention: the direction of migration flows of the more educated population (>12.00 years) and that of the population with low educational attainment (<12.00 years).

A general movement of people with higher educational attainment from the central clusters (Centre and Peri-Centre) towards the urban periphery (Periphery and Extra-Periphery) has been taking place since 1992. Three specific development paths can be identified:

- The first refers to the central cluster. Although characterized by negative net migration of inhabitants with higher educational levels, positive values were registered for and particularly after the period 1997–2002. As a result of this recent trend, the city centre raised its educational level by migration, unlike in previous periods.
- The second refers to the Eastern Peri-Centre. Although this cluster was affected by negative net migration of the more educated residents, the results of Encuesta CASEN indicated positive values. With the exception of Vitacura, however, the municipalities of the eastern peri-central cluster registered a positive impact of intra-urban migration on their already high educational levels for the periods 1987–1992 and 1997–2002 (especially Lo Barnechea). For the period 2002–2006, the Eastern Peri-Centre (notably Ñuñoa and Las Condes) registered very high positive changes in the average number of years spent in education due to migration. The impact of intra-urban migration on the eastern peri-central cluster is evident from the ongoing increase in the high average number of years its inhabitants spend in education. Thus for educated families it remains an attractive place to live.
- The third refers to the peripheral and extra-peripheral clusters. Both clusters registered very high positive net migration of residents with higher educational attainment. These values are decreasing over time, but maintain the level of the Eastern Peri-Centre. As a result, municipalities of the peripheral cluster such as Huechuraba and Peñalolén – two historically emblematic, poor municipalities – registered major increases in terms of the number of average years of education as a result of intra-urban migration. Nevertheless, recent migration trends (2002–2006) failed to repeat these results. On the contrary, there is evidence of a reverse tendency, in particular in Huechuraba, where the average number of years spent in education among people of 25 and over decreased by 12% as a result of intra-urban migration. Only El Bosque, La Florida, Peñalolén and Pudahuel continued to increase their values. Similar to the peripheral cluster, the five municipalities of the extra-peripheral cluster have been widely favoured by intra-urban migration in the periods 1987–1992 and 1997–2002, produced an increase in the average number of years spent in education of between 1% and

5% in the course of 5 years. Again, recent migration trends (2002–2006) have failed to show similar results. As a sole exception, Pirque continues to register positive increases. In summary, the fact that immigrants were for the most part well-educated families seems to have had a significant influence on the educational levels of these municipalities. Nonetheless, recent trends tend to impeach the optimistic vision of the heterogeneity and social ascent of the Periphery and Extra-Periphery.

People with low educational attainment have also been migrating from the central clusters (Centre and Peri-Centre) towards the urban periphery (Periphery and Extra-Periphery) since 1992. The Centre, the Eastern Peri-Centre and to some extent the Peri-Centre are habitually characterized by the continuous expulsion of low-skilled residents. Of particular interest are the peripheral and extra-peripheral clusters. Both have been recipients of residents with low educational attainment. As a result, the concentration of low educational qualifications in several municipalities (primarily La Pintana, Puente Alto and San Bernardo) reduced the overall educational level of the peripheral cluster, in particular in the periods 1997–2002 and 2002–2006.

Against this background, residential mobility in Santiago de Chile has displayed three general tendencies. Firstly, residential intra-urban mobility plays a key role in the expansion of GMAS and impacts on socio-spatial differentiation processes. Nevertheless, the importance of the extra-peripheral cluster for future expansion could not be verified. Secondly, more highly skilled residents tend towards greater residential mobility than those with lower education and could have a major influence on processes of socio-spatial differentiation. Thirdly, the Centre, the Eastern Peri-Centre and the Periphery are the municipalities most affected in terms of modification of educational levels as a result of intra-urban mobility. Over time, however, highly divergent paths of development were identified. These three tendencies question the role of land markets and housing policies in residential mobility.

8.5 Drivers of Socio-spatial Differentiation: Housing Construction and Land Prices

The function of the land market is to allocate land resources according to supply (constructors, real estate companies) and demand (residents, business companies, industries). The relationship between supply and demand operates within a system of regulating factors, such as norms and standards, providing incentives and disincentives for investment decisions and land use (e.g., housing subsidies, property tax, zoning and land-use plans, infrastructure investments). In Santiago de Chile the land market for residential purposes (houses, condominiums) is both a key driver of socio-spatial differentiation processes (Sabatini 2000) and an influencing factor in decisions on housing construction locations (supply) as well as the location or moving of households (demand).

This section considers two sets of variables, both of which relate to the 'supply side' of residential land and housing: housing construction (housing stock, building permits) and land offer (quantity and price). It analyses the temporal and spatial trends of these variables to define general 'patterns'. According to previous studies (Sabatini and Salcedo 2007), such trends and patterns are the result of efforts made by proprietors and real estate developers to realize profits from land and the corresponding housing market. Real estate developers have begun to extend their residential land investments, no longer confining themselves to high-income neighbourhoods. Instead they locate development in areas with low land prices and convert them into 'high value' locations, thereby maximizing their profits. In this real estate market logic the location decision is subordinated to the development of 'cheap' land, which in recent decades has corresponded to areas in the Periphery and Extra-Periphery of Santiago with low-income residents. Consequently, areas that previously had low land prices and a strong presence of low-income residents have over time exhibited a larger presence of upper-income residents (Sabatini 2000). Furthermore, socio-spatial differentiation processes are influenced by social housing programmes, a prominent feature of housing construction in Santiago (see the section on housing policy below). These programmes are likewise closely linked to land prices. To maintain an affordable range of housing for beneficiaries, programmes primarily target peripheral locations where land is 'cheap' (Brain and Sabatini 2006; Brain et al. 2009a).

Turning to housing construction as the first variable under consideration, available census data shows that the overall residential housing stock in Santiago increased from 1.2 million units in 1992 to 1.5 million units in 2002. This increase of roughly 300,000 units (an annual rate of 2.4%) slightly exceeds the population growth rate for the same period.

Manifestation of this increase differs across the municipal clusters. As shown in Table 8.3, the peripheral cluster has seen a strong increase in both numbers and proportion. This cluster alone absorbed three out of four new housing units, amounting to an annual increase of more than 6%. In contrast, the housing stock in the central and peri-central clusters came close to stagnation. Interestingly, the

Table 8.3 Change in housing stock 1992–2002

Cluster of municipalities	Housing stock 1992	Relative share (%)	Housing stock 2002	Relative share (%)	Change 1992–2002	% Change 1992–2002
Centre	60,029	5.25	64,167	4.56	4,138	0.69
Peri-centre	535,008	46.83	531,231	37.77	−3,777	−0.07
Eastern Peri-centre	181,043	15.85	230,221	16.37	49,178	2.72
Periphery	337,472	29.54	544,441	38.70	206,969	6.13
Extra-Periphery	28,843	2.52	36,606	2.60	7,763	2.69
Aggregated GMAS	1,142,395	100.00	1,406,666	100.00	264,271	2.31

Source: The authors, based on INE 1992, 2002

Table 8.4 Number of building permits for residential dwellings aggregated by municipal cluster

Cluster of municipalities	Average annual building permits (dwellings) 1992–1996	Average annual building permits (dwellings) 1997–2001	Building permits (dwellings) 2006
Centre	2,286	3,966	15,427
Peri-centre	7,447	6,659	9,608
Eastern Peri-centre	11,820	7,097	13,733
Periphery	34,103	25,162	19,909
Extra-Periphery	1,077	1,864	2,444
Aggregated GMAS	56,733	44,747	61,121

Source: The authors, based on Arriagada 2009

eastern peri-central municipal cluster registered a significant increase despite a stagnation in population trends for the same period.

The number of building permits for residential dwellings in GMAS generally backs this trend but shows some interesting specifications for more recent years.

With the exception of the peripheral cluster, all clusters show a higher number of permits in 2006 compared to the annual average between 1992 and 2001 (see Table 8.4). The Centre registered a significantly higher number of building permits in 2006 compared to annual averages for the 1990s. This dramatic increase is due to the trend towards high density and high rise development, the comparatively large proportion of use change (e.g., residential to commercial) and the availability of subsidies for construction provided by the MINVU urban renewal programme (renovación urbana) (Arriagada 2009). In the adjacent eastern peri-central and peri-central clusters, the number of permits issued likewise exceeds averages for previous years, particularly in Providencia and Ñuñoa. The sole exception is the peripheral cluster, where the number of building permits issued in 2006 fell below the average of previous years. Nonetheless, the total number continued to be the highest of all clusters (with some degree of concentration in the municipality of Maipú, with almost 8,500 permits). Finally, the number of permits issued in the extra-peripheral municipalities displayed a slight increase but remained low in absolute terms.

The review of trends in land offer (quantity and price), as the second set of variables, also reveals substantial differences between the individual municipal clusters.

As evident from Table 8.5, the largest share of land offered for residential development is found in the peripheral and extra-peripheral clusters. This is not unusual considering that the municipalities in these clusters, particularly in the Extra-Periphery, have far greater land reserves for potential urbanization than those located in the more central clusters. Some extreme inter-annual variations can be observed in the Extra-Periphery, where the quantity of land offered in 1994 and in 2000 exceeded that of other years substantially. The explanation for these inconsistencies is linked to legislation. New laws on the 'Zona de Urbanización Condicionada' (ZODUC) and 'Proyecto de Desarrollo Urbano Condicionado' (PDUC) were introduced in both periods, putting preconditions in place for the development of large-scale residential estates.

Table 8.5 Land offered for transaction (in m²)

Cluster	1992	1994	1996	1998	2000	2002	2006	2007
Centre	95,013	77,170	115,960	93,654	89,103	106,247	77,893	78,078
Peri-centre	471,150	371,388	616,052	234,504	303,702	209,487	293,356	335,003
Eastern Peri-centre	1,804,677	1,201,540	1,531,338	683,840	735,717	825,158	635,534	613,457
Periphery	3,498,253	2,760,106	3,330,754	1,551,212	1,066,704	883,774	1,213,823	1,073,586
Extra-Periphery	3,067,953	42,465,772	4,062,475	1,778,814	16,536,557	1,717,125	1,031,924	2,126,685

Source: Boletín del Mercado de Suelo and El Mercurio; aggregated volume for the last quarter of the years shown[1]

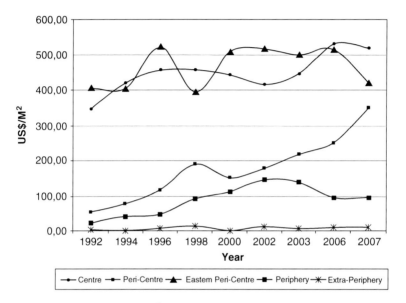

Fig. 8.4 Average land prices (US$/m²) aggregated by municipal cluster (Source: The authors, based on Arriagada 2009)

In the remaining three municipal clusters, the amount of land offered for residential development was substantially lower, with intra-annual differences over the period under review. While it declined steadily in the eastern peri-central cluster, the amount of land offered in the central and peri-central clusters remained fairly stable over the years.

The trends in aggregate land prices by municipal cluster shown in Fig. 8.4 point to major differences between the municipal clusters of the Centre and Eastern Peri-Centre, on the one hand, and the Extra-Periphery, on the other, whereby in 2007 prices in the former were 50 times in excess of those in the latter. Not unexpectedly,

---

[1] Data sources: Boletín del Mercado de Suelo, published by P. Triveli, for the 34 municipalities of AMGS, and the Real Estate Supplement (Suplemento de Propriedades) of the *El Mercurio*; prices for land (undeveloped) ready for development; data shows aggregate average of the last quarter of each year under study; values are inflation adjusted to prices in 2007.

the range of prices is consistent with the 'relative distance' to the city centre: land prices tend to fall from the central clusters outwards to the extra-peripheral cluster.

Overall, the trend suggests three general phases:

First phase (1992–1996/1998): a 'convergence' of prices across clusters largely due to the rapid expansion of the urban land market with vast land offers in the peripheral and extra-peripheral municipalities.

Second phase (1996/1998–2002/2003): a 'consolidation' phase with minimal adjustments across all sectors. The Periphery and Extra-Periphery failed to uphold the trend towards incorporation into the urban land market with the intensity of earlier years.

Third phase (2002/2003–2007): a 'differentiation' phase in which the central and peri-central clusters in particular show price increases. While the former can be explained by the urban renewal policies and incentives that began to take effect at this time, the explanation for the gradual incorporation of the peripheral cluster into the high price segment is less evident.

The figures allow for several interesting and more specific observations. The two traditional 'high-price' clusters (Centre, Eastern Peri-Centre) maintain their position, although prices progress at comparatively low average annual rates. In the central cluster, prices increased moderately from 421 US$ in 1994 to 521 US$ in 2007. The average price level in the Eastern Peri-Centre increased from 404 US$ in 1994 to 515 US$ in 2006, but dropped below central cluster prices to 420 US$ in 2007. This can be attributed to construction incentives and major urban renewal projects in the area.

The Peri-Centre in contrast saw a more notable rise in land prices. With an increase from 77 US$ in 1994 to 350 US$ in 2007, the cluster shows a tendency towards convergence with the Centre and the Eastern Peri-Centre. This is particularly true for the municipalities close to the central municipality of Santiago, such as Macul, Estación Central, Recoleta, San Joaquin and San Miguel. Here land prices of more than 400 US$ in 2007 indicate that the high price segment is expanding from the Centre to its neighbouring municipalities.

The municipalities in the peripheral cluster show a somewhat ambivalent trend. While they had some of the highest price increases from the 1990s up until 2002, trends have meanwhile reversed and prices have remained stagnant (e.g., Huechuraba, La Florida, Puente Alto) or even dropped (e.g., Pudahuel). Only a few locations showed evidence of further price increases (e.g., Peñalolén, Maipú).

Finally, absolute prices in the extra-peripheral cluster of municipalities showed no significant change. The early 1990s saw a slight development from 5 US$ in 1992 to around 11 US$ in 2000, with substantial variations over this period. Since then, prices have been more stable at between 8 and 11 US$.

The combined view of the variables considered in this section (housing construction, land offers) suggests a variety of patterns across the municipal clusters. The Centre and the Eastern Peri-Centre are defined by high prices and growing construction. In the Peri-Centre, land prices are rising but building activity is low. The lion's share of residential development took place in the Periphery, albeit in small quantities. Likewise, the trend in prices for development land did not follow the upward progression of the more central clusters. The Extra-Periphery is the cluster with the highest amount of land offered for development, although building activity has not yet picked up. This cluster continues to be a low-price area. Thus the notion that the real estate sector in Santiago is shifting its focus to incorporation of the (low-price) extra-peripheral municipalities into the urban land market has not materialized so far, partly due to the political, legal and administrative barriers that limit this expansion to some extent.

As shown, there is some indication that housing policy and housing programmes provide incentives that explain the trend in housing. The following section on housing policies focuses on gaining deeper insights into the interlinkages and consequences of the socio-spatial differentiation process.

8.6 Drivers of Socio-spatial Differentiation: Social Housing as the Housing Policy Focus

Housing policy plays an important role in steering supply and demand in the housing market. The aim is to provide affordable dwellings for all socio-economic groups and avoid housing deficits. In this context, social housing programmes are vital to poorer households. Social housing competes with 'traditional housing' and other demands for land use on the market, but requires 'cheap' land if construction and maintenance are to remain affordable. It is mostly directed to the peripheral locations of the city, where low land prices prevail. This frequently leads to socio-spatial homogenization and the concentration of poorer households in specific areas. Social housing is therefore seen as a key component of the housing policy and the promotion of socio-spatial differentiation processes. This section deals with Chilean housing policy, a centralized system with the National Ministry of Housing (MINVU) as the main actor for policy development. The goal is to gain a deeper understanding of its role as driver of socio-spatial differentiation.

The establishment of the MINVU in 1965 constituted the official inauguration of the Chilean housing policy. At this time the public sector played a prominent role, enacting laws and regulatory plans, and acting as the principal source of housing for poor and middle-class households (housing supply). It also regulated the land market. The private sector concentrated exclusively on housing construction under the supervision of the MINVU. While in-migration from the rural regions to Santiago exhausted the housing supply and led to housing shortages, 'auto construction' became extremely popular. The outcome was a high degree of illegal land occupation at the beginning of the 1970s (Hidalgo 2005).

The period between 1974 and 1989 was deeply affected by a radical shift in ideology to the more neoliberal policy of the military dictatorships, paving the way for a housing policy focused on the enabling market. The public sector continued to be responsible for housing construction and implemented a strategy to promote demand with a voucher system, targeting the poorest households exclusively in the

1980s. The main objective of this programme was to exercise more efficient control over supply and demand in the real estate sector in the wake of the economic crisis. Families were treated as numbers on a voucher list and aspects such as preservation of social networks or location preferences ignored. The standard housing (vivienda básica) provided by the state to the poorest of the poor was perceived as state 'welfare' rather than a 'right'.

For the Greater Metropolitan Area of Santiago de Chile, statistics confirm that from 1974 up to 1989 social housing programmes increased in all municipal clusters, apart from the Centre. Most of the construction in the Peri-Centre took place in the municipality of Renca (~7,800). In the Eastern Peri-Centre construction was highest in the municipality of Lo Barnechea (~1,800), albeit less than in other clusters. No data was available for the Extra-Periphery. The number of social housing units in the Periphery is conspicuous, with an exceptionally high number in the municipalities of La Florida and La Pintana (>13,000). Other municipalities showed values above 6,000 new units: Puente Alto, San Bernardo, El Bosque and Pudahuel, the main bulk of which were built in the 1980s. These values bear witness to the high concentration of social housing projects in peripheral locations (see Fig. 8.5).

Although a considerable number of housing units did materialize, on the whole a welcome development, the outcome was not always positive. This period exemplifies the close link between social housing, and by extension the housing policy in general, and socio-spatial differentiation processes. Massive relocation

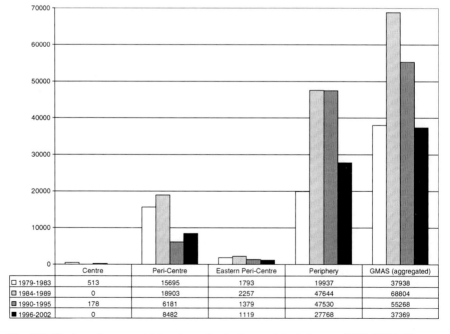

Fig. 8.5 Number of new social housing units in the municipal clusters (1979–2002) (Source: The authors, based on Hidalgo 2007 (no data for the Extra-Periphery))

activities were set in motion, pushing intra-urban campamentos (illegal settlements) to the urban fringe. Even today, these areas are hotspots of segregation, stigmatization and social homogeneity with reference to poorer households (Hidalgo 2004) (e.g., the municipality of La Pintana carries social housing units to 95%, Brain et al. 2009b).

The years between 1990 and 2005 saw the return of democracy to Chile and vigorous efforts to reduce the housing deficit and improve the steering of the growing demand for housing. MINVU amplified the voucher system to include middle-class residents in the facilitated access to housing. Based on subsidized demand, this policy culminated in large-scale segregation in the Periphery, where land prices were generally lower. A key housing policy measure was the introduction in 1997 of the Chile Barrio programme (known today as Línea de Atención a Campamentos (LAC)), the aim of which was to abolish the remaining campamentos. The potential of these illegal settlements to sustain social capital and social networks was acknowledged for the first time. The relocation of households took place within a municipality or in others among specific groups, thereby enhancing the persistence of social networks. Other programmes such as Fondo Solidario de Vivienda and Vivienda Dinámica sin Deuda were established to assist low-income families (first quintile). Social integration gradually gained currency. Subsidy financing was traditionally threefold: (a) the subsidy itself (39%), (b) savings (6%), and (c) loans (56%) (Brain et al. 2009b). New programmes required conditions that could be met easily by the poorest households: (a) organization in groups, (b) purchase of a subsidy covering 97% of the housing value, and (c) possession of a savings minimum corresponding to 3% of the overall housing value (ibid.).

All of this led to the construction of new social housing units (mainly multi-storey housing) on the urban fringe. While the total number of newly built social housing units in GMAS, the Centre and the Eastern Peri-Centre declined up until 2002, the Peri-Centre witnessed an increase between 1996 and 2002 (see Fig. 8.5) as a result of new social housing units (>2,000) in the municipalities of Cerrillos and Renca. Growth rates in the Periphery remained high. The increase in social housing units in the municipality of Puente Alto was exceptionally high, with more than 23,500 units erected between 1990 and 2002. Housing also expanded in San Bernardo, albeit to a lesser extent. Major increases were observed in Quilicura and Peñalolén, while La Florida received no new housing units at all. The growing demand for social housing units once again pressured the private sector. As a result, the housing shortage fell by almost half in quantitative terms from 888,681 units in 1992 to 450,000 in 2000 (Hidalgo 2007).

Again, these developments led to a concentration of poorer socio-economic groups in the Periphery and subsequently to problems such as criminality, unemployment, greater distance to job opportunities and lack of social infrastructure (Sabatini 2000; Brain et al. 2009b). In the mid-1990s, the inadequacy of the overall situation could no longer be ignored. The outcome was a reduction in the number of social housing units in a specific area to a maximum of 300. At the same time, the number of subsidies awarded increased. Unfortunately, these programmes failed as soon as niches were found to avoid these restrictions. On the other hand, the poor quality of the houses built began to show.

The principal objectives of MINVU have undergone significant change since 2005 and led to new priorities in the public sector. Housing policy now seeks (a) to diminish the housing deficit for the first and second income quartiles, (b) to upgrade housing standards, including an increase in size and the improvement of construction materials. Importantly, for the first time in history, efforts were made (c) to advance social inclusion by improving middle-class housing. To achieve these goals, MINVU created several programmes oriented to:

- The construction of housing areas with mixed socio-economic groups;
- The promotion of housing projects in central locations;
- The protection of family heritage;
- Further amplification of existing housing, and
- 'Best practice projects' for social integration such as the Parque Bicentenario.

There is likewise a trend towards 'second-hand' housing, which gives families greater freedom to choose their housing location. The innovative Quiero mi Barrio programme applies a bottom-up strategy combined with strong citizen participation and seeks to improve the social and physical aspects of social housing. The private sector, on the other hand, continues to concentrate on new social housing projects. On the whole, it can be stated that citizens' involvement in housing policy has grown substantially. Citizens are represented by the Entidad de Gestión Inmobiliaria Social (EGIS) promoted by MINVU. The task of this organization is to coordinate the demand for social housing in the municipalities on their behalf and to assist with applications for government subsidies and public housing.

Two programmes were introduced to revitalize neighbourhoods and upgrade the quality of housing: the Programa de Mejoramiento de la Gestión/PMG (Management Improvement Programme) in 2002 and the Plano de Aseguramiento de la Calidad/PAC (Quality Insurance Plan) in 2006 (Jiménez-Cavieres 2009). The initiatives reflect a new governance approach that recognizes how citizens' needs and expectations change over time.

In order to advance social inclusion, more subsidies are now being assigned to middle-class families. One such programme is the D.S.40 offered for the first three quintiles in possession of a minimum of 50 UF^2 savings. It can be allocated in urban or rural areas and includes new or second-hand homes. Subsidies for poorer households are based on the overall programme Fondo Solidario de Vivienda (FSV), which determines the allocation of funds for construction of new and second-hand homes, and their purchase. FSV is divided into FSV I and FSV II, with FSV I dedicated to the poorest families of the first quintile for housing up to 650 UF with a minimum size of 38 m (max. 55 m). Homes supported by the FSV I

---

[2] UF (Unidad de Fomento) refers to a *unit of account* used in *Chile*. The exchange rate between the UF and the *Chilean peso* is constantly adjusted to inflation, allowing the value of the Unidad de Fomento to remain constant.

Table 8.6 FSV I and FSV II acquisition in GMAS between 2006 and 2009

	2006	2007		2008		2009	
	FSVI	FSVI	FSVII	FSVI	FSVII	FSVI	FSVII
Centre	0	4	1	10	4	19	4
Peri-centre	100	943	73	2,884	481	1,729	162
Eastern Peri-centre	0	1	1	18	6	17	5
Periphery	384	2,575	371	2,717	2,008	3,206	588
Extra-Periphery	9	200	36	386	69	171	13

Source: MINVU observatorio habitacional

includes family savings of at least 10 UF, a state subsidy between 330 and 470 UF for new homes and 280–420 UF for second-hand homes. Other public and private grants were also available. The FSV II targets the 40% of the most 'vulnerable' in the population and is exclusively responsible for the purchase of new or second-hand housing. Savings of 30 UF are a prerequisite, state support amounts to between 280 and 420 UF, and households may apply for loans.

Statistics show that FSV I subsidies correspond to the concentration of social housing in the Periphery (see Table 8.6). Values increased steadily between 2006 and 2009. The two municipalities with the highest acquisition of homes are Puente Alto and San Bernardo. The Peri-Centre shows comparatively high values, peaking in 2008. The Extra-Periphery also displays a significant amount of FSV I supported housing with the highest value in Colina. Regarding FSV II, again the Periphery presents the highest number of subsidies, albeit under the values of FSV I. Outstanding municipalities here are Maipú and, yet again, San Bernardo. It is worth noting that in the case of FSV I for the years 2007–2009 almost the entire housing in the Metropolitan Region was second-hand, in contrast to the year 2008 with only 28% (Brian et al. 2009b). Overall, the location of second-hand homes is more advantageous to those of the more recently built homes, a circumstance due in the main to existent social networks.

Today's Chilean housing policy of financing and allocating homes is basically a system of demand. This model of subsidizing the housing demand was the first of its kind worldwide and influenced housing policies in many other countries of Latin America, and indeed of other continents (Gonzáles Arrieta 1997; Gilbert 2002). From this follows that the housing supply is to a great extent determined by the private sector. Housing subsidies are no longer exclusively allocated to poor households (first quartile) but, based on the Ficha de Protección Social point system, also to the second quartile and to middle-class households. It should be emphasized that subsidized households receive ownership of their homes. As a result of the quantitative deficit in housing supplies, the real estate sector and the land market have a strong influence on housing policies. Furthermore, coherence between housing and urban limits, a potent factor for both housing policies and urban development, is likewise the responsibility of MINVU.

To sum up, the quantitative approach of the Chilean housing policy in many aspects disregards quality. In other words, Chile's housing policy has a 'dark side'. Despite its achievements this 'successful' housing policy has reinforced 'social and spatial exclusion', as evidenced by its contribution to dividing the city according to social classes (Jiménez-Cavieres 2009), making it a key driver of socio-spatial differentiation. Although this trend can be identified in other Latin American cities, the responsibility here lies with the government, since Chile's comprehensive public housing policies in the past exacerbated urban 'social' problems. Jiménez-Cavieres (2009) argues that these quality deficits are, among other things, the product of a policy that relied on the market, ignored people's expectations and the housing procedures of low-income families, and located housing projects in zones already marked by the absence of infrastructure, social services and employment.

8.7 Synthesis: Hotspots, Risks and Opportunities

The three major drivers of processes of socio-spatial differentiation analysed above expose development paths that are both separate and interlinked. This section provides a trend synthesis for each municipal cluster and defines 'socio-spatial hotspots'. Hotspots are locations within GMAS, where a combination of trends in migration, social housing and the land market has produced 'extreme' types of socio-spatial conditions (see Sect. 8.2 above). This section seeks to synthesize processes and changes by evaluating their positive consequences or opportunity for inclusion and their undesired negative consequences or risk of exclusion. It combines an overview of the principal trends in each of the five clusters with some considerations of selected 'hotspot' municipalities, and assesses the potential risk of exclusion and the opportunity for social inclusion. The results can be summarized as follows:

The Centre is characterized by a moderately positive change in housing stocks and a marginal proportion of social housing. The population loss is significant although a strong decline in negative intra-urban net migration can be observed. Since 1997 migration has contributed positively to the educational level average, which is constantly rising in this cluster. These tendencies are linked to the immigration of young people. Land prices, already in the upper bracket, have increased. Since the early 1990s, the majority of real estate activities in the central cluster have concentrated on residential use. The recent sharp increase in building permits can be explained by the availability of construction subsidies offered within the frame of the MINVU urban renewal programme (renovación urbana) (Arriagada 2009). Consequently, land prices have soared yet again, impeding the construction of a large number of social housing units in the cluster.

Coupled with the immigration of young families and the renewal of the built environment, which bears signs of gentrification, this overall tendency has the capacity for potential risks as well as opportunities. If it continues, it may, on the one hand, lead to an improvement of facilities in the area, such as public space. On the other hand, it could culminate in a major shift in population composition and in the displacement of the current residents. Whether and to what extent such tendencies will impact on social networks, either positively or negatively, requires further study.

The Peri-Centre displays a decline in its housing stock, despite an upsurge in social housing. It likewise suffers from a significant population loss, although average educational levels are rising. Negative net migration is vast and on the increase, while migration continues to have a negative impact on educational levels. In other words, the educated outmigrate at disproportionately high levels, and land prices, although considerably lower than those in the neighbouring central cluster, are constantly rising.

Far from bright, the prospects for municipalities in this cluster point to either decay or renewal. Several locations (e.g., Lo Espejo) exhibit a tendency towards large-scale agglomeration of lower-income households and stigmatization (social isolation). Other municipalities, typically those close to the Centre (e.g., San Miguel), are affected by urban renewal projects and may therefore face the same risks and opportunities as the Centre.

Similar to the Centre, the Eastern Peri-Centre has reconfirmed its attractiveness, realizing a relatively positive net migration since 2002, despite exorbitant land prices. This traditional 'high-price' cluster has maintained its position, with prices continuing to rise, albeit at comparatively low average annual rates. In addition, and similar to the Centre, the enormous price of land has thwarted efforts to build a large number of social housing units in this cluster. Moreover, resettlement policies frustrated social housing construction, and the comparatively low level of public housing decreased over time. The initially high average educational levels in this cluster also increased. All of these trends confirm the persistent attractiveness of the area, leading to a concentration of the urban 'rich' (e.g., Las Condes).

This trend of the Eastern Peri-Centre to act as a magnet for educated households begs the question of whether or not a process of 'elitism' peters out of its own accord. Can we assume that this concentration will ultimately lead to auto-exclusion?

The Periphery in general, is characterized by a major change for the better in the housing stock and a moderate proportion of social housing units. Population is growing and the already high average educational levels are stable. The cluster shows a significant reduction in positive net migration, but a positive contribution of migration to the average educational level. Land prices increased up until 2002, but since then have declined.

Apart from such general trends, the cluster reveals two 'extremes'. On the one hand, it contains municipalities with a large-scale agglomeration of less-educated households with a heavy increase in numbers and proportion. Outstanding examples of municipalities with a tendency towards marginalization are La Pintana and Puente Alto. On the other hand, the cluster includes locations with a 'new' spatial mix of households from different socio-economic groups in close proximity.

Moreover, this cluster has been the main absorber of construction for residential purposes and is characterized as having the lowest access to services and commercial use (as evident from the share of construction in these branches). This trend could be related to the consolidation of the periphery as a low-price location.

The process of concentrating the poorest families on the urban periphery, where land prices are low and links to social services, public transport, jobs and education rudimentary, makes abundantly clear the influence of the land market and housing policies, especially social housing, on socio-spatial differentiation processes. As shown elsewhere (Orellana 2007), municipalities with a large concentration of poor families and social housing programmes tend to have a comparatively low tax revenue base and therefore a low budget. On the other hand, these municipalities are reported to be likewise confronted with growing fiscal needs for public services such as schools and hospitals, which they are unable to meet due to budget constraints. These interwoven processes evidence the hazard of living in these areas: the risk of social exclusion.

With regard to locations characterized by a 'new' spatial mix of households from different socio-economic groups in close proximity, the analysis of residential mobility differentiated by the average number of years spent in education indicates that educated families display a growing interest in living on the periphery and beyond. This tendency is perhaps closely linked to real estate development in this area and the marketing of mega-urban residential projects. Again activities in the private housing market of the peripheral cluster are linked to the relatively low land prices that prevailed at the beginning of the 1990s but have risen over time as a result of land originally bought at a low price being sold for a great deal more. This is particularly evident in the municipalities of Huechuraba, La Florida and Peñalolén. The influx of educated people to these areas was a major factor in the more recent increase in the average number of years spent in education in the municipalities of the peripheral cluster. This relocation of households characterized by high educational attainment to areas designated as 'poor' and the new spatial proximity of various socio-economic groups increases the potential opportunities for social inclusion.

The Extra-Periphery is still in the early stages of being incorporated into the city. It has seen a moderate increase in housing stocks but boasts a similarly high share of social housing. There is a strong increase in population as well as in average educational levels. Those who migrate in have a positive influence on levels of education. Based on extremely low unit prices, the price of land has spiralled. The Extra-Periphery has recently become the location of a rapid increase in the private housing sector and new forms of residential development other than social housing. Since the early 1990s, the Plan Regulador Metropolitano de Santiago (PRMS) has faced tremendous pressure to 'open' the rural areas. One instance was the deviation in 1994 from a then 'compact' vision to one that permitted urban uses in hitherto non-urban rural locations. Another example was the introduction of new land-use instruments, e.g., the development of specific Zonas de Desarrollo Urbano Condicionado (ZODUC) in 1997, aimed at a 'harmonious' expansion that included the planning of basic services and green space. The establishment of Áreas Urbanas de Desarrollo Prioritario (AUDP) additionally encouraged the urbanization of areas close to the cities. In 2003, the PRMS further expanded the idea of converting rural land to urban use with the Proyectos de Desarrollo Urbano Condicionado (PDUC), based on consideration of minimum requirements with regard to infrastructure and social housing. In response to these instruments, several residential mega-projects are emerging in the Extra-Periphery, delivering further evidence of the crucial role of private housing developments and the real estate sector. An assessment of housing developments in the Extra-Periphery must of necessity be left to future study.

8.8 Conclusions

The combination of tendencies in migration, land prices and (social) housing programmes has led to individual trends and conditions in socio-spatial differentiation across the city. Overall, the extremes (agglomeration of households from 'homogenous' groups) seem to have become more distinct (for rich and poor alike), while in several cases/locations, this aggregation of groups occurs in close proximity.

With respect to risks of exclusion and opportunities for inclusion, no clear pattern of consequence could be identified. While the contemporary debate on these aspects focuses to a large extent on the agglomeration of lower-income groups in social housing in the Periphery, our results suggest that the Peri-Centre should be given greater attention. This applies both to locations in decay as well as to those with the potential to follow the gentrification trend already observed in the Centre.

Migration trends and housing construction/reconstruction react strongly to incentives set by the state. In three aspects the public actor has been fundamental in determining conditions for operations in the housing construction sector: (1) social housing in the Periphery, (2) incentives for private real estate investments in the Periphery and Eastern Peri-Centre, and (3) construction associated with the 'urban renewal programme' in the Centre. Further investigation of the socio-spatial effects of state housing policy on a smaller geographical scale should contribute to a thorough understanding of its potential and constraints in guiding socio-spatial development.

References

Arriagada, C. (2009). The role of land markets in patterns of socio-spatial differentiation. Working paper.

Bolt, G., & van Kempen, R. (2010). Ethnic segregation and residential mobility: Relocations of minority ethnic groups in the Netherlands. *Journal of Ethnic and Migration Studies, 36*(2), 333–354.

Brain, I., & Sabatini, F. (2006). Los precios del suelo en alza carcomen el subsidio habitacional, contribuyendo al deterioro en la calidad y localización de la vivienda social. *Revista ProUrbana, 4*, 2–13.

Brain, I., Celhay, P., Prieto, J. J., & Sabatini, F. (2009a). Living in slums. Residential location preferences in Santiago, Chile. Land lines October 2009, Lincoln Institute of Land Policy.

https://www.lincolninst.edu/pubs/dl/1696_908_Oct%2009%20Article%203.pdf. Accessed 9 Oct. 2009.

Brain, I., Mora, P., Rasse, A., & Sabatini, F. (2009b). Report on social housing in Chile. RHM working paper.

Cáceres, G., & Sabatini, F. (2004). *Barrios cerrados en Santiago de Chile: entre la exclusión y la integración residencial*. Santiago de Chile: Lincoln Institute of Land Policy, Pontificia Universidad Católica de Chile.

De Ramón, A. (1978). Santiago de Chile 1850–1900. Limites urbanos y segregación espacial según estratos. *Revista Paraguaya de Sociología, 15*(42/43), 253–276.

Friedrichs, J., & Nonnenmacher, A. (2008). Führen innerstädtische Wanderungen zu einer ethnischen Entmischung von Stadtteilen? In F. Hillmann & M. Windzio (Eds.), *Migration und städtischer Raum. Chancen und Risiken der Segregation und Integration* (pp. 31–48). Budrich: Opladen/Farmington Hills.

Gilbert, A. (2002). Power, ideology and the Washington consensus: The development and spread of the Chilean housing policy. *Housing Studies, 17*(2), 305–324.

Gonzáles Arrieta, G. (1997). Acceso a la vivienda y subsidios directos a la demanda: Análisis y lecciones de las experiencias latinoamericanas. Serie Financiamiento del Desarrollo N° 63. CEPAL. Santiago de Chile.

Guilford, J. (2000). *Making the case for social and economic inclusion*. Halifax: Population and Public Health Branch, Atlantic Regional Office, Health Canada.

Hidalgo, R. (2004). La vivienda social en Santiago de Chile en la segunda mitad del siglo XX: Actores relevantes y tendencias espaciales. In C. De Mattos, M. E. Ducci, A. Rodríguez, & G. Yáñez Warner (Eds.), *Santiago en la globalización: ¿Una nueva ciudad?* (pp. 219–241). Santiago de Chile: Ediciones SU.

Hidalgo, R. (2005). La vivienda social en Chile y la construcción del espacio urbano en el Santiago del siglo XX, Santiago. *EURE, 31*(93), 108–112.

Hidalgo, R. (2007). ¿Se acabó el suelo en la gran ciudad? Las nuevas periferias metropolitanas de la vivienda social en Santiago de Chile. *EURE, 33*(98), 57–75.

Ihlanfeldt, K. R., & Sjoquist, D. L. (1998). The spatial mismatch hypothesis: A review of recent studies and their implications for welfare reform. *Housing Policy Debate, 9*(4), 849–892.

INE – Instituto Nacional de Estadísticas. (1982). *Censo de Población y Vivienda 1982*. Santiago de Chile: Instituto Nacional de Estadísticas.

INE – Instituto Nacional de Estadísticas. (1992). *Censo de Población y Vivienda 1992*. Santiago de Chile: Instituto Nacional de Estadísticas.

INE – Instituto Nacional de Estadísticas. (2002). *Censo de Población y Vivienda 2002*. Santiago de Chile: Instituto Nacional de Estadísticas.

Jiménez-Cavieres, F. (2009). Housing and urban planning in Chile. TU INTERNATIONAL, 64 (August), 15–17

Kabisch, S., Kindler, A., & Rink, D. (1997). *Sozialatlas der Stadt Leipzig*. Leipzig: Helmholtz Centre for Environmental Research – UFZ.

Kain, J. (1992). The spatial mismatch hypothesis: Three decades later. *Housing Policy Debate, 3*(2), 371–460.

Kleve, H. (2005). *Inklusion und Exklusion: Drei einführende Texte*. Potsdam: Fachhochschule Potsdam, University of Applied Sciences, Fachbereich Sozialwesen.

Kuhn, K. M. (2000). Message format and audience values: Interactive effects of uncertainty information and environmental attitudes on perceived risk. *Journal of Environmental Psychology, 20*(1), 41–51.

Massey, D. S., Gross, A. B., & Shibuya, K. (1994). Migration, segregation and the geographic concentration of poverty. *American Sociological Review, 59*(4), 425–445.

MIDEPLAN – Ministerio de Planificación y Cooperación (2006). Encuesta CASEN 2006. Santiago de Chile. http://www.mideplan.cl. Accessed 12 Jan 2008.

Orellana, A. (2007). La Gobernabilidad metropolitana: nuevos escenarios para el desarrollo urbano y territorial del área metropolitana de Santiago. In C. de Mattos, R. Hidalgo, C. de

Mattos, & R. Hidalgo (Eds.), *Santiago de Chile: Movilidad Espacial y Reconfiguración Metropolitana* (pp. 189–206). Santiago de Chile: GEOlibros 8, Insitituto des Estudios Urbanos y Territoriales, Pontificia Universidad Católica de Chile.

Rodriguez, J. (2001). Segregación residencial socioeconómica: ¿qué es?, ¿cómo se mide?, ¿qué está pasando?, ¿importa? Santiago de Chile: CEPAL/CELADE – SERIE Población y Desarrollo, N° 16.

Rodriguez, J. (2008). Dinámica sociodemográfica metropolitana y segregación residencial: ¿qué aporta la CASEN 2006? *Revista de Geografia Norte Grande, 41*, 81–102.

Rodriguez, J., & Arriagada, C. (2004). Segregación Residencial en la Ciudad Latinoamericana. *EURE, 30*(89), 5–24.

Sabatini, F. (2000). Reforma de los mercados de suelo en Santiago, Chile: efectos sobre los precios de la tierra y la segregación residencial. *EURE, 26*(77), 49–80.

Sabatini, F., & Salcedo, R. (2007). Gated communities and the poor in Santiago, Chile: Functional and symbolic integration in a context of aggressive capitalist colonization of lower-class areas. *Housing Policy Debate, 18*(3), 577–606.

Sabatini, F., Cáceres, G., & Cerda, J. (2001). Segregación residencial en las principales ciudades chilenas: Tendencias de las tres últimas décadas y posibles cursos de acción. *EURE, 27*(82), 21–42.

Tokman, A. (2006). El MINVU, la política habitacional y la expansión excesiva de Santiago. In A. Galetovic (Ed.), *Santiago: Dónde estamos y hacia dónde vamos* (pp. 489–522). Santiago de Chile: Centro de Estudios Públicos.

Chapter 9
Energy Systems

Sonja Simon, Volker Stelzer, Luis Vargas, Gonzalo Paredes,
Adriana Quintero, and Jürgen Kopfmüller

Abstract Almost entirely dependent on energy imports from outside, cities are the key driving force behind the demand for energy, which is an essential resource for industries, households and services. Up to now, the Chilean energy system has met Santiago's needs satisfactorily. However, development trends in the current energy system pose significant risks to its future. Using selected energy indicators and a distance-to-target approach, a detailed sustainability analysis of the energy sector in the Metropolitan Region of Santiago, and of Chile as a whole was conducted. Risks to the sustainable development of the energy sector were detected, such as increasing concentration in the energy sector, import dependency on fossil fuels and rising CO_2 emissions due to energy consumption. Alternative options were assessed for a more sustainable development of the megacity Santiago within the frame of the national Chilean energy system, such as enhancement of energy efficiency and greater use of renewable energies.

Keywords Efficiency • Energy system • Non-conventional renewable energies • Renewable energies

9.1 Introduction

Unsustainable energy supplies in modern societies are commonplace. Only with tremendous effort will we be able to satisfy the constantly growing energy needs of current generations without risking at the same time the livelihood of future generations. While almost completely dependent on energy imports from outside to provide industry, households and services with the necessary energy for life and

S. Simon (✉) • V. Stelzer • L. Vargas • G. Paredes • A. Quintero • J. Kopfmüller
Institute of Technical Thermodynamics, German Aerospace Center (DLR), Pfaffenwaldring 38-40, 70569 Stuttgart, Germany
e-mail: sonja.simon@dlr.de

production, cities are the driving force behind energy demands and will be even more so in the future. Megacities like Santiago epitomize the culmination of this development. They demand an increasing amount of energy as a result of population growth, economic activity and the associated infrastructure requirements.

This chapter analyses the most glaring deficits in the Santiago energy supply and, moreover, asks how it will develop in the future: is it possible to decelerate energy demands? And how can the sustainable use of renewable resources for electricity, heat and fuel production be enhanced?

In the wake of mounting demands for energy, risks arise to the sustainable development not only of the system *Megacity* itself, but also of the respective country's energy supply system. Megacities are highly vulnerable to energy shortages. This is particularly the case in Santiago, not least as a result of Chile's massive dependence on imported fossil fuels, the confinement of import operations to two harbours only and the radial characteristics of the electricity grid. Megacities are furthermore affected by other problems that emerge in the aftermath of an explosive energy demand, such as greenhouse gas emissions or air pollutants.

This tremendous demand for energy places megacities centre stage when it comes to strategies for more sustainable development. The core issue here is how to manage the fast-growing demand for energy services. In this regard, both the size and the concentration of megacities have, in principle, vast potential for efficient energy consumption.

For two primary reasons, however, the national level needs to be taken into account. First of all, a look at the supply side must integrate a broader view of the national energy system, with its large variety of energy sources vital to the megacity. Hence greater sustainability in the energy sector of the Santiago Metropolitan Region implies a view beyond the region towards an integrated solution to Chile's energy problems. Secondly, the political framework conditions for the energy sector are set predominantly by national ministries. Energy policy is not yet an institutionalized topic at regional level.

The analysis of the energy sector in this chapter focuses on the national perspective, since the available data for the Metropolitan Region is still limited.

Sections 9.1 and 9.2 of this chapter give an overview of energy demand and supply in Chile today, including information relevant to the description of the Metropolitan Region. This is followed in Sect. 9.3 by an assessment of the sustainability performance according to selected indicators. Based on these findings, Sect. 9.4 discusses two essential strategic approaches towards more sustainability: to increase energy efficiency and to intensify energy source diversification, especially by building up the share of non-conventional renewables. Conclusions are drawn in Sect. 9.5.

9.2 The Energy System Today

Chile's energy sector profoundly resembles that of an industrialized country, with its high growth in energy demand and strong dependence on fossil fuels. The total energy use in Chile (final consumption) increased from 460 PJ (Petajoule) in 1990 to 1,013 PJ in 2007 (CNE 2010a), an annual growth off 13%. This surge in energy consumption is for the most part a result of population growth, a highly dynamic economic development and deficiencies in the effective use of energy resources. In the household sector, growing incomes lead to higher living standards and a rise in the demand for energy.

Table 9.1 below shows the development of the energy carrier mix for Chilean secondary energy provision between 1990 and 2007. Currently, oil accounts for about 55%, natural gas and derivates for about 5%, coal and derivates for about 3%, electricity 19% and firewood for 18% (CNE 2010a).

While electricity consumption per capita in the Metropolitan Region of Santiago (MRS) increased to the same degree as the national level, the figures point to sharp differences between individual municipalities (Fig. 9.1). The municipality of Vitacura

Table 9.1 Final energy use (in PJ)

	Oil derivates	Electricity	Coal	Coke, tar and blast furnace gas	Natural gas, methanol and city gas	Wood	Final consumption
1990	252	56	22	10	13	107	460
1992	293	69	26	13	15	131	548
1994	345	78	18	14	17	126	599
1996	399	97	22	15	18	141	692
1998	436	111	26	16	31	147	766
2000	446	132	22	13	53	164	830
2002	432	147	22	13	64	162	841
2004	451	169	22	13	70	162	887
2006	500	182	19	15	64	177	957
2007	560	191	17	15	44	187	1,013

Source: CNE 2008e, 2010a

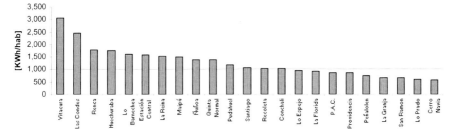

Fig. 9.1 Electricity consumption per capita in selected municipalities of Santiago for 2002 (Source: Vargas 2008)

displays the highest consumption in kWh/cap, i.e., six times higher than in Cerro Navia.

The principal consumers of final energy are industry and mining, on the one hand, and transport, on the other, both with a share of almost 30%. Electric power claims a share of 23%, while household, commercial and public use combined amounts to 20% of the final energy consumption (see Fig. 9.2).

Chile's energy dependency has grown considerably in recent decades. The percentage of primary energy imports increased from 50% in 1990 to 70% after 1999. The root cause of this rising energy import dependency was a combination of soaring energy demands and Chile's own scant conventional energy resources. This led to an upsurge in natural gas imports from Argentina up until 2007, when Argentina reduced its natural gas exports to Chile dramatically. Natural gas imports to Chile were replaced in many instances by diesel imports, which increased from 104 PJ in 2006 to 207 PJ in 2007, representing between 55% and 64% of total energy imports. Additionally, oil and coal imports have risen (see Table 9.2). Chile invested one billion USD in two large sea terminals (Quintero and Mejillones) with a capacity to handle 16 million m^3 of imported liquid natural gas (LNG) per day. The *BG Group* has its own production and supply contracts with Trinidad and Tobago, Egypt, Nigeria, Equatorial Guinea and other gas producing countries (Company of Liquid Natural Gas 2010). Several gas pipelines were installed from the Chilean coast to the eastern border in order to compensate speedily for the missing gas from Argentina.

Chile has very few fossil energy resource deposits. Existing reserves are estimated at about 45,000 million m^3 of gas and 30 million barrels of oil, located for the most part in Patagonia. In January 2006 the journal Oil and Gas reported that Chile owned 3.5 trillion cubic feet (Tcf) of proven natural gas reserves and 150

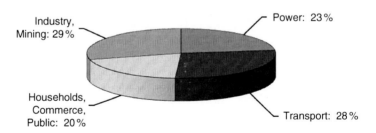

Fig. 9.2 Share of final energy consumption 2007 (Source: CNE 2007a)

Table 9.2 Import of energy resources (PJ)

	Natural gas	Oil	Diesel	Coal
1999	148	419	33	166
2001	204	435	24	81
2003	234	461	30	83
2005	259	461	79	117
2007	98	489	207	168

Source: CNE 2008a

million barrels of proven crude oil reserves (Langdon 2008). Coal reserves are located in the southern region and the corresponding estimates amount to approximately 83 MMtons (million tons) in Arauco Area and 734 MMtons in Magallanes (Industria carbonífera Chile 2010). Due to the rather low quality and remote localization, particularly of the coal reserves, very few of them have been exploited so far.

In the southern area of Chile, some of the domestic gas resources are used locally. Due to geographical conditions, however, gas extraction here is unprofitable and production therefore has declined. Since 2006, the California Oil & Gas Corporation and March Resources Corporation have conducted prospective drillings in the south of Chile, where they own gas exploration rights to an area of 10,000 km^2. Native resources currently supply only 10% of the Chilean fossil energy use. Hydropower has the biggest share in the power mix of the country (see Sect. 9.2.1 below).

9.2.1 Electricity Supply

A vital component of the energy system is the use of electricity. It is currently indispensable to industry, mining, the public domain, households, and potentially in the future in the transport sector. Due to the particular geographical situation, electricity production in Chile boasts some special characteristics. The power supply consists primarily of two large electricity grids (see Fig. 9.3) used by several electricity production and distribution companies, i.e., the northern grid SING (Northern Interconnected System) and the central grid SIC (Central Interconnected System) that covers the Metropolitan Region of Santiago (MRS). Coordination of the SIC and SING nets is carried out by Load Economic Dispatch Centres (CDEC). These are autonomous administrations composed of representatives of the transmission, generating, and distribution companies. With the enactment of the law *Decreto Supremo 291* in August 2009, large clients and sub-transmission companies became part of these CDEC.

Furthermore, two regional and vertically integrated grids known as the Aysen and Magallanes systems exist in the south east. Owned by the Edelaysen group (Aysen system) and Edelmag (Magallanes system), they are operated by their own CDEC (CNE 2010b).

Almost three-quarters of the national installed capacity operate in the SIC (72.6%), which includes the Metropolitan Region of Santiago (MRS) and provides electricity for 92% of the Chilean population. In 2007, the installed power production capacity amounted to 12,847 MW (megawatt), with 7,720.4 MW of thermal power plants, 4,907 MW of hydropower plants and 20 MW of wind power plants (CNE 2008b). An electric power auto-production capacity of 1,064 MW (including industrial Combined Heat and Power – CHP – plants) was also installed (CNE 2007b). A total production capacity of 9,910 MW was installed in the SIC grid in 2008 (CDEC-SIC 2008b), 53% of which was based on hydropower, 46% on thermal and 0.02% on wind energy.

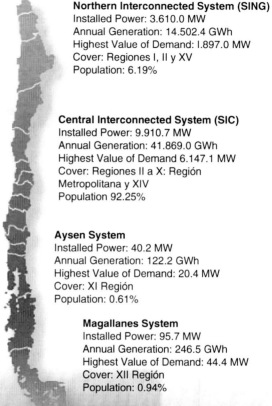

Fig. 9.3 Location and data (2008) of the four power grids (Source: CDEC-SIC 2008a)

In 2007, the total electricity production in Chile amounted to 60,138 gigawatt hours (GWh), 6.3% of which is generated by auto-producers, mainly the mining companies in the copper industry (CNE 2007c). Power generation in the SIC system increased by 11% from 37,965 GWh in 2005 to 41,971 GWh in 2008 (CNE 2008c).

Within this time frame, the energy carrier mix for power production in the central SIC grid changed radically (see Table 9.3), in particular with respect to gas and diesel. Natural gas consumption dropped from 18% to 3%, while the use of diesel increased from 4% to 23%. In 2008, only 56% of the 41,971 GWh of electricity production was based on hydropower compared to 67% in 2005.

9.2.2 Organization of the Energy Supply

Chile began to privatize its electricity sector in the 1980s. Today, all generation, transmission and distribution activities are managed privately. The National Energy

9 Energy Systems

Table 9.3 Electric system (SIC) generation by source

	Coal (%)	Diesel (%)	Natural Gas (%)	Hydraulic (%)	Others (%)
2005	10	4	18	67	1
2006	12	2	15	70	1
2007	15	23	7	53	2
2008	15	23	3	56	2

Source: Chilectra 2008a

Commission (CNE) and the Ministry of Economy and Energy (MEE) are principally responsible for the regulation of the electricity sector.

The energy sector in Chile is to a large extent privately organized. A total of 72 companies control approximately 90% of the installed electricity production capacity in the country, i.e., 31 power production companies, five transport companies and 36 other power-related companies. The remaining 10% of the installed capacity belongs in the main to the state-owned copper companies. With a market share of 40%, the biggest electricity producer is the Spanish company Empresa Nacional de Electricidad (Endesa). Along with the distribution company Chilectra, it belongs to Enersis Holding, which in turn is part of the Spanish trust Endesa.

Endesa runs almost all the hydropower plants in Chile and, with 1.35 million clients, supplies close to 50% of Chilean households with electricity. The second largest electricity producer is AES Gener, with a market share of almost 30%, while the Chilean Colbún company comes in third. For electricity transmission, 80% of the grid is run by the private company Transelec.

In Chile, 85% of combustibles (with the exception of firewood) are traded from the state-owned Empresa Nacional de Petroleo (ENAP), which is controlled by the mining ministry (Minminería). ENAP plays a leading role in the market and trades 40% of primary energy.

This picture of the Chilean energy sector is thus one of a highly concentrated market with a pronounced dependency on imported fossil fuels. How this setting constitutes a risk to sustainable development is illustrated in the next section.

9.3 Risk Analysis and Sustainability

Up to now the energy system has on the whole been able to satisfy the needs of the Metropolitan Region of Santiago. However, the future development of the Region poses considerable challenges to the energy system, exposing it to a variety of risks, as demonstrated in the following.

9.3.1 Methodology

The risk analysis of the energy sector is based on the distance-to-target methodology and adheres to the *Helmholtz Integrative Sustainability Concept* described in

Chap. 4. It consists of three steps. Building on the status analysis presented in the previous section, a first step defined a set of sustainability indicators and, where possible, target values for these indicators. A second step comprised a current sustainability performance analysis of the selected indicators. The analysis, which was based on the framework scenarios outlined in Chap. 4 focusing on the business-as-usual scenario, also included an exploration of future trends and prospects. Due to Santiago's tight integration in the national energy system and the absence of regional data, the analysis was conducted at the national level. A third step outlines the most urgent sustainability problems in the energy sector, concentrating in this chapter on current performance.

9.3.1.1 Selection of Sustainability Indicators

The risk analysis of the energy system began with the implementation of the sustainability rules contained in the *Helmholtz Integrative Sustainability Concept*. Its objective was to select appropriate indicators for the Metropolitan Region of Santiago.

For this purpose, the international literature, e.g., UN (2007a, b), Southeast False Creek Steering Committee (2005), CEPAL (2007), OECD (2007), Atkinson et al. (2004), Berman and Phillips (2000), Böhnke (2001), Hacking (2005), Shookner (2002), Van der Maesen and Walker (2006), and Arancon (2009), was reviewed and a comprehensive set of 44 indicators assembled. 16 core sustainability indicators were selected from the set, based on criteria such as the appropriate description of sustainability in accordance with the *Integrative Concept*, the possibility to determine target values, and data availability. Section 9.3.2 outlines historical, current and future data for these indicators. The non-availability of data at the metropolitan level for a number of these indicators was a limiting factor.

Finally, the following indicators were selected for the sustainability evaluation:

1. Share of rural households with no access to electricity in the MRS;
2. Duration of electricity supply interruptions in MRS;
3. Total primary energy consumption per capita;
4. Energy intensity (i.e., energy per GDP);
5. Share of non-conventional renewable energies in electricity production;
6. Energy-related CO_2 emissions;
7. Energy import dependency (percentage of final energy use based on imported energy), and
8. Degree of economic concentration in the energy sector.

Some of these indicators were used at national level, given that most of the energy supply system is organized at the national (CNE activities) or regional level, such as the SIC electricity grid (see Sect. 9.2.1).

In the next step, target values based on existing regulations, scientific debates or international experiences were determined at local, regional and national level.

9.3.1.2 Model Used for the Scenario Analysis

To gain a perspective on the future development of the energy system, a model was integrated into the analysis process. Based on an existing energy system model, a MESAP/PlaNet (Modular Energy System Analysis and Planning) model was adapted for the case of Chile. MESAP/PlaNet is a quantitative energy system model originally designed at the University of Stuttgart, Germany, and currently being refined by *Seven2one Informationssysteme GmbH* (Seven2one 2006). This model is used for scenario analysis at global and regional levels at the German Aerospace Center (DLR) (for details, see Krewitt et al. 2007). MESAP relies on the principle of Reference Energy Systems. This allows for structuring and delineating of energy systems with processes (e.g., electricity generation plants, heating plants and transportation processes) and commodities (e.g., electricity, heat and fuel demand, primary energy consumption and CO_2 emissions) that are linked by flows and calculation rules. Particular equations describe each process. The backbone of the model is an extended data base for a variety of energy supply technologies, such as power plants, heat production, combined heat and power production (CHP production) and fuel production.

Fuel efficiency, full load hours and emission factors are some of the technical specifications the data base provides to feed equations. The specifications are adapted to regional characteristics. The model is calibrated on the basis of national energy balances derived from the International Energy Agency (IEA) energy statistics for 2009 (IEA 2009a). Projection of the gross domestic product (GDP) and population growth are key drivers behind the energy demand, which is supplied by the energy system in line with technical restrictions as defined in the data base. The model is used for scenario analysis in Chile at the national level. Basic technical data development is adopted for Chile from previous scenario analysis at the German Aerospace Center (see Greenpeace International and EREC 2009).

The analysis outlined in this chapter considers a business-as-usual scenario (BAU) for the energy system, taking the year 2030 as a time horizon. Its basic *philosophy*, which consists in the persistence of strong market forces and the current social security system was adopted in the model. The GDP will rise at an average of 3.5% per year and the Chilean population increases from 16.3 in 2005 to 19.6 million in 2030. Changes in energy intensity were adopted from the projection from a previous study (see Greenpeace International, and EREC 2009) and lead to a 26% improvement by 2030. In the transport sector, electric vehicles will gradually enter the market and make up a share of 6% of the vehicle stock in 2030. For heat production, gas and biomass will persist, while the increase in CHP will be slow. For the power sector, a vast expansion of hydropower is estimated. In particular, the currently planned and hotly debated *HidroAysen* hydropower project will be realized and a heavy increase in coal-fired power plants is expected.

9.3.2 Sustainability Analysis

The analysis of the current performance of the selected indicators presents a heterogeneous picture. Some trends are positive, while other indicators point to deterioration in terms of set targets. Such cases call for immediate action.

9.3.2.1 Share of Rural Households with No Access to Electricity in the Metropolitan Region of Santiago

In the modern world access to electricity is the precondition for satisfaction of basic needs such as healthy food (refrigerator, stove), security (light), communication (computer, fax, telephone), and information (television, radio). All of the houses in the urban areas of the Metropolitan Region of Santiago have electricity. Only a small number of buildings in the rural periphery of the region are not connected to the power grid. However, in comparison to the beginning of the 1980s, when over 20% of buildings had no grid access, the situation has altered significantly over time, with the percentage reduced by half in the 1990s, finally dropping to 1% of houses in recent years. Many of these, however, are located in the mountains far from the urban centre, making the grid connection extremely expensive. The best option for access to the electricity supply in the future is the use of *island systems* of renewable energy (windmills, solar panels, batteries, small hydro, and biogas) (Fig. 9.4).

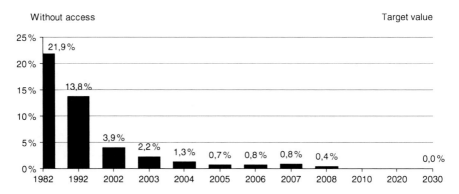

Fig. 9.4 Share of rural households in the MRS without access to electricity (Source: CNE 2008d, 2010c)

9.3.2.2 Duration of Electricity Supply Interruptions in the Metropolitan Region of Santiago

One of the chief problems of the energy supply is its temporary disruption. Long periods of disruption lead to interruptions in production, to cooling difficulties, both

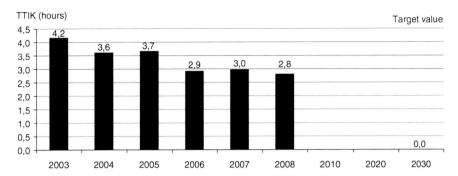

Fig. 9.5 Duration of interruption of electricity supply in MRS Data (Source: Chilectra 2005a, 2006a, 2007a, 2008b)

in private homes with failing refrigeration and in the production sector, e.g., in warehouses.

The standard used in Chile to measure power supply interruptions is *TTIK*, i.e., a unit of less than 13 h of interruption in 12 months in an urban area and 18 hours in a rural area. Chilectra, the company that supplies most of the Metropolitan Region of Santiago, is currently (2010) operating with a TTIK average of 18.46. The method of calculating TTIK altered in 2009, although past values do reveal improvements in the security of the grid (see Fig. 9.5). This is a laudable development. However, to avoid substantial damage to food, machines and other goods, the target should be lower than established by law and aim for zero interruption.

9.3.2.3 Energy Intensity Per GDP

Energy intensity is the reciprocal value of energy efficiency and indicates the amount of energy required to produce 1 unit of GDP. As energy consumption equates with high costs, reducing energy intensity is not merely a question of sustainable development but of economic rationale. Historically, in accordance with the degree of technological progress, Chile's energy intensity has declined over the years, as shown in Fig. 9.6. Since 1990, for example, energy intensity in

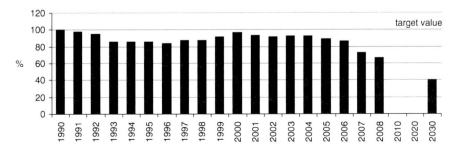

Fig. 9.6 Amount of energy per GDP (relative, 1990 as 100) Data (Source: INE 2010)

Chile has dropped by about 35%. This trend, however, was not linear. Between 1997 and 2006, energy intensities even increased compared to 1996. To meet the target of a 60% reduction of the 1990 energy intensity value by 2030 will require monumental effort.

9.3.2.4 Share of Energy from Renewable Sources

For the indicator *Share of non-conventional renewable energies in power production* (excluding hydropower), the Chilean government established a quota model to increase the use of non-conventional renewable energy. From 2010 to 2014, all agents of the power system (SIC or SING) that sell electricity to distribution companies or to final clients must provide a 5% share of Non-Conventional Renewable Energy (NCRE). From 2015 onwards, this 5% quota will increase by an annual 0.5% points, ultimately reaching a quota of 10% by 2024. A subsequent annual increase of 1% points is assumed up to 2030 (see Fig. 9.7).

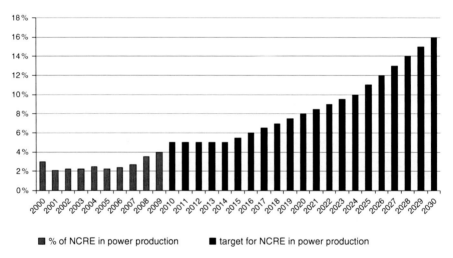

Fig. 9.7 Share of power production from NCRE in SIC in relation to the requirements of Law 20.257 for the development of NCRE and other targets (Source: CDEC-SIC 2008a)

9.3.2.5 Energy-Related CO_2 Emissions

The Metropolitan Region of Santiago, home to 40% of Chileans and the production site of over 40% of the national GDP, is a major energy consumer and thus a prime source of CO_2 emissions. In essence, the rise in gas consumption and particularly of coal, is responsible for mounting greenhouse gas emissions. Considering the long investment cycles in the energy sector, Chile risks being locked in with even greater CO_2 emissions in the future, if the country does not invest more heavily in

9 Energy Systems

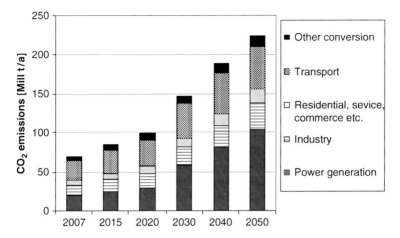

Fig. 9.8 Development of CO_2 emissions in Chile under BAU assumptions (Source: The authors, based on calculations with MESAP model)

renewable energy technology. Up to now, Chile has not been obliged to mitigate CO_2 emissions within the framework of the Kyoto Protocol. Although new targets on CO_2 reduction are not foreseeable, international agreements beyond Kyoto are likely to impose more pressure, at least on newly industrializing countries like Chile. The scenario calculations show that a *Business-as-usual* development of the energy sector will lead to a doubling of energy-related CO_2 emissions by 2030 (see Fig. 9.8).

Even if plans for vast new hydropower plants such as HidroAysen are implemented, power production will still be responsible for a growing share of CO_2 emissions. Current policy directed at establishing new sources of energy will only partly alleviate the risk of low diversification and emission increase within the energy sector.

9.3.2.6 Energy Import Dependency: Percentage of Final Energy Consumption Based on Imported Energy

Another sustainability deficit is the substantial dependence on fossil fuel imports. For many years now, Chile has imported more than two-thirds of its energy supply (see Table 9.4).

Table 9.4 Chilean dependency on imports for energy provision

Origin	1999 (%)	2000 (%)	2001 (%)	2002 (%)	2003 (%)	2004 (%)	2005 (%)	2006 (%)	2007 (%)
National	31.0	32.0	34.0	34.0	30.0	27.5	30.3	29.8	28.6
Imported	69.0	68.0	66.0	66.0	70.0	72.5	69.7	70.2	71.4

Source: IEA 2009a

A share of over 80% of fossil fuel-based primary energy is commonplace in industrialized countries. Chile imports more than 90% of fossil fuels (IEA 2009a). Compared to the situation in 1990 (see Table 9.1), the energy demand is now largely covered by imports. This exposes the energy system to the risk of energy supply insecurity, given the expected increase in energy prices on the world market (IEA 2007).

The fact that the power sector is underdiversified and depends on Argentinian gas exacerbates the problem. Three-quarters of Chile's power production is currently based on hydro and gas, especially in the Metropolitan Region of Santiago and its highly specialized SIC grid electricity supply. Santiago de Chile, the country's largest consumer of electricity, is therefore almost completely dependent on two fuel sources: gas imports from Argentina and hydropower, with the latter providing almost 50% of the installed capacity (IEA 2009b). The risk associated with an underdiversified energy system is not a new phenomenon in Chile. Much more crucial is its 100% dependence on gas imports from Argentina, which are indispensable to industry and the transport sector. Due to border conflicts and the geographical situation, Bolivia refused to deliver gas to Chile. Moreover, the gas supply contract between Bolivia and Argentina explicitly interdicts Argentina from transferring gas from Bolivia to Chile. Since 2004, the Argentinian energy ministry has given strong preference to using national gas reserves rather than exporting them. As a result, Argentina temporarily cut back 90% of its gas deliveries contracted to Chile (Fig. 9.9), leaving Chile to cope with the ongoing deficit in gas supplies. These energy supply shortages will and already have had an impact on economic growth.

The high percentage of electricity generated from national hydropower exposes the energy system to an additional shortage risk: supply bottlenecks in the past occurred in years with little rainfall, e.g., in 2008. As a result of diminished gas supplies from Argentina, climatic variability and the predicted 6% increase in the

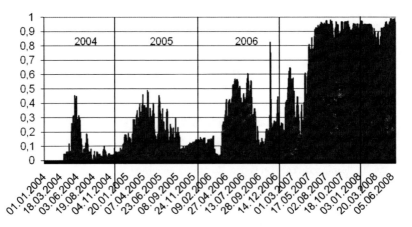

Fig. 9.9 Reduction of gas supply from Argentina (in percent of contracted delivery) (Source: CNE 2009a)

annual demand for power over the next 10 years, securing the energy supply remains a key challenge.

Energy consumers currently compensate power shortages, as in 2007 and 2008 (see Fig. 9.9), with additional diesel aggregates. In the short term, Chile addresses its deficits in energy security by importing Liquified Natural Gas (LNG). Under the current framework conditions, this dependency will only gradually diminish, for instance, with the growing capacity of coal-fired power plants, to be added over the next few decades. CNE predicts a 26% share of coal in the power supply by 2020. Currently, 2.2 gigawatt (GW) of coal power plant capacity is under construction, another 4.7 GW are under consideration (IEA 2009b).

9.3.2.7 Degree of Economic Market Concentration in the Energy Sector

Market concentration on the supply side carries the risk of either a monopoly or an oligopoly and thus of reduced competition. This could result in higher prices for private households and industry or a tendency to reduce service quality.

As Table 9.5 shows, the energy sector is highly concentrated. For several years, Chilectra Metropolitana had a share of approximately two-thirds of electricity sales and along with Cia. General de Electricidad (CGE) and Cia. Eléctrica del Maipo covered 94% of sales up to 2004. In 2005, Cia. Eléctrica del Maipo was replaced by Emelectric S.A., which was third in the ranking. Today, these three companies together cover almost 99% of the market.

With regard to electricity distribution, the concentration is slightly lower but still high in relation to other commodity markets. In the last 10 years, however, market concentration has increased to a dramatic 98%. Liquid fuel distribution is one example of a market decline in concentration in recent years. Here, the market share of the three largest companies fell from 96% in 1997 to 84% in 2006. This notwithstanding, the energy sector remains a highly concentrated domain. Disentanglement is advisable in the interests of greater competition.

Table 9.5 Percentage of the three most important branches in the energy sector: electricity sales, electricity purchase, and market distribution of liquid fuels

	1997	1998	1999	2000	2001	2002	2003	2004	2005	2006	2007	2008
Electricity sales organizations	95.1	94.7	94.6	94.9	94.1	94.2	94.0	94.0	98.5	–	–	–
Electricity purchase organizations	–	–	–	71.3	81.4	83.5	85.5	95.1	94.6	95.0	95.0	98.2
Liquid fuel distribution organizations	96.0	94.8	92.6	92.1	91.0	–	91.5	–	–	84.4	–	–

Sources: CNE 2007d; Chilectra 2000, 2001, 2002, 2003, 2004, 2005b, 2006b, 2007b, 2008b; Republic of Chile 2002; Fuentes et al. 1994; Agostini and Saavedra 2009; Copec 2006; Sepúlveda and Miranda 2001

9.4 Policy Options for More Sustainability

Chile has already taken various steps to augment the sustainability of the energy system. Chile Sustentable, a national non-governmental sustainability initiative, explicitly addresses new energy policies, including long-term considerations, internalization of external costs, efficiency improvement and increased use of renewable energies (Chile Sustentable 1997). The CNE guidelines (CNE 2006; CNE, CONAMA et al. 2006) also display a growing awareness of environmental issues. Moreover, improving energy efficiency, a neglected aspect for the last 20 years, has now become a priority.

Yet another energy policy focus in Chile is non-conventional renewable energy sources (NCRE). In the last 10 years, various steps have been taken to increase the use of NCRE. In 2005, an amendment to the energy law facilitated the access of small producers to the power grid (CNE 2006) and 2008 saw the introduction of an NCRE quota. This last obliges power suppliers to increase the share of NCRE power to 10% by 2024 or face a penalty for non-compliance. While 0.6 GW of NCRE installations are currently undergoing environmental impact assessments (CNE 2009b), CNE projections suggest that an additional capacity of 2.2 GW of NCRE could be installed by 2020. The potential currently amounts to 12 GW. In 2009, CNE announced plans for solar thermal (10 MW) and photovoltaic (0.5 MW) power plants (IEA 2009b). Figures from CNE indicate a rapidly expanding market for renewable energies. In 2008, a total of 1,240 MW underwent environmental impact assessment (Sistema de Evaluación de Impacto Ambiental, SEIA), including almost 900 MW of wind projects. By 2009, the number had increased to 2,500 MW, containing over 2,000 MW of wind generation (CNE 2008f; CNE 2009b). The potential could and should be tapped, not only in the power sector, but also in the heating and transport sectors. The latter's current biofuel policy proposes a share of 10% of biofuels in vehicle fuel consumption by 2020 (IEA 2009b).

However, the positive effects of the projected increase in renewable energy capacities and biofuels will be compensated by the increase in conventional fossil capacity and fuel consumption – a major reason for the persistence of the sustainability problems outlined above. Tackling existing and future sustainability deficits calls for an integrated approach that covers two key elements: energy efficiency improvement and an increased share of renewable energies.

9.4.1 Energy Efficiency Improvement

Forced by the energy crisis, the Chilean government set about developing energy efficiency initiatives in 2005, beginning with programmes for the residential sector. These included activities such as light bulb exchange programmes, energy labelling of electrical devices and energy certification of residential buildings, as well as insulation retrofit programmes for existing buildings (IEA 2009b). In 2008, CNE proposed new energy policy guidelines to address an institutional framework for energy efficiency, the development of a more advanced knowledge data base, the broad promotion of energy efficiency measures and regulations for overall efficiency improvement, especially in the energy sector. In terms of energy demand, the guidelines contain a target proposal for a 20% efficiency increase by 2020 (CNE 2008g).

If the available (technical) efficiency potential is to be exploited, however, more ambitious energy efficiency targets will be necessary. Results of Chilean estimates indicate that efficiency gains could reduce up to 50% of the increase in energy demand over the next 20 years (Larraín et al. 2003). Modern insulation technologies applied to existing buildings, for instance, would reduce today's average heating and cooling demands by 50%. This could impact significantly on final energy demands, since buildings and agriculture account for 30% of the total energy demand in Latin America (Krewitt et al. 2009).

Likewise in the industrial sector, a large variety of efficiency measures could be applied, e.g., in the petrochemical industry and the production of metals and minerals, which are associated with high energy intensities. Based on currently available best practice technologies, an annual technical efficiency potential of 2.5% can be assumed for Latin America, e.g., in industrial processes and including mounting recycling rates. At 2.9% per year, the technical efficiency potential of the transport sector in Latin America is even higher (Krewitt et al. 2009).

Past experience shows, on the other hand, that market barriers to efficient technologies still prevail and that current policy mechanisms might not be ambitious enough to overcome them, calling urgently for an appropriate response.

9.4.2 Implementation of Renewable Energies

Although about 50% of the electricity supply in Chile is currently produced by water, for primarily economic reasons the Chilean government actually supports the construction of coal and oil-fired thermal power plants. The objective is to base the future energy system on a diversity of energy sources, focusing on renewable energies. Due to its exceptional geographical location along the Andes and its long coastline, Chile has a remarkable potential for hydro, wind energy and ocean power. Areas like the Atacama Desert in the north could provide masses of energy from solar radiation. Furthermore, Chile is situated in the Pacific Ring of Fire, which has huge geothermal potential. According to estimates of the Universidad de Chile (UCH) and the Universidad Técnica Federico Santa María (UTF), the renewable energy potential in 2025 will even exceed the growing power demand in Chile: the potential for hydro amounts to 20 GW, for wind 40 GW, for geothermal 16 GW, for solar between 40 and 100 GW, for photovoltaic 1 GW, for biomass 13.6 GW, leading to a total potential of 190 GW (UCH, and UTF 2008).

Even more conservative estimations show that Chile has barely started to use its outstanding NCRE potentials. The National Petroleum Company (ENAP) estimates a potential of 3.4 GW for electric generation (IEA 2009b). Several case studies by the Deutsche Gesellschaft für Technische Zusammenarbeit (GTZ) have assessed considerable biomass potential, concluding that currently available residues from agriculture, the food industry and forests would be able to supply between 0.6 and 0.8 GW. Agricultural and food industry residues could account for 4.8–6.5 TWh (terrawatt hours) from biogas, with forest residues supplying 310–470 MW (Chamy and Vivanco 2007; Spichiger and Verdugo 2008).

All in all, numerous studies conclude that non-conventional renewable energies (NCRE) in Chile have enormous potential for the future of the energy sector, by far exceeding current policy targets. UCH and UTF estimate that by 2025, non-conventional renewables could account for up to 30 TWh produced by almost a 6 GW installed capacity and provide more than 30% of the total power generation (UCH, and UTF 2008).

While most studies focus on the power sector, the German Aerospace Center has analysed the implementation potential of renewables in power, heat and transport in a scenario study for Greenpeace. The Greenpeace *Energy [R]evolution* report estimates NCRE coverage of more than 35 TWh in 2020. Figure 9.10 shows how a fitting combination of renewable energy technologies and efficiency measures can lead to a share of more than 50% of renewables in 2030. Exploiting foremost the huge wind and solar potential would subsequently accelerate this increase. *Energy [R]evolution* also addresses the role of renewables for both heat and transport. While biomass could magnify its position as the most important renewable heating source, new technologies that use geothermal and solar energy could contribute to meeting higher demands for heating and cooling or process heat. If Chile's transportation sector was geared to electro mobility in the future, its vast potential for renewable power could be exploited in the interests of a more sustainable system of transport (Greenpeace International and EREC 2009).

NCRE are currently discussed for Chile's power sector in particular. However, the heating and transport sector should not be neglected when it comes to looking

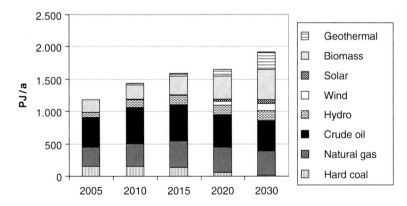

Fig. 9.10 Potential development of primary energy demand in the "Energy [R]evolution" scenario (Source: Greenpeace International and EREC 2009)

for suitable paths to a sustainable energy future. Although Chile's potential to supply first generation biofuels is small, if tapped electro mobility could prove to be a solution for the transport sector as well. Geothermal, solar and biomass resources will be vital to the future provision of heating and/or cooling services. Thus, current activities to augment the share of NCRE in the power market must take future demands for heating and transportation into account if consumption competition and its negative impact on all sectors is to be avoided.

For the Metropolitan Region of Santiago, integration of renewable energy into the existing grids is essential, given the large distances between generation and demand in the city. Long-distance power transmission, such as high voltage direct current transmission, is already in the planning stage for remote hydropower projects such as HidroAysen. Its urban counterpart, intelligent load management, will be necessary to harmonize demand and supply over time, a hitherto unsolved challenge associated with the massive introduction of renewables into the grids. To realize the power potential for mobility, new technologies such as plug-in hybrids and electro vehicles will have to be introduced, with the added benefit of meeting transport and emission-related targets in the city.

The Chilean government is fully aware that reorientation of the energy supply towards sustainability criteria and diversification is inevitable. This growing scientific, political and even public relevance of renewable resources notwithstanding, it will be a rocky road ahead if challenges are to be met. Privatization processes and the high market concentration have created barriers to renewable energy generators, which are frequently small companies on the market fringe. The fact that renewables are forced to compete with other energy sources along purely economic lines is one such barrier.

Hence the setting of appropriately modified framework conditions is a precondition for achieving ambitious renewable energy targets. One approach, already successfully implemented in countries like Germany, is for the government to exert more influence in the form of subsidizing or initial-funding renewable technologies and their implementation, of increasing the obligatory quota for renewables in the main grid and of establishing renewable quotas in the heating and transport sector. Sharing the financial risks involved in developing innovative renewable energy technologies or introducing them to the market could promote investment in renewable energies and the development and implementation of new projects in this field.

9.5 Conclusions

This chapter presented the main features and sustainability-related risks of Chile's energy system: (1) the need to satisfy the fast-growing demand for fossil fuels heightens the country's import dependency; (2) current and planned investments in the fossil fuel-based power capacity – especially in coal-fired power plants – cancel out the positive effects of using more renewable energy; (3) certain institutional and organizational framework conditions, e.g., a high economic concentration of the energy sector, inhibit the efficacy of market mechanisms, for instance, by preventing companies from entering the market.

On the other hand, Chile has tremendous potential for renewable energies: high solar radiation in the north, strong and continuous wind along the coast, biomass prospects in the centre and geothermal potential throughout the country. Nevertheless, these renewable energy prospects call for complementary efforts to exploit the existing potential to increase energy efficiency and meet ambitious sustainability targets for the energy sector. This furthermore requires a variety of far-reaching measures and structural changes. Chile has taken first steps towards a more sustainable energy system by implementing regulations associated with energy efficiency and renewable energy sources, by establishing an energy round table and setting up an energy ministry. These activities clearly are of both national and Metropolitan Regional relevance. The prime challenge is to develop a suitable combination of topic-specific measures, to bring the relevant actors on board, and to design and implement an overall policy that will enable the just distribution of the resultant burdens and benefits across social groups.

References

Agostini, C., & Saavedra, E. (2009). *The oil industry in Chile*. http://fen.uahurtado.cl/wp/wp-content/uploads/2010/07/inv219.pdf. Accessed 21 Apr 2011.

Arancon, S. (2009). *Grounding sustainable development in urban planning: A framework of sustainability indicators for the metropolitan region of Santiago, Chile*. Master thesis, Pontificia Universedad Católica, Santiago de Chile.

Atkinson, A., Marlier, E., & Nolan, B. (2004). Indicators and targets for social inclusion in the European union. *Journal Common Market Studies, 42*(1), 47–75.

Berman, Y., & Phillips, D. (2000). Indicators of social quality and social exclusion at national and community level. *Social Indicators Research, 50*(3), 329–350.

Böhnke, P. (2001). *Nothing left to lose? Poverty and social exclusion in comparison. Empirical evidence on Germany*. Berlin: Veröffentlichungen der Abteilung Sozialstruktur und Sozialberichterstattung des Forschungsschwerpunktes Sozialer Wandel, Institutionen und Vermittlungsprozesse des Wissenschaftszentrums Berlin für Sozialforschung.

CDEC – SIC – Central interconnected electric system/Load economic dispatch center (2008b). *SIC Map*. https://www.cdec-sic.cl/contenido_en.php?categoria_id=4&contenido_id=000028. Accessed 14 June 2010.

CDEC-SIC – Central interconnected electric system/Load economic dispatch center (2008a). *Operation Statistics 1999/2008*. (Work document). http://www.cdec-sic.cl/datos/anuario2009/. Accessed 14 June 2010.

CEPAL – Economic commission for Latin America. (2007). *Risk, habitat and megacities Latin American metropolitan panorama*. Santiago de Chile: Draft.

Chamy, R., & Vivanco, E. (2007). *Identificación y clasificación de los distintos tipos de biomasa disponibles en Chile para la generación de Biogás*. Santiago de Chile: Comisión Nacional de Energía (CNE), Deutsche Gesellschaft für Technische Zusammenarbeit (GTZ) GmbH, Ecofys GmbH.

Chile Sustentable (1997). *Programma Chile Sustentable*. http://www.chilesustentable.net/. Accessed 10 Aug 2008.

Chilectra (2000). *Yearbook of Chilectra, 2000*. http://www.chilectra.cl/wps/wcm/connect/ngchl/ ChilectraCl/La+Compania/Gobiernos+Corporativos/Memoria/. Accessed 15 June 2010.
Chilectra (2001). *Yearbook of Chilectra, 2001*. http://www.chilectra.cl/wps/wcm/connect/ 4bc2fc80450f37bf8c669fab9584e337/Memoria_Chilectra_2001.pdf? MOD=AJPERES&Tipo=DOC. Accessed 21 Apr 2011
Chilectra (2002). *Yearbook of Chilectra, 2002*. http://www.chilectra.cl/wps/wcm/connect/ 933c8f00450f37a58c509fab9584e337/MemoriaAnual2002publicado2002.pdf? MOD=AJPERES&Tipo=DOC. Accessed 21 Apr 2011
Chilectra (2003). *Yearbook of Chilectra, 2003*. http://www.chilectra.cl/wps/wcm/connect/ f2dc6f80450f37908c3a9fab9584e337/Memoria_Chilectra_2003.pdf? MOD=AJPERES&Tipo=DOC. Accessed 21 Apr 2011.
Chilectra (2004). *Yearbook of Chilectra, 2004*. http://www.chilectra.cl/wps/wcm/connect/ 9752b780450f37798c249fab9584e337/MEMORIAANUAL2004PUBLICADA2005.pdf? MOD=AJPERES&Tipo=DOC. Accessed 21 Apr 2011
Chilectra (2005a). *Sustainability reports of Chilectra 2005*. http://www.enersis.cl/enersis_web/ sostenibilidad/Informes_Sostenibilidad/Chilectra/Chilectra_Sostenibilidad05.pdf. Accessed 21 Apr 2011.
Chilectra (2005b). *Yearbook of Chilectra, 2005*. http://www.chilectra.cl/wps/wcm/connect/ ffe26a00450f37628c0e9fab9584e337/MEMORIAANUAL2004PUBLICADA2005.pdf? MOD=AJPERES&Tipo=DOC. Accessed 21 Apr 2011.
Chilectra (2006a) *Sustainability reports of Chilectra 2006*. http://www.enersis.cl/enersis_web/ sostenibilidad/Informes_Sostenibilidad/Chilectra/Chilectra_Sostenibilidad06.pdf. Accessed 21 Apr 2011.
Chilectra (2006b). *Yearbook of Chilectra*, 2006. http://www.chilectra.cl/wps/wcm/connect/ 26e07200450f374a8bf89fab9584e337/Memoria_Chilectra_2006.pdf? MOD=AJPERES&Tipo=DOC. Accessed 21 Apr 2011.
Chilectra (2007a) *Sustainability reports of Chilectra 2007*. http://www.enersis.cl/enersis_web/ sostenibilidad/Informes_Sostenibilidad/Chilectra/Chilectra_Sostenibilidad07.pdf. Accessed 21 Apr 2011.
Chilectra (2007b). *Yearbook of Chilectra, 2007*. http://www.chilectra.cl/wps/wcm/connect/ 49649200450f372e8be29fab9584e337/Memoria_Chilectra_2007.pdf? MOD=AJPERES&Tipo=DOC. Accessed 21 Apr 2011.
Chilectra (2008a). *Sustainability reports of Chilectra 2008*. http://www.enersis.cl/enersis_web/ sostenibilidad/Informes_Sostenibilidad/Chilectra/Chilectra_2008.pdf. Accessed 21 Apr 2011.
Chilectra (2008b). *Yearbook of Chilectra*, 2008. http://www.chilectra.cl/wps/wcm/connect/ ed3f4500450f37108bcc9fab9584e337/Memoria_Anual_2008.pdf? MOD=AJPERES&Tipo=DOC. Accessed 21 Apr 2011.
CNE – National Energy Commission. (2006). *Las Energías Renovables no Convencionales en el Marco de la Diversificación Energética*. Santiago de Chile: Comisión Nacional de Energía.
CNE – National Energy Commission (2007a). National balance of energy of 2007. Sectorial Consume. Table 21-C. http://www.cne.cl/cnewww/opencms/06_Estadisticas/Balances_Energ. html. Accessed 13 June 2010.
CNE – National Energy Commission (2007b). National balance of energy of 2007. Sectorial consume. Table 22-B. http://www.cne.cl/cnewww/opencms/06_Estadisticas/Balances_Energ. html. Accessed 13 June 2010.
CNE – National Energy Commission (2007c). National balance of energy of 2007. Table BALANCE_ELECT. http://www.cne.cl/cnewww/opencms/06_Estadisticas/Balances_Energ. html. Accessed 13 June 2010.
CNE – National Energy Commission (2007d). History of sales distribution concessionaire. period 1997–2005, Systems: SING, SIC, Aysen and Magallanes. Personal communication.
CNE – National Energy Commission (2008a). National balance of energy of 1999, 2001, 2003, 2005 and 2007. Primary energy balance. http://www.cne.cl/cnewww/opencms/ 06_Estadisticas/Balances_Energ.html. Accessed 13 June 2010.

CNE – National Energy Commission (2008b). *National balance of energy of 2008*, http://www.cne.cl/cnewww/opencms/06_Estadisticas/Balances_Energ.html. Accessed 13 June 2010.
CNE – National Energy Commission (2008c). National balance of energy of 2007. National Gross generation by system. http://www.cne.cl/cnewww/opencms/06_Estadisticas/Balances_Energ.html. Accessed 13 June 2010.
CNE – National Energy Commission (2008d). *Energy sector statistical yearbook Chile. National coverage of rural electrification*. http://anuario.cne.cl/anuario/electricidad/php_electricidad-5.01.php. Accessed 16 June 2010.
CNE – National Energy Commission (2008e). National balance of energy of 2008 Santiago de Chile 2008, Table 21. http://www.cne.cl/cnewww/opencms/06_Estadisticas/Balances_Energ.html. Accessed 13 June 2010.
CNE – National Energy Commission (2008f). Wind energy in Chile and the Renewable Energy Law 20.257. Berlin. Comisión Nacional de Energía.
CNE – National Energy Commission. (2008g). *Transformando la crisis energética en una oportunidad*. Santiago de Chile: Comisión Nacional de Energía (CNE).
CNE – National Energy Commission (2009a). Gráfico Restricciones de Gas desde Argentina.
CNE – National Energy Commission. (2009b). *Renewable energy portfolio and wind energy in Chile from GTZ*. Berlin: Comisión Nacional de Energía.
CNE – National Energy Commission (2010a). Energy sector statistical yearbook Chile. Series of secondary energy balance. National Energy Commission. http://anuario.cne.cl/anuario/balance/php_balance-2.3.php. Accessed 13 June 2010.
CNE – National Energy Commission (2010b). Medium systems menu, Aysen y Magallanes. http://www.cne.cl/cnewww/opencms/07_Tarificacion/01_Electricidad/otros_sistemas_medianos/Sistemas_Medianos/sistemas_medianos_aysen_magallanes.html. Accessed 14 June 2010.
CNE – National Energy Commission (2010c). *Rural electrification program (1995–1999)*. http://www.cne.cl/cnewww/opencms/06_Estadisticas/Programas/index.html. Accessed 17 June 2010.
CNE – National Energy Commission, CONAMA – National Comission of Environment, & GTZ – Deutsche Gesellschaft für Technische Zusammenarbeit GmbH. (2006). *Guía del Mecanismo de Desarrollo Limpio para Proyectos del Sector Energía en Chile*. Santiago de Chile: Comisión Nacional de Energía, Comisión Nacional del Medio Ambiente.
CNE – National Energy Commission, & GTZ – Deutsche Gesellschaft für Technische Zusammenarbeit GmbH (2009). *Las Energías renovables non convencionales en el mercado eléctrico chileno*. Santiago de Chile.
Company of Liquid Natural Gas (2010). *Homepage, natural gas supply*. http://www.gnlquintero.cl/GNL/Abastecimiento.htm. Accessed 14 June 2010.
Empresas Copec (2006). *Yearbook of Copec 2006*. http://www.empresascopec.cl/copec.asp?idq=3&con=22. Accessed 21 Apr 2011.
Fuentes, R., Paredes, R., & Vatter, J. (1994). Deregulation and competition in the gasoline market. *Journal Public Surveys Estudios Públicos*, (56)25. http://www.cepchile.cl/dms/archivo_1391_173/rev56_fuentes.pdf. Accessed June 2010
Greenpeace International, & EREC. (2009). *Energy [R]evolution – A sustainable Chile energy outlook*. Amsterdam: Greenpeace International/European Renewable Energy Council (EREC)/Deutsches Zentrum für Luft- und Raumfahrt (DLR)/Ecofys.
Hacking, S. (2005). *European network indicators of social quality – ENIQ – Social quality. The British National Report*. Sheffield: University of Sheffield. Department of Sociological Studies.
INE – National Statistics Institute of Chile (2010). *Intensidad energética*. http://centralenergia.cl/actores/distribution-chile/. Accessed 17 June 2010.
IEA – International Energy Agency. (2007). *World energy outlook 2007*. Paris: IEA.
IEA – International Energy Agency (2009a). *Energy balance of non-OECD countries* (2009 ed.). IEA Energy Statistics (Beyond 20/20). Paris.
IEA – International Energy Agency. (2009b). *Chile energy policy review 2009*. Paris: IEA.

Krewitt, W., Simon, S., Graus, W., Teske, S., Zervos, A., & Schaefer, O. (2007). The 2 degrees C scenario – A sustainable world energy perspective. *Energy Policy, 35*(10), 4969–4980.

Krewitt, W., Nienhaus, K., Kleßmann, C., Capone, C., Stricker, E., Graus, W., et al. (2009). *Role and potential of renewable energy and energy efficiency for global energy supply*. Stuttgart/Berlin/Üttrecht/Wuppertal: German Aerospace Center e.V., Ecofys Germany GmbH, Ecofys Netherlands BV, Wuppertal Institute for Climate, Environment and Energy GmbH, German Federal Environment Agency.

Langdon, D. (2008). *Energy profile of Chile. Encyclopedia of earth*. In C. J. Cleveland (Ed.), (Washington, D.C.: Environmental Information Coalition, National Council for Science and the Environment). 4 March, 2007, last revised 4 Sept 2008. http://www.eoearth.org/article/Energy_profile_of_Chile. Accessed 16 June 2010.

Larraín, S., Karím P., & Aedo, M. P. (2003). *Chile Sustentable – Propuesta Ciudadana para el Cambio*. Santiago de Chile. http://www.chilesustentable.net/nweb_portal/site/propuesta/introCS.pdf. Accessed 10 March 2009.

OECD – Organisation for Economic Development and Co-Operation, Environmental Directorate. (2007). *OECD key environmental indicators 2007*. Paris: OECD Environment Directorate.

Republic of Chile (2002). *National Economic Fiscal. Legal department research*. Resolution of order N° 211 of November 5 2002. http://mailweb.fne.gob.cl/db/jurispru.nsf/.../Combustibles_liquidos.pdf. Accessed June 2010

Sepúlveda, M., & Miranda, M. (2001). Oil company of Chile S.A., industrial sector, risk analysis. *Journal Fitchrating*. http://www.fitchratings.cl/pdf/Inf-emp.pdf. Accessed June 2010

Seven2one (2006). *MESAP*. http://www.seven2one.de/mesap.phpS. Accessed 8 July 2009.

Shookner, M. (2002). *An inclusion lens workbook for looking at social and economic exclusion and inclusion*. Halifax: Atlantic Regional Office, Population and Public Health Branch.

Southeast False Creek Steering Committee. (2005). *Sustainability indicators, goals, targets for south east false creek adopted by the Vancouver City Council*. Vancouver: Southeast False Creek Steering Committee.

Spichiger, J. B., & Verdugo, E. M. (2008). *Potencial de generación de energía por residuos del manejo forestal en Chile*. Santiago de Chile: CNE – National Energy Commission, & GTZ – Deutsche Gesellschaft für Technische Zusammenarbeit GmbH.

UCH, & UTF. (2008). *Estimación del aporte potencial de las Energías Renovables No Convencionales y del Uso Eficiente de la Energía Eléctrica al Sistema Interconectado Central (SIC) en el período 2008–2025*. Santiago de Chile: Universidad de Chile, Universidad Técnica Federico Santa María.

UN – United Nations (2007b). Commission on Sustainable Development. *CSD indicators of sustainable development* – 3rd ed.

UN – United Nations (2007a). Secretary-General. *Millennium development goals*. New York. http://www.un.org/millenniumgoals/index.html. Accessed 26 Sept 2007.

Van der Maesen, L., & Walker, A. (2006). Indicators of social quality: Outcomes of the European scientific network. *European Journal of Social Quality, 5*(1–2), 8–24.

Vargas, L. (2008). *Field of aplication energy system. Presentation at the first status conference of the Risk Habitat Megacity Project*. Santiago de Chile.

Chapter 10
Santiago 2030: Perspectives on the Urban Transport System

Andreas Justen, Francisco Martínez, Barbara Lenz, and Cristián Cortés

Abstract This chapter summarizes recent developments in Santiago's urban public and private transport system and outlines perspectives for the year 2030. The analysis is conducted with a set of indicators: motorization rate, current and expected congestion levels, level of service, modal split and accessibility levels. For each indicator the expected values for 2030 are estimated and analysed to determine the feasibility of a more sustainable urban transport system in Santiago. Assumptions regarding economic and demographic growth for the Metropolitan Area of Santiago, as well as infrastructural projects and operational adjustments are considered. Indicator quantification is achieved with mathematical economic models for transport and land use. Based on the indicator values, conclusions are drawn about the perspectives on Santiago's urban transport system. Finally, recommendations for further policy action in the field are discussed.

Keywords Santiago de Chile • Transport indicators and policies • Transport modelling • Urban development

10.1 Introduction

The transport systems of large urban agglomerations such as Santiago de Chile are an integral part of the highly dynamic urban environment. The dynamic refers to changes in land use, the construction of new infrastructure or alterations in the socio-economic situation of households, all of which influence the interaction between transport supply and demand in the city. It also impacts on other city sectors, such as energy consumption, emission levels and health.

A. Justen (✉) • F. Martínez • B. Lenz • C. Cortés
German Aerospace Center (DLR), Institute of Transport Research, Rutherfordstr. 2, 12489 Berlin, Germany
e-mail: Andreas.Justen@dlr.de

The subject of this chapter is the urban transport system of the Metropolitan Area of Santiago (AMS).[1] The overall aim is to summarize recent developments in Santiago's public and private transport and to outline perspectives for the year 2030. The potential future of the transport system in 2030 is designed, given demographic and economic growth rates and assumed developments in infrastructure or vehicle technology. Certain risks associated with elevated levels of motorization, congestion and increased average travel times are analysed.

The examination is guided by several questions:

- How will the transport system perform in 20 years?
- What policies are required to maintain the system at the current or improved level of supply quality?
- To what extent will motorization and car use increase?
- What is the role of public transport in the future?

Closely linked to these questions is the issue of the type of city – and consequently the life style – that will evolve, and how it will react to the policies implemented in the transport sector. The analysis is conducted with the use of sustainability indicators that allow monitoring of the transport system and its development. The indicators evolve positively or negatively over time according to the desired future, one that aims, for example, to reduce energy consumption and achieve an efficient transport system that guarantees a high level of mobility for the citizens of Santiago. For the assessment of these indicators, mathematical economic transport and land-use models were applied to quantify the impact of demographic and economic change, as well as policies assumed to be implemented between now and 2030.

The chapter is structured as follows: the next section gives an overview of Santiago's current transport system, summarizing recent developments and current challenges. Section 10.3 describes the methodology, introduces the risk concept and its associated sustainability indicators, and concludes with a description of the models applied. Section 10.4 introduces the main assumptions. This is followed by an illustration and discussion of the analysis findings. Finally, Sect. 10.5 returns to the driving questions and, based on the obtained results, draws conclusions.

10.2 Recent Developments in Public and Private Transport in Santiago

Urban infrastructural developments and operational changes in Santiago in the last 10 years are impressive. Between 2000 and 2010, the metro network expanded from 40 to 101 km of track. In the 3 years between 2004 and 2006, urban highways were constructed with a total length of 155 km. In February 2007, the public transport

---

[1] The Metropolitan Area of Santiago comprises the 32 municipalities of Santiago province and the six municipalities of San Bernardo, Puente Alto, Pirque, Calera de Tango, Colina and Lampa.

system Transantiago introduced an integrated fare scheme for bus and metro transfers, while simultaneously the organization and operation of the bus system was remodelled. Research and evaluation of the transport system in Santiago must be carried out against this backdrop. Expensive infrastructure cannot be rebuilt easily and – as evident in Santiago – the reorganization of public transport requires substantial financial resources and planning efforts. Although these facts do not lessen the need to develop strategies to improve the environmental or economic performance of the system, they do set the basic – somewhat restrictive – parameters for analysis. Figure 10.1 illustrates the Santiago road and metro network, the cornerstone for this analysis.

10.2.1 Private Transport and Motorization

Reliable information about the travel behaviour of citizens can be gained from large-scale Origin Destination Surveys. The most recent Santiago Household and Travel Survey, *Encuesta Origen Destino (EOD)*, dates back to 2001 (DICTUC 2001).

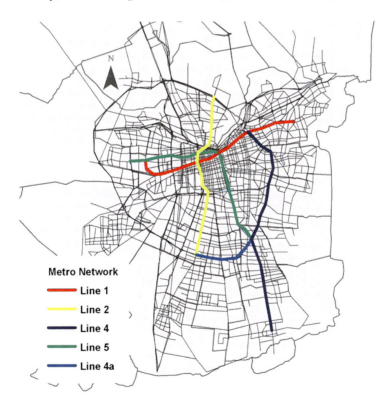

Fig. 10.1 Santiago road and metro network (Source: ESTRAUS transport and network model for Santiago)

Table 10.1 Modal split in Santiago, 1991 and 2001

	EOD 1991	EOD 2001
Bus	47.1	30.4
Metro	6.7	5.0
Car	14.6	27.4
Taxi / shared taxi	2.8	4.1
Walk	21.1	26.6
Other	7.7	6.5

Source: SECTRA 2002

In comparison to its predecessor, the EOD 1991, this survey identified trends indicating a change in the city's modal split. Table 10.1 illustrates in percentage the dramatic increase in the use of private transport for a representative average working day at the notable expense of public transport.

It is likely that this tendency persisted after this date. Despite political declarations to improve public transport and discourage the use of private cars, large investments were made in urban highways. With an investment of two billion USD, the extension of the highway network is historically the most important component of Santiago's new infrastructure. Car use is also a result of the rising motorization that goes hand in hand with the continuous growth in income levels. Between 2000 and 2008, the GDP per capita in Chile rose from 4,900 USD to 10,100 USD (IMF 2009). Additionally, it can be assumed that failure to implement the Bus Rapid Transit (BRT) system Transantiago in 2007 obliged users of public transport to buy cars. Table 10.2 summarizes population growth, the number of motorized vehicles and motorization rates between 2001 and 2009.

The motorization rates measured in 2001 were distributed unevenly across the population. Based on data from the EOD, the highest motorization rate in that year was found in the municipality of Vitacura, with 429 vehicles/1,000 inhabitants and the lowest in the municipality of La Pintana with 52 vehicles/1,000 inhabitants (Gschwender 2007). Given that socio-spatial differentiation processes in Santiago de Chile determine the spatial pattern of household localization, the same can be assumed for the distribution of car ownership rates. The highest income and, thus, the highest motorization rates are concentrated in the northeastern area of the city (the municipalities of Las Condes, La Reina, Providencia and Lo Barnechea), while the lowest rates are found in the southwestern municipalities (Lo Espejo, San Ramón, La Granja and Pedro Aguirre Cerda). The phenomenon that rising motorization in emerging countries is tied to higher income levels is well known. Nevertheless, this development is crucial to understanding the visible shift in Santiago from public to private transport in recent years. It is also worth noting that emerging car ownership tends to lead to strong car-oriented travel behaviour, aspect in which Santiago is no exception.[2]

---

[2] Qualitative interviews conducted in the municipality of La Florida (Santiago) with 56 people who had bought their first car between 2003 and 2008 indicate that trips to work are now made to 80% by car once a vehicle has been purchased (Keil 2008). La Florida had a motorization rate of 159 vehicles/1,000 inhabitants in 2001, placing the municipality slightly above average (148 vehicles/ 1,000 inhabitants).

Table 10.2 Motorization in the metropolitan area of Santiago between 2001 and 2009

	2001	2002	2003	2004	2005	2006	2007	2008	2009
Population (a) in millions	5.72	5.79	5.86	5.92	5.98	6.04	6.10	6.16	6.22
Motorized vehicles (b) in thousands	850	852	860	920	973	1,042	1,115	1,183	1,206
Motorization rate	149	147	147	156	163	172	183	192	194

Source: Instituto Nacional de Estadisticas de Chile (INE), (a) Proyecciones de Población 1990–2020; (b) Parque de Vehículos en Circulación, 2001–2009
Note: The data represents the 38 municipalities of the AMS. Motorized vehicles include private cars, motorcycles, buses, taxis and trucks.

The construction of Santiago's urban highway system is partially financed by private companies and the Chilean government within the framework of public-private partnerships. Private companies recover their infrastructure investments with 30-year concessions that permit them to obtain toll-generated revenues. The Electronic Toll Collection system introduced in Santiago is an electronic payment system whereby car-owner accounts are debited in a fraction of a second, as cars are not required to decelerate. The system received wide acceptance by its users, since concessionaires were required to distribute the necessary electronic devices (*tags*) free of charge. Predicted revenues generated by the system were underestimated by the Chilean state, such that the companies concerned now expect to recover their investments much earlier than planned in the concession contracts. Apart from the above-mentioned failure of the BRT system, this unforeseen increase in revenue was an immediate result of increases in family incomes and the corresponding low price elasticity of car travel demand.

10.2.2 Public Transport and Transantiago

The metro network forms the backbone of Santiago's public transport system. Its first line (Line 1) along the city's main transport axis *Avenida Libertador Bernardo O'Higgins (Alameda)* was inaugurated in 1975. As the population increased and the spatial expansion of the city progressed, further lines were constructed, linking the densely populated southern parts of the city with the centres of employment in the historic city centre and the commercial and financial districts in the eastern part of the city (Lines 2 and 5, respectively). Between 2000 and 2010, the metro network more than doubled in length, with extensions of existing lines to the north and south, and the construction of a partial metro ring (Line 4a) parallel to the Americo Vespucio highway. Additionally, the fifth line is currently under expansion to linkage the southwestern municipality of Maipú, a densely populated municipality that will most likely experience further population growth in the coming years. The construction of a new metro line (Line 6) was announced at the end of 2009. It will connect the municipalities of the centre-south (Cerrillos, Pedro de Aguirre Cerda) with the eastern districts of Providencia and Las Condes. One of the driving forces for the construction of this line is the attempt to reduce current demand levels on Line 1, which increased significantly when changes to the bus system were introduced within the context of implementing Transantiago in 2007.

Planned around the structure of trunk and feeder services for buses, the Transantiago integrated public transport system caters for transfers between metro and bus with no additional cost and enables payment via a touchless smart card. This new system was inspired by other experiences with BRT in the region, particularly in Bogotá and Curitiba (Muñoz and Gschwender 2008). Transantiago replaced the former bus system, which was characterized by small private entrepreneurs whose salary depended directly on the number of passengers carried, creating severe competition for passengers on the street. The system now in operation eliminates street competition; the bus fleet has been renewed and metro network operations adjusted. Contrary to the successful examples of Bogotá and Curitiba, where implementation of BRT lines was gradual, the ambitious goal of Transantiago was to begin operating immediately and simultaneously alter existing bus operations and ticketing throughout the entire city. There is a consensus view that the implementation of Transantiago in February 2007 was a failure, albeit the reasons are disputed. Too few buses on the streets, an underestimated demand for the metro, incentive problems in operator contracts and the malfunction of GPS monitoring of buses and their schedules are some of the causes. Significantly, poor public opinion of the system in the first few months took a heavy toll on the government. Although Transantiago has meanwhile made giant strides, it still suffers deficits. These are gradually being corrected by the authorities. More importantly, implementation consumed greater financial resources than originally expected, leaving financial autonomy unattainable. The ambitious project and its malfunctions generated extensive discussion on political and planning responsibilities and errors made in its technical design, from which lessons were learned and reported (Quijada et al. 2007).

The last 10 years has seen an increase in the use of bicycles. Prior to this, bicycles were rarely seen in the urban environment of Santiago, and the respective infrastructure was sparse. Propelled by the Transantiago system, cycling has recently become popular, and some municipalities have begun to construct new or extend bicycle lanes. This progress notwithstanding, bicycles as a means of transport only play a minor role, especially during peak hours, when competing with cars for road space can be a dangerous sport. However, cycling is now an established transport option within Santiago's urban transportation system and recognized as a planning issue by public authorities.

In summary, the challenges facing Santiago's transport system today are to some extent not unlike those of large urban agglomerations elsewhere in the world, with growing populations and urban territory expansion. Huge investments were made in road and metro infrastructure and the bus system was completely redesigned. The total number of private cars continued to rise, acting as both a source and a product of the simultaneous urban expansion. From a long-term city perspective, the surge in dependency on car-based travel and the ongoing loss of public transport in the modal split (despite an overall increase in ridership) tend to reduce sustainability as a result of high energy consumption levels and the attendant emissions. Mitigation of these tendencies requires further policy action, which is where the analysis and development of a potential future for Santiago's urban transport system presented in the forthcoming sections can contribute.

10.3 Methodological Considerations

The methodology applied in this chapter adopts a stepwise approach. Initially, the contextual framework for the analysis was set and refers to issues of risk and sustainability. All policy interventions and subsequent changes to Santiago's transport system were assessed against this framework. Indicators are defined to allow an estimation of how the transport system evolves over time. Quantification of the indicators is accomplished by applying transport and land-use models. Their functionalities and role in the assessment is briefly described. Finally, the issue of policy definition and prioritization is raised and, in combination with the application of the models, allows for a portrayal of Santiago's transport system in 2030.

10.3.1 Understanding Risk and Sustainability in the Context of the Urban Transport System

Risk in the context of this chapter is defined as the tendency of the transport system to function inefficiently, to consume a vast amount of natural resources and to fall short in achieving high accessibility levels throughout the urban area. In practical terms, risks are associated with elevated levels of congestion and air pollution, and a system that excludes segments of users either in physical or monetary terms, or both.

A distance-to-target approach is therefore applied, taking the current situation of the system as described in Sect. 10.2 as the starting point. Based on this 'bottom line,' it is analysed how selected transport system indicators evolve over time and whether or not they demonstrate an improvement in the system's efficiency or accessibility levels. The targets remain qualitative and descriptive but are useful in conjuring up an image of desired and undesired developments in the urban transport system in Santiago. Rising congestion levels, increased pollution or less accessibility in 2030, for example, are interpreted as deviation from the objective to maintain or even upgrade the situation prevailing in 2010. With the application of land-use and transport models, the relevant indicators are quantified and can perform either negatively i.e., constitute a sustainability deficit or even risk, or positively according to the bottom line, and thus act as signposts for a situation that is either more or less sustainable.

Most large cities agree that a resource-efficient transport system that provides a high degree of accessibility can be associated with an advanced public transport system that balances out the increase in the use of cars. Public transport is also evidence of a more compact city and less car dependency, promoting a greater interaction between different sections of the population. Thus, a sharp increase in car use in the city-wide modal split would indicate a negative development of the transport system in terms of resource consumption, emissions or congestion levels.

The terms risk and sustainability refer to the current state in 2010 and the predicted future situation in 2030. Factors beneficial to the individual user could in fact accelerate the emergence of risks at city level. The unpunctuality of the bus service might be perceived initially as a nuisance, but frequent occurrences could drive users to alter their mode choice as a protest against an unsatisfactory situation. The individual decision to use a private car rather than public transport becomes a risk to the overall system when such a decision is taken by the user majority, leading as it does to high congestion levels and low patronage of public transport. In this context, congestion levels and the associated air pollution remain at the top of the city's policy agenda. At the same time, investment in public transport tends to become less affordable as the demand for this mode declines. The above example demonstrates the interaction and interdependency of reduced sustainability and growing risk levels, where less revenue for public transport will most likely aggravate supply quality and further enhance the attractiveness of car-based travel. In sum, numerous risks are related to how public and private transport and their interrelationship evolve over time, and how land-use changes affect the distribution of residential and economic activity, which, in turn, partly depends on the car versus public transport balance.

Quantifying indicators helps to understand the direction a complex system is taking and, in the context of risk analysis, allows for evaluation of whether or not a set of policy measures is perceived as adequate in terms of risk mitigation. Given the expected development of demography, the economy, infrastructure investment and technology, the indicators support the description of the urban transport system in light of sustainability and risk. The analysis makes reference to the following indicators analysed for the years 2010 and 2030:

- Motorization rate (number of vehicles per 1,000 inhabitants);
- Congestion levels (saturation levels on major transport links);
- Level of service (travel times and velocities by mode);
- Modal split;
- Accessibility levels, and
- Emission levels.

The indicator values for vehicle usage in private and public transport, the development of motorization rates and the modal split, for instance, allow for estimates of energy consumption intensity and likewise serve as a concrete evaluation of environmental efficiency. Congestion levels and the level of service variables indicate system-wide time gains or losses. If time saving is converted into monetary units, gains and losses will indicate economic impacts. With regard to health and the environment, the amount of transport-related emissions helps to evaluate the extent to which cleaner technologies, i.e., advanced vehicle emission standards, are nullified by increased motorization and car use. The estimation of transport-related emissions is based on the analysis conducted here and described in more detail in Chap. 11 of this book.

10.3.2 Transport and Land-Use Models for Santiago

In order to address the research questions, the existent modelling instruments for Santiago with respect to transport and land-use are applied. As stated earlier, the analysis covers two points in time: the base year 2010 and the future year 2030. In general terms, models can reproduce the behavioural relationships of supply and demand in the transport and land-use sector, given the conditions of, for example, demography, prices or network quality, and the response to policy interventions. Comprehensive data sets and an established software infrastructure are available for both land use and transport. The development of the land-use model MUSSA and the transport model ESTRAUS was financed by the Chilean government. Today planning entities make use of these models to analyse infrastructure projects or predict the city's land market (de Cea et al. 2003; Martínez 1996; Martínez and Donoso 2001; SECTRA 2008). As the models represent the chief instruments used for the calculations of indicators, their functionalities are briefly described.

MUSSA (Land-Use Model of Santiago) is a software and refers to a mathematical model designed to describe, predict, simulate and carry out analysis of the urban real estate market and allow planners to forecast and simulate its economic equilibrium under different demographic, macroeconomic and regulatory scenarios. The user simulates the real estate market for these scenarios and assesses the economic impact on the city. According to the model, urban real estate market consumers, whether in households or economic activities, are located where they are the highest bidders in the market, with real estate assigned to a particular use according to 'auction' rules. This approach sees the rent of each property defined by the highest bid submitted. The socio-economic household attributes are included here in order to consider household life cycles, while zone quality indices, for instance, reflect accessibility to the transport system. Urban management policies designed to stimulate (subsidies) or restrict (taxes) the location of activities in the city can be added to the analysis.

In line with the aim of this chapter, the MUSSA land market model is seen as input for the ESTRAUS urban transport model. Localizing households differentiated by car ownership and income class is key to calculating trip generation and the attraction for each zone of the transport system. Likewise, the transport model passes information to MUSSA with respect to spatial measures dealing with accessibility and the attractiveness of zones. Consumers (households or firms) regard these access measures as variables in the process of choosing a location. Strategic projects with impact on the transport system are considered in the land-use system, i.e., the localization of households and firms adjusted with MUSSA in turn affects the transport demand calculated by ESTRAUS (SECTRA 2008). ESTRAUS is a classic four-step transport demand model that estimates trip generation (number of trips by household and zone), realizes trip distribution according to the attractiveness of each zone in the area, and partitions modes by dividing calculated traffic flows by the modes in question. In a fourth step, the estimated demand is iteratively assigned to public and private transport networks via Origin Destination Matrices until equilibrium between supply (networks) and demand (users) is achieved. Figure 10.2 outlines the components of each model and their interaction.

Work with these types of models as described in this chapter requires diverse and comprehensive input data. Figure 10.3 gives an overview of the land-use and transport model input and output for quantifiable indicators. The overall population is divided into 13 socio-economic user groups, representing different motorization levels and the associated travel behaviour (e.g., by different mobility rates, defined as the number of trips per household unit).

The spatial scale of the models covers the entire urban transport system and all transport modes within the boundaries of the AMS and the 38 municipalities. For the application, this area is further divided into 618 Traffic Analysis Zones (TAZ). As a result of land use modelling with MUSSA, households by socio-economic

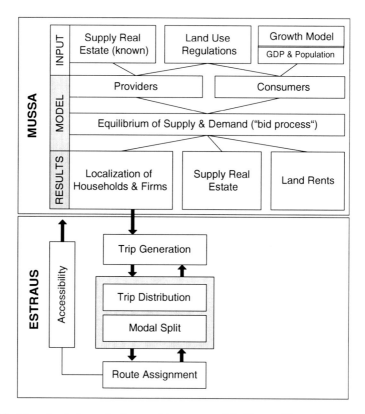

Fig. 10.2 Interaction between land-use and transport models (Source: The authors)

10 Santiago 2030: Perspectives on the Urban Transport System

MUSSA LAND-USE MODEL		ESTRAUS TRANSPORT MODEL	
INPUT & REQUIREMENTS	**OUTPUT & INDICATORS**	**INPUT & REQUIREMENTS**	**OUTPUT & INDICATORS**
• number of firms by economic activity • number of households by car ownership and income: 13 user groups • average income by zone • already constructed area by economic activity	• located number of firms by economic activity; number of household user groups and housing type • location by different geographic areas: TAZ (618), municipalities (38)	• number of households by 13 user groups • location of households by TAZ • trip generation rates by household user groups	• OD traffic flows by transport modes and travel purposes • number of trips by household user group, travel purpose and mode • vehicle miles travelled by mode
• accessibility indices	• constructed area by residential use • constructed area by non residential use	• behavioural model: utility functions for the joint estimation of mode and destination choice	• saturation levels of networks
• real estate market: number of buildings or departments (incl. size), lot size, constructed area, age of construction	• constructed lot size by residential use • constructed lot size by non residential use	• networks: private transport (highways, streets), public transport (bus corridors, frequencies, metro lines)	• average travel times by mode • average travel velocities by mode
• behavioural model considering: willingness to pay, consumer characteristics and preferences, real estate market, accessibility indices	• land rent by TAZ and construction type • (predicted) average income by zone		• traffic flows for estimation of emissions
• land use regulations: e.g. construction area by zone, minimum/maximum elevation of buildings, permits of land use types, taxes, subsidies			

Fig. 10.3 Input and output of transport and land-use models (Source: The authors)

category[3] are localized in the 618 Traffic Analysis Zones of the city. The number of households by category and area constitutes the input for the ESTRAUS transport model and the subsequent estimation of trips per household, the spatial distribution of trips and the selection of transport modes. The focus in this chapter is primarily on selected results (indicators) provided by the ESTRAUS transport model. However, it is important to note that the coupling of both models is indispensable and meaningful to considering the influence of land-use on transport and vice versa.

10.3.3 Priori Among Transport Policies

Prior to application of the models, the appraisal of experts in the field in Santiago was sought to broaden the discussion on measures that are both useful and to be expected on the path to a more sustainable transport system in 2030. This was vital, since the indicators analysed with the models alter in response to the policies and assumptions considered. The results of an expert survey conducted in 2008 revealed that the most urgent policies were the introduction of a congestion charge in the centre of the city, the construction of more bus lanes, restrictive parking management, and the expansion of bicycle lanes and pedestrian areas (see Fig. 10.4).

The survey results were considered in the selection of policies to be integrated into the analysis. As shown in the following section, particularly policies involving

---

[3] The 13 household types were segmented as follows: **01** (0 cars, 0–150,000 Chilean Pesos (CHP)), **02** (1+ cars, 0–150,000 CHP), **03** (0 cars, 150,001–300,000 CHP), **04** (1+ cars, 150,001–300,000 CHP), **05** (0 cars, 300,001–600,000 CHP), **06** (1 car, 300,001–600,000 CHP), **07** (2+ cars, 300,001–600,000 CHP), **08** (0 cars, 600,001–1,200,000 CHP), **09** (1 car, 600,001–1,200,000 CHP), **10** (2+ cars, 600,001–1,200,000 CHP), **11** (0 cars, > 1,200,000 CHP), **12** (1 car, > 1,200,000 CHP), **13** (2+ cars, > 1,200,000 CHP).

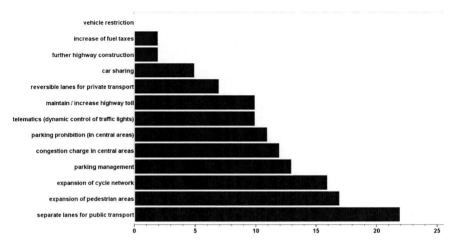

Fig. 10.4 Prioritization of transport policies in Santiago 2008 (Note: A total of 54 questionnaires were sent to transport experts in government, universities, the private economy and non-governmental organizations in Santiago. The 22 respondents were asked to evaluate the effectiveness of the described policy measures: *From your point of view – do the following measures contribute to a more sustainable urban transport system in Santiago, defining sustainability as a high level of accessibility [social component], with reduced traffic emissions [ecological component] and less congestion, and shorter travel times [economic efficiency component].* Multiple response was possible; thus the overall number of mentions was 127. Source: Klementz 2008)

improvements to public transport and non-motorized modes were included. Some measures needed to be 'translated' so that the model could quantify their effects, e.g., in the case of telematics that could indicate a slight increase in road capacities. The analysis focuses nonetheless on the impact of processes and policies that lead to irreversibility, e.g., the type of transport infrastructure developed, the type of buildings and their location, and demographic and economic growth.

10.4 Santiago 2030: Principal Assumptions and Main Results

In order to describe Santiago's transport system in 2030, several assumptions are made taking the state of the urban transport system in 2010 as described in Sect. 10.2 as a point of reference. Exogenous economic growth in annual GDP for the AMS is assumed, as well as an overall population figure and average household size for 2030.[4] Based on GDP and population data, another sub-model used by the Chilean government to predict the city's vehicle fleet is applied, and the

---

[4] Assumptions about GDP, population and households were jointly developed by Chilean and German researchers during the *Risk Habitat Megacity* Research Initiative, and were adopted for calculations undertaken in this analysis.

Table 10.3 Economic and demographic growth and expected motorization rate

	2010	2030
Population (in million)	6.28	7.30
Average household size	3.6	3.0
GDP growth rate per year	5% (2010–2019)	4% (2020–2030)
Number of vehicles (in million)	1.26	2.37
Motorization rate	201	325

Source: The authors (Assumptions for household size and GDP, modeling results for vehicles and motorization rate)

motorization rate in the city for the year 2030 estimated[5]. The estimated population, household, economic growth and motorization figures constitute the necessary input to predict the localization of households with the MUSSA land-use model, and their distribution into socio-economic categories. Table 10.3 above summarizes assumptions about economic growth and demography, as well as the expected motorization rate in Santiago by 2030.

Further assumptions about road and metro network extensions, public transport frequencies and the relevance of bicycle use are made and integrated in the models. In combination with the figures shown in Table 10.3 this reflects the policy package expected to influence Santiago's transport system up to 2030. The following set of projects and operational adjustments are bundled and form the backdrop for an estimate of transport demand in Santiago in 2030:

- Division of transport demand in the morning peak hour due to the introduction of flexible working hour regimes and school start times.[6]
- Three per cent modal share of bicycle trips in the morning peak hour (7% throughout the day). The bicycle mode is not yet established in Santiago's transport model, so that its development had to be assumed externally. Investment in bicycle infrastructure in the recent past and the upcoming use of bicycles in Santiago suggests greater use in the future.
- Capacity increase in urban road network: 5% due to pavement and alignment improvements of roads, adjustment of traffic signals and other traffic management measures.
- Capacity increase by one extra lane in existing urban highway networks.
- Capacity increase at critical nodes, i.e., congested crossings or roundabouts.

---

[5] The 'Vehicle Prediction Model' (Chumacero and Quiroz 2007) is applied by the Secretaría de Planificación de Transporte (Sectra) and partly based on historical information from different countries on the interaction of GDP and motorization.

[6] The ESTRAUS transport model estimates the transport demand for the morning peak hour between 06.30 and 08.30. Two demand matrices are calculated, for each hour. The overall demand is currently divided, given the empirical observation that about 75% of all trips occur between 07.30 and 08.30 and 25% between 06.30 and 07.30. This was altered to a ratio of 65% (07.30–08.30) to 35% (06.30–07.30).

- Construction of urban highways along current roads of *Avenida Isabel Riquelme* and *Vicuña Mackenna*.
- Construction of metro Line 6 as announced by the Chilean Government in December 2009.
- A 15% frequency increase on all Transantiago bus lines, reflecting the expected increase in velocities following the construction of more bus lanes and the provision of additional buses.

Naturally, the assumptions presented affect the overall quality and functionality of the transport system. The results reflect the impact of these assumptions without presuming substantial changes in economic circumstances or user behaviour. Thus, it is an attempt to show a likely development path, should internal and external urban processes perform as they have done in the past.

10.4.1 Localization of Households

The increase in the number of households by municipality between 2010 and 2030 calculated by MUSSA is shown in Fig. 10.5. The percentage categories were calculated by first defining the absolute increase in households, i.e., approximately 0.7 million between 2010 and 2030. Furthermore, the growth share that each municipality in the AMS is expected to experience was calculated in percentage.

For 2010, MUSSA localized 6.28 million people living in 1.74 million households in the 38 municipalities (3.6 persons per household). For 2030, it is assumed that 7.30 million people in 2.4 million households (3.0 persons per household) will be living in the urbanized space of the Metropolitan Area. The map indicates a concentration of the expected increase in the municipalities close to the centre municipality of Santiago and on the periphery. In particular, the southeastern municipality of La Florida and the southwestern municipality of Maipú are expected to show a significant increase in the number of households. This is explained by the large amount of vacant land still available in these areas with development permits for residential use.

As shown in Table 10.3, a total of 1.26 million vehicles are in use in 2010, representing a motorization rate per 1,000 inhabitants of 201. It is predicted that about 2.37 million motorized vehicles will be on the streets by 2030. Given the prediction for population numbers in the 38 municipalities of the AMS, we can expect a motorization rate of 325. This represents an 88% increase in the total number of vehicles and a 62% increase in motorization. Based on these predictions and the spatial localization of households by income and car ownership, the ESTRAUS transport model was applied to estimate the potential transport demand.

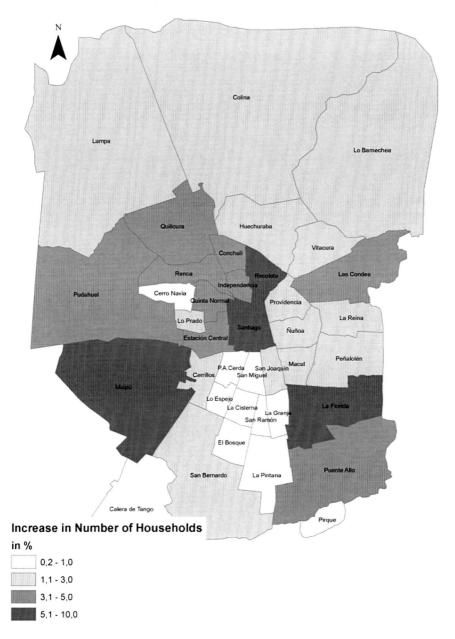

Fig. 10.5 Percentage increase in the number of households by municipality in the Metropolitan Area of Santiago between 2010 and 2030 (Source: The authors, based on MUSSA land use model for Santiago)

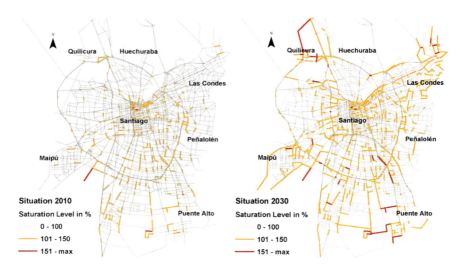

Fig. 10.6 Saturation on major transport links 2010 (*left*) and 2030 (*right*) (Source: The authors, based on ESTRAUS transport and network model for Santiago)

10.4.2 Congestion and Level of Service

The congestion level measured in the saturation of major road links is shown in Fig. 10.6. Each map is based on the morning peak hour demand, represented by the hour between 07.30 and 08.30. In 2010, a total of 1.4 million interzonal trips (0.51 million car trips) are made during the morning peak, a figure that will increase to 2.1 million trips (0.83 million car trips) by 2030. Although primarily due to the absolute increase in population and in the number of households, it is also a result of a greater share of households with higher income, more cars and higher rates of mobility. It is evident that future capacity bottlenecks will occur most on the periphery, where urban road infrastructure capacities have not adjusted to absorb the growing number of households and the concomitant increase in cars on the streets in 2030. For this reason the relatively sparse road network coverage is likely to be denser than imagined in the model. Although an increase in road capacity was assumed, the historic city centre in particular will suffer in 2030 from high congestion levels and a decrease in vehicle travel times. Most of the centre and the connecting roads to the eastern districts, the hub of the financial and commercial activities, are expected to be oversaturated by 2030. This is also the case for the roads that connect the southern and southwestern municipalities of, for example, Maipú, La Florida and Puente Alto with the main employment centres.

The saturation of transport links, the road network for cars and buses, directly affects the level of service indicators such as average velocities and travel times in public and private transport. The following Table 10.4 indicates the expected values of velocities and travel times for trips taken by car and by public transport:

Table 10.4 Average velocities and travel times 2010 and 2030

	Car		Public transport	
	2010	2030	2010	2030
Velocities in km/h	29.4	23.8	23.6	21.7
Travel times in min/trip	24.4	31.7	37.1	52.3

Source: The authors, based on ESTRAUS transport and network model for Santiago

The results clearly indicate that the rise in population and motorization leads to an increase in the overall demand for transport in the morning peak hour, which in turn impacts negatively on the overall performance of travel time levels in 2030. There is a velocity decline of almost 20% for car trips and almost 10% for public transport. At the same time, the average travel time per trip increases by 30% for car trips and 40% for public transport trips. The vast increase in travel time for public transport is also related to the huge number of vehicles on the road by 2030. Buses and private cars in competition produce greater congestion and decrease flow velocities for buses, at least for those not yet running on corridors with separate bus lanes. This is reflected in the velocity decline in both car and bus-related traffic. These findings indicate that despite the assumption of significant improvement in network capacities in both public and private transport, a service level comparable to the level observed in 2010 cannot be guaranteed for 2030. In other words, the implemented policy package does not suffice to mitigate the negative effects generated by the anticipated increase in motorization and mobility in general in Santiago.

10.4.3 Modal Split

Another important indicator for the evaluation of the urban transport system is the modal split. In Table 10.5, modal split values for 2010 and 2030 are compared. The shares in Table 10.5 are based on morning peak hour traffic (07.30–08.30), whereas Table 10.1 reflects the average mode choices throughout an entire day. Interestingly, for example, the share of walking trips in the morning peak hour is less than whole day values. The share of public transport trips in 2030 remains high in the morning peak hour. This phenomenon is due to the fact that the key destinations in the morning peak hour (work and education) are farther away from home than those related to other trip purposes and, hence, reached by motorized modes rather than slower, non-motorized modes.

In 2030, over 40% of all trips in the morning peak hour are still made by public transport. The reason for this is twofold. Firstly, the imposed improvements regarding increased frequencies and high congestion levels for cars have led to the conclusion that for certain origin–destination pairs, public transport is the best mode option. Secondly, a significant number of households are still not in possession of a car in 2030. Nevertheless, the increase in motorization and car availability

Table 10.5 Modal split in Santiago, 2010 and 2030, morning peak hour 07.30–08.30

	2010	2030
Bus+metro	49.0	44.2
Car	36.6	39.7
Taxi/shared taxi	3.2	2.9
Walk	11.2	10.2
Bicycle	–	3.0

Source: The authors, based on ESTRAUS transport and network model for Santiago

induces car use, reflected in the prediction that about 40% of all trips in the morning hour are made by car (as car driver or car passenger). Bicycle-related trips are responsible for 3% of all trips. It is worth noting that for the whole day mode share values, an even higher share of bicycle trips can be expected, as this mode will play a more vital role in leisure or private errand-oriented trips of shorter distances made during the day.

10.4.4 Accessibility

The definition of accessibility in the context of model application depends on the spatial units, transport modes and travel purposes considered. Thus, a specific analysis can be conducted for a selected Traffic Analysis Zone, e.g., identifying the number of shopping locations accessible by car in 30 min. Taking into account that 618 TAZ form the background for the analysis presented in this chapter, a more aggregated representative of the accessibility indicator was drawn up. Since the objective is to provide a system-wide indicator, however, accessibility was defined here as the average travel time by mode (public transport, car travel) from one municipality to another with reference to all 38 municipalities. Figure 10.7 shows the estimated increase and decrease in minutes of travel times by transport mode between 2010 and 2030.

Figure 10.7 shows, for instance, that the average travel time from the centre municipality of Santiago to any other municipality increases by 3.7 min for cars and 1.5 min for public transport between 2010 and 2030. In general terms, the figures confirm phenomena addressed by the analysis of congestion levels and velocities, bearing in mind that system average velocities showed a higher increase for private cars than for public transport. Furthermore, a major increase in congestion levels was detected, which leads to higher travel times for cars system wide. In accordance with the accessibility values, higher average travel times are expected for car travel in all municipalities in 2030; thus, a decline in the level of accessibility is confirmed for this mode throughout the city. On average, levels of accessibility – in line with the definition used here and due to increased travel time – will drop by 26% for car travel and by 6% for public transport between 2010 and 2030. Given the improvement in bus frequencies, deterioration in the level of accessibility is less

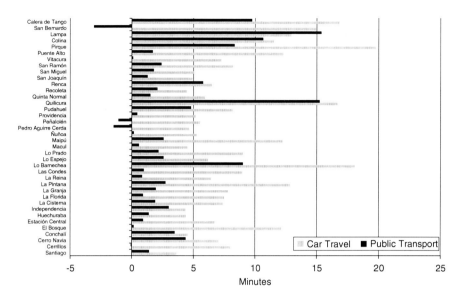

Fig. 10.7 Increase and decrease in average travel time between 2010 and 2030 by mode and municpality (Source: The authors, based on ESTRAUS transport and network model for Santiago)

pronounced for public transport. A slight improvement in public transport accessibility was even detected for some municipalities.

10.5 Conclusions

The study of urban transport systems faces a number of challenges; the interplay between transport supply and demand is complex, and influencing factors are numerous. The speed of change in these systems is high, making ongoing research a sine qua non, since urban development is essentially based on infrastructure that endures and thus conditions the life style of future generations. This type of study harbours a level of uncertainty, as several assumptions have to be made. However, the long-term perspective provides valuable insights into how the system might evolve under certain circumstances. In this sense, an extended analysis can open up a space for different scenario alternatives and create awareness among policymakers in terms of the economy, demography and specific transport policies.

The results obtained confirm some expectations about the future of the city of Santiago but also deliver some news. Urban development and its derived transport needs are heavily fuelled by a combination of demographics – with population growth coincident with a reduction in the average household size – and improvements to system economics. This blend detonates a huge increase in the total number of vehicles by a factor of 1.9 in 20 years and in daily trips realized in the morning peak hour by a factor of 1.5. The consequence is enormous pressure on the transport system with potential risk to quality of life as a result, i.e., higher congestion levels and urban sprawl. Returning to the original research questions, it can be concluded that system performance will have slowed down by 2030 in the case of both private and public transport. In order to relieve the pressure, several measures that reflect the tendencies observed in recent decades were considered, i.e., a combination of public transport and road infrastructure adjustments in a free market oriented system. In this scenario the models predict a mixture of sprawl and density increase in developed areas, with a higher consolidation of the primary and secondary city centres. In sum, this yields slower travel speeds, especially of cars, and thus an increase in average travel times. A similar indication arises with the accessibility factor for car and public transport travel, which performs negatively, i.e., the average travel time for both modes increases between 2010 and 2030. In other words although substantial network and system management progress was assumed, the indicators tend to perform negatively, leading to increased pressure on the urban transport system. As indicated in the respective section, motorization is expected to rise continuously with up to 325 vehicles per 1,000 inhabitants by 2030 and hence gains further relevance in the modal split. The role of public transport will be nonetheless a vital one in 2030; in the morning peak hour, the majority of trips will continue to be made by bus or metro. It can be concluded that the trend in both public and private transport will be a reduction in quality due to greater congestion and longer travel times. In a direct comparison, public transport users will be less affected by this scenario than car users, since overall accessibility is less likely to deteriorate.

The main conclusion is that Santiago's urban system is still experiencing an increase in travel demand and expansion of its spatial coverage. This calls for urgent studies of further scenarios to mitigate impacts and quantify the gap between a desirable future and expected realities. Despite efforts to introduce mitigating policies, the likelihood of Santiago becoming a car-based city is strong; an irreversible tendency given the assumed adjustments to the infrastructure for roads and buildings. Hence, the study conducted indicates that further policy action is required. Other explorative studies could include a pricing scheme for the city centres or a toll increase on urban highways. At the same time, more investment is needed for the introduction of bus lanes to reduce public transport travel time and make them independent of car-based congestion effects. Additionally, the expansion of information technologies, especially in public transport (real-time information at bus stops and on the buses, schedule control), is pending. Although difficult to reproduce with a classic transport model, measures of this kind seem promising in terms of adding to the reliability and attractiveness of Transantiago.

In summary, the methodology applied supports the approach of quantitative modelling to predict the future of the urban transport system in Santiago. Both the land-use and the transport model indicated trends, taking into account predicted developments and assumptions made. Although final discussions and conclusions for policy action were, of course, complemented by a discursive and qualitative analysis, it was possible to showcase the ability of the modelling tools to shed light on the complex processes of urban development.

References

Chumacero, R., & Quiroz, J. (2007). *Modelos de proyección del parque vehicular*. Santiago: Informe Final.

de Cea, J., et al. (2003). *ESTRAUS: A computer package for solving supply-demand equilibrium problems on multimodal urban transportation networks with multiple user classes*. Washington, DC: TRB.

Dirección de Investigaciones Científicas y Tecnológicas de la Pontificia Universidad Católica de Chile – DICTUC. (2001). *Encuesta Origen Destino 2001*. Santiago de Chile: Informe Final.

Gschwender, A. (2007). *A comparative analysis of the public transport systems of Santiago de Chile, London, Berlin and Madrid: What can Santiago learn from European Experiences?* Ph. D. thesis. Wuppertal University, Germany.

Instituto Nacional de Estadisticas de Chile (INE), (a) Parque de Vehículos en Circulación, 2001–2009; (b) Proyecciones de Población 1990–2020

International Monetary Fund (2009). World economic outlook database. Country level data, Chile (2010). http://www.imf.org/external/pubs/ft/weo/2009/02/weodata/index.aspx. Accessed 14 June 2010.

Keil, M. (2008). Gründe für den Autokauf und Auswirkungen des Autobesitzes auf das individuelle Mobilitätsverhalten in Santiago de Chile. Unpublished diploma thesis, München.

Klementz, F. (2008). Auswirkungen von verkehrspolitischen Maßnahmen auf die Verkehrsnachfrage. Unpublished diploma thesis, Berlin.

Martínez, F. (1996). MUSSA – Land use model for Santiago city. *Transportation Research Record, 1552*, 126–134.

Martínez, F., & Donoso, P. (2001). MUSSA – A land use equilibrium model with location externalities, planning regulations and pricing policies. *7th International Conference on Computers in Urban Planning and Urban Management (CUPUM 2001)*, Hawaii.

Muñoz, J., & Gschwender, A. (2008). Transantiago: A tale of two cities. *Research in Transportation Economics, 22*, 45–53.

Quijada, R., et al. (2007). *Investigación al Transantiago: Sistematización de Declaraciones hechas ante la Comisión Investigadora, Resume de Contenidos de los Principales Informes Técnicos*. Santiago: Información de Documentos Públicos Adicionales y Comentarios Críticos.

Secretaría Interministerial de Planificación de Transporte – SECTRA (2002). Mobility Survey EOD 2001. Executive Report. Santiago, Chile.

Secretaría de Planificación de Transporte – SECTRA. (2008). EFFUS – Efectos urbanos futuros en Santiago: Exploración aplicada de los cambios en el uso de suelo generados por Transantiago en los años 2010 y 2015, estimados con los modelos MUSSA y ESTRAUS. Santiago de Chile.

Chapter 11
Air Quality and Health: A Hazardous Combination of Environmental Risks

Peter Suppan, Ulrich Franck, Rainer Schmitz, and Frank Baier

Abstract This chapter addresses the link between Santiago's transport sector and its corresponding emissions, on the one hand, and the impact on the health of its citizens, on the other. The dispersion of pollutants at the micro and meso scale is based on an integrated approach that incorporates different platforms – satellite data, in situ measurements and emission data. Specific attention is paid to traffic emission estimates based on the traffic flow, fleet composition and emission factors in the Metropolitan Area of Santiago de Chile. Health-related indicators such as PM_{10}, O_3 and NO_x are discussed and their dispersion throughout the city analysed in order to arrive at a comprehensive health impact assessment. The possible development of PM_{10} levels forms the basis of a detailed discussion on adverse health effects in the future.

Keywords Adverse health effect • Air quality assessment • Micro/meso scale modelling • Particles • Traffic emissions

11.1 Introduction

Air pollution in megacities and densely populated areas in general is recognized as a major environmental issue (e.g., Molina and Molina 2004; Gurjar and Lelieveld 2005; Gurjar et al. 2010). Due to the exceptionally high emissions generated by anthropogenic activity, people living in megacities are more likely to be affected by air pollution than those in rural areas. It is furthermore widely accepted that particulate matter (PM), nitrogen oxides (NO_x), ozone (O_3), carbon monoxide (CO) and sulfur dioxide (SO_2) impact on mortality and/or morbidity.

P. Suppan (✉) • U. Franck • R. Schmitz • F. Baier
Karlsruhe Institute of Technology (KIT), Institute for Meteorology and Climate Research,
Kreuzeckbahnstr. 19, 82467 Garmisch-Partenkirchen, Germany
e-mail: peter.suppan@kit.edu

The World Health Organization (WHO) ranks Santiago among the world's most polluted cities (WHO/UNEP 1992; Krzyzanowski and Cohen 2008). Pollution here is the result of two factors: anthropogenic emissions and the city's geographic location in a basin surrounded by high mountains, above which meteorological phenomena force stable conditions, causing high levels of air pollution. This scenario finally led to implementation of an air quality management programme (Plan de Prevención y Descontaminación Atmosférica – PPDA) in the 1990s, which has contributed substantially to the decline in levels of air pollution. In 1997, the air quality management programme declared the Metropolitan Area of Santiago de Chile a saturated zone for carbon monoxide (CO), particulate matter smaller than 10 μm in diameter (PM_{10}), and ozone (O_3).

Despite a major advance in the reduction of air pollution levels and efforts to grasp the problem from a systematic perspective, little progress has been made since the beginning of the twenty-first century (CONAMA 2004, 2006). Continuing high levels of pollution pose a major risk to the population in the Metropolitan Area of Santiago in terms of health and quality of life. Health impacts have been reported, for example, by Ostro et al. 1995, who attribute more than 1,000 deaths per year to particulate matter (PM). PNUMA (2003) and even estimate an increase of up to 4,800 premature deaths per year as a result of PM_{10} concentrations. These figures indicate that air pollution – and with it the implied health impact on Santiago's population – remains one of the city's (and probably the country's) most challenging environmental burdens.

This dramatic scenario calls for the development of efficient mitigation strategies for the reduction of air pollution levels in the Metropolitan Area. Emissions must be characterized as precisely as possible and atmospheric processes that culminate in high pollution levels understood. Major anthropogenic contributors to air pollution are traffic emissions, domestic activities (e.g., heating, cooking, use of solvents) and industrial sources (CONAMA 2007).

The work discussed in this chapter concentrates on analysis of the current state of air quality in the Metropolitan Area of Santiago and the risks associated with air pollution. To gain a more thorough understanding of the driving factors behind the high pollution levels observed for the Metropolitan Area, an integrated approach was applied in an effort to chart the pathway from traffic and other emissions to their concrete impacts on health. Specific tasks of the 'chain' (emissions – air quality – health impact) are discussed, with special attention given to traffic emissions as a key source of low air quality and its adverse health effects. A bottom-up approach is used to calculate traffic emissions (see Chap. 10 in this volume). Due to their particularly serious effect on people's health, PM_{10} concentrations and their impact on health are the core of the analysis.

11.2 Methodological Approach

The strong interdependency between human activity and air quality, beginning with transport demands and the need to provide products and services, requires an integrated modelling of city traffic and its emissions, and a link to the corresponding health risks. Creating this integrated system (integrated approach) is crucial to the dispersion modelling of air pollution in Santiago de Chile (see also Chap. 10).

When a large quantity of air pollutants are released near the ground by motor-vehicle exhaust and subsequently trapped within the urban street canyon walls, the interaction between emissions and meteorology leads to higher variability in terms of risk and the impact on human health, especially at the local scale. As densely populated regions, urban street canyons are in fact high-risk zones in terms of health, particularly in the case of megacities like Santiago de Chile. The population health risk is not determined by the risk increase to the individual alone, but also by the total number of people exposed to that risk. The combination of high population density and high traffic volumes in densely populated areas implies a greater degree of risk compared to that in less populated urban areas, resulting in an increased mean health risk for the entire megacity population.

In the context of this chapter, risk is defined as the vulnerability of humans in relation to their level of exposure to air pollutants. The principal air pollutants NO_2, PM_{10}, and O_3 are used as indicators for the vulnerability analysis. Human vulnerability varies and depends, for instance, on age, gender, socio-economic state, and prior diseases/pre-existing conditions, and can be modified by individual behaviour (Pope et al. 1991, 1992, 1995; Wietlisbach et al. 1996; Wichmann et al. 2000; Schwartz 2000; Schwartz 2001; Brajer and Mead 2004; Boldo et al. 2006; Breitner et al. 2009).

The total risk for the population of Santiago de Chile is linked to an increase in per capita risk, the number of elements at risk (exposed population), the increase in per capita risk per unit of pollutant, and the actual concentration of the pollutant itself. In the present context, per capita risk is defined as the risk of morbidity or mortality. The per capita risk here is the mean risk of the total population.

While the use of total population risks is a valid approach for public health protection, the need to give greater attention to vulnerable groups is gradually gaining ground. As a rule, children are regarded as a group vulnerable to air pollution (WHO 2005b). On the other hand, environment-related morbidity has increased dramatically in the elderly, who tend to suffer from pre-existing diseases. Furthermore, total population risks strongly depend on the risks of such vulnerable sub-groups. With respect to mortality in Santiago de Chile, the elderly exhibit the highest risks (Cakmak et al. 2007).

The methodological approach presented in this chapter is to integrate and develop a chain of models covering transportation ($\rightarrow$ traffic modelling), traffic emissions ($\rightarrow$ traffic emission modelling) and air quality simulations ($\rightarrow$ at the

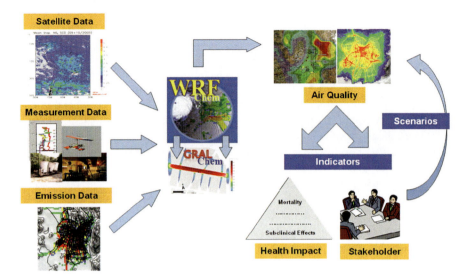

Fig. 11.1 Methodological approach of combining satellite information with measurements and emission data (*left*) as input for coupled meso- and micro-scale modelling tools (*centre*), a prerequisite for health impact assessment studies, stakeholder involvement and scenario development (*lower right*) (Source: The authors)

micro[1] and meso[2] scale), and carry out a health impact assessment with a high temporal and spatial resolution ($\rightarrow$ epidemiological modelling) (Fig. 11.1). This approach allows for a holistic view of the link between traffic emissions and adverse health effects. Identifying links between sectors and different scales facilitates the implementation of comprehensive assessment studies on the current air quality and future trends, e.g., based on the future development of traffic emissions (Suppan 2010).

11.2.1 Transport, Emission and Air Quality Models

The ESTRAUS transport model developed by SECTRA (Secretaria de Planificacion de Transporte, Corvalán et al. 2002) calculated the traffic flow for 2010, based on the number and location of households, car ownership and mobility generation. A more detailed description of the model can be found in Chap. 10. Calculation of the traffic flow in the Metropolitan Area of Santiago (AMS/34 municipalities) was performed by SECTRA in close cooperation with the transport

---

[1] Micro scale defines the spatial scale at which air quality simulations are performed, in this case from a few metres to a few kilometres.

[2] Meso scale: from a few kilometres to several 100 km.

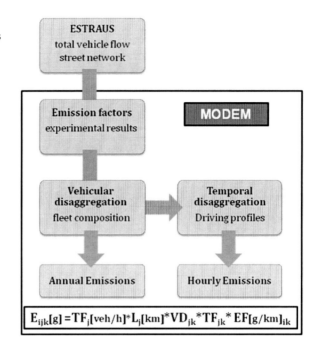

Fig. 11.2 Methodology of traffic emission modelling as a basic requirement for the assessment of current and future emission states (Source: Nogalski 2010)

sector (see Chap. 10). Traffic emissions were calculated with the MODEM emission model (Modelo de Emisiones vehiculares, SECTRA 2001) for the air quality state in 2010 (Fig. 11.2), based on the flow of traffic in the 34 municipalities of the AMS.

In order to consider the traffic situation for the reference year 2010, MODEM was adapted to current fleet composition and consequently current valid emission factors (Nogalski 2010). The model considers the emission factors of 61 vehicle categories and driving profiles, and calculates four categories of emissions (cold, hot, evaporation, re-suspension) for six lumped pollutants (PM_{10}, SO_2, NO_x, VOCs, CO, CO_2). Calculations are performed for all streets (approx. 15,000) in the Metropolitan Area of Santiago de Chile.

In a further step in the modelling chain, the dispersion of emissions throughout the Metropolitan Region of Santiago was calculated with the micro-scale air quality Grazer Lagrangian Model (GRAL, Öttl et al. 2003). The model was designed to calculate the dispersion of non-reactive pollutants on a grid resolution of 10 m to 25 m within a region of 35 km × 35 km. Apart from traffic emissions, wind direction, wind speed, atmospheric stability classes and air temperature from the regional scale (meso-scale) served as additional model input. These parameters were provided by the state of the art Weather Research and Forecasting (WRF) Model (http://www.mmm.ucar.edu/wrf/users/document.html).

The meso-scale dispersion model WRF/chem. was used to assess air quality in the Greater Metropolitan Region of Santiago de Chile (Grell et al. 2005). WRF/chem. is a three-dimensional fully coupled meteorology/chemistry model and, depending on the region, designed to simulate meteorology and chemistry with a high temporal (1 h) and horizontal spatial resolution (in the case of Santiago de Chile) of 4 km × 4 km. The chemistry modul of the model considers the gas phase (e.g., O_3, NO_x, SO_2, VOCs), the particle phase (e.g., PM_{10}, $PM_{2.5}$) of the atmosphere, and their interaction. The modelling region is at a scale of 450 km × 500 km.

11.2.2 Boundary Conditions and Further Requirements

Additional information on emissions from industrial (e.g., power plants), domestic (e.g., heating, cooking, use of solvents) and biogenic sources (e.g., parks, forests) are needed for reliable air quality simulations at micro and meso scale. Recently available emission inventories in Santiago de Chile, however, need constant improvement for reliable air quality simulations, as demonstrated by Schmitz et al. (2010).

Adequate boundary conditions for meteorological and chemical parameters are also crucial to regional air quality simulations. Large-scale information on air quality as provided by satellite data is paramount for regions with little or no available in-situ observations, such as Latin America, and particularly the Metropolitan Region of Santiago de Chile. Technically speaking, the chief objective of using satellite data is to improve air quality modelling with a more thorough description of background conditions and assistance in model verification.

A key characteristic of the air quality simulation approach in this chapter is the coupling of processes on different scales, i.e., from the micro scale, where emission sources are located, e.g., traffic emissions, to the meso scale where emissions (air pollutants) have their strongest impact on health (exposure levels). In order to link these scales, the exchange of data with high spatial and temporal resolution must be ensured. In the present case, meteorological parameters (wind direction, wind speed and temperature) from meso-scale simulations based on a grid resolution of 4 km × 4 km were used as input for the micro-scale simulation (grid resolution of 25 m × 25 m). For future research, the results of these micro-scale simulations (e.g. PM_{10}, NO_2, CO) can be used for comprehensive health impact studies.

11.2.3 Health Impact

High concentrations of airborne pollutants (e.g., O_3, NO_x, other gases, particles) are associated with the development of environment-related diseases and the exacerbation of selected diseases. In the time scale of hours to several years, respiratory and cardiovascular illnesses are associated with airborne pollutants. Concentrations of these pollutants differ within one city due to on-site variations in emissions, topography and building type, and density (Alastuey et al. 2004; Beelen et al.

2009; Berner et al. 2004; Duzgoren-Aydin et al. 2006; Harrison and Jones 2005; Houthuijs et al. 2001; Penard-Morand et al. 2006; Wehner and Wiedensohler 2003; Beelen et al. 2009; Bell et al. 2007; Lianou et al. 2007; Tuch et al. 2006). High traffic densities result in higher emissions, but also in greater exposure of individuals, particularly in street canyons and valleys. Emission sources in non-built-up areas, on the other hand, result in the dispersion of airborne pollutants. Typically, industrial emission sources are not equally distributed across cities but concentrated in urban areas where industries predominate. These variations play a leading role in megacities as they represent spaces with a high land-use mix. Furthermore, concentrations of pollutants vary over time including seasonal, weekly, diurnal, and other variations, for the most part caused by changing weather conditions and diverse human activity.

Pollutant effects must be quantified and ranked if environment-related diseases are to be prevented effectively and analysis carried out with respect to time and spatial scales. Mean concentration levels should be considered if pollutant concentration varies little over space and a linear dose–response relationship is assumed (dependent on the pollutant).

The definition of air pollution indicators helps to describe the scale or dimension of the health impact. In line with WHO Air Quality Guidelines (Krzyzanowski 2008), the following indicators were selected:

- Threshold limit values of specific pollutants (e.g., NO_2, O_3, PM_{10}, $PM_{2.5}$, VOCs), and
- Mortality rates due to cardio-respiratory disease.

Threshold limit values for air pollutants were defined for factual or pragmatic reasons. Since a factual threshold level for adverse health effects did not emerge in epidemiological investigations with regard to PM exposure, a linear dose–response relationship was generally assumed. For practical reasons, the WHO also defined threshold values for PM (WHO 2005a). These values can be adopted by governments formulating policy targets. Compliance with these target values helps to protect the resident population from the adverse health effects of PM. A gradual tightening of regulations would further prevention.

To meet the requirements for adverse health effects, threshold values for air pollutants are also defined for different time intervals. Typically, a 24 h or a 1 year limit is considered a threshold, although an 8 h running mean value is also considered (Table 11.1).

On the whole, emphasis should be on the development and evaluation of emissions from the transport sector, as traffic emissions have a strong impact on the overall quality of air due to their characteristic proximity to the surface (street canyons) and their emergence in residential areas.

Other sources apart from those related to traffic contribute to air pollution with the same or similar pollutants. Small-scale urban industry and individual heating can play a major role. Long-distant transport adds to the concentration of several air pollutants (e.g., airborne particulates). Nevertheless, traffic emissions are the predominant contributors to small-scale differences in the concentration of air

Table 11.1 Threshold values for specific pollutants in Chile compared with the air quality guidelines of the World Health Organization

Pollutant	Chile	WHO	Time interval
	$\mu g/m^3$		
O_3	120	120	8 h mean
NO_2	100	40	Annual mean
PM_{10}	150	50	Daily mean
	50	20	Annual mean

Source: WHO 2005a

pollutants in the inner city, and as such are relevant to inner-city impacts on human health.

A first step with regard to the impact on human health was to assess the potential to mitigate the PM_{10} concentrations associated with risk of premature death from respiratory and cardiovascular disease. As the simulation process is ongoing, a first-guess health assessment was performed on the basis of time series (at least 1 year) using PM_{10} measurements from the Parque O'Higgins monitoring station, located in the park of the same name.

Applied methods for health risk assessment yielded quantitative data. The necessary quantitative input data for these assessments can be provided by past and current measurements or simulation methods for future developments.

Typically, a linear dose–response relationship between pollutant concentrations and health effects is assumed for particulate air pollution. Hence an immediate, quantitative prognosis of changes to health risks will be possible if pollution concentrations can be predicted.

For the assessment of adverse health effects, a non-linear regression time series analysis was carried out, with time series based on daily mean values for exposure concentrations obtained at the Parque O'Higgins measuring site. Due to its location in the heart of Santiago, this monitoring station is representative for a vast region between the ridges of the Andes and its foothills. Exposure concentrations were related to daily case-specific death counts. The analysis distinguished between hypertensive disease, ischemic heart disease, other forms of heart disease, influenza and pneumonia, and chronic lower respiratory disease. A total of 20,918 deaths were included in the assessment. These cursory assessments were carried out for PM_{10}, only. Thus, an additional, more detailed analysis is required (see conclusions).

11.3 State of the Art Knowledge of Air Quality in Santiago de Chile

According to the emission inventories for the Metropolitan Area of Santiago (CONAMA 2007), traffic is the most dominant emission source of all (Fig. 11.3). To perform reliable air quality assessments and health impact studies, however, exploration of other emission sources is equally important. Schmitz et al. (2010)

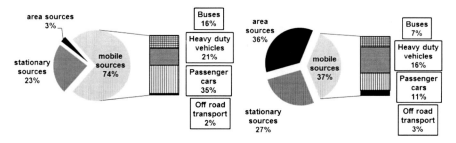

Fig. 11.3 Emission sources in the greater region of Santiago de Chile: NO_x emissions (*left*) and PM_{10} emissions (*right*) (Source: CONAMA 2007)

have recently shown that although emission inventories give some insight into emission characteristics, a great deal more work has to be done if the quality of emissions is to be improved.

Several studies have been carried out to identify the atmospheric characteristics responsible for air pollution in the Santiago Basin (e.g., Rutllant and Garreaud (1995)). Schmitz (2005) used a numerical air quality model to describe dispersion patterns in the Basin. Rappenglück et al. (2005) conducted a photochemical campaign defining the fundamental features of ozone chemistry dynamics in Santiago. Although the basic knowledge required for applied research is in place, there is still plenty of scope for further scientific activity. Applied research, however, calls for a systemic approach that includes the assessment of traffic, air quality and health, and their links to other disciplines such as transportation (see Chap. 10) and energy (see Chap. 9).

Knowledge of the temporal evolution of specific air pollutants in the past is a prerequisite for the overall assessment of current air quality in the Metropolitan Area of Santiago de Chile and its potentially adverse human impact both now and in the future. The development of the NO_2 and PM_{10} annual means, and maximum 8 h mean value of O_3 for the time period 1998–2007 is presented in Fig. 11.4. The data set consists of the MACAM II network (Red de Monitoreo Automática de Contaminantes Atmosféricos), which has operated with three stations in Santiago de Chile since 1988 and since 1997 with an additional four stations (CONAMA 2003). Subsequent to the introduction of the air quality management plan for a cleaner Santiago (PPDA), air pollution levels dropped slightly. It is obvious, however, that the sum total of strategies and mitigation processes implemented in the past have failed to reduce the mean values below Chilean limit values or WHO guidelines (dotted lines). PM_{10} is still above the Chilean limit value (50 µg/m³) and way above the WHO guideline of 20 µg/m³. O_3 values are above the WHO guideline 8 h mean value of 100 µg/m³ and the Chilean limit value of 120 µg/m³. Only the NO_2 value has come close to the WHO guideline annual mean of 40 µg/m³ and lies well below the Chilean threshold of 100 µg/m³.

In particular, the annual mean values of PM_{10} have decreased within the space of 10 years from almost 100 µg/m³ in 1998 to around 70 µg/m³ in 2008. Concentrations

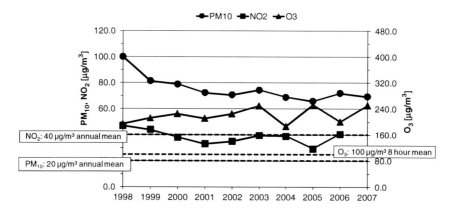

Fig. 11.4 Annual mean time series of NO_2, and PM_{10} and 8 h annual maximum of O_3, based on the MACAM II monitoring network of the Metropolitan Area of Santiago de Chile (Source: The authors, based on CONAMA 2006). Dotted lines illustrate the WHO guidelines for NO_2, PM_{10} (annual mean) and O_3 (8 h mean value)

Table 11.2 Statistics on PM_{10} concentrations of all monitoring stations in the Metropolitan Region of Santiago de Chile for 2006

PM_{10} [μg/m³]	La Paz	Pudahuel	La Florida	Parque O'Higgins	Las Condes	Cerrillos	El Bosque
Min	19.7	10.8	19.1	16.8	13.4	13.0	16.0
Median	69.7	58.4	67.4	65.3	52.5	67.3	70.2
Mean	77.8	71.2	75.5	74.7	53.8	72.2	79.0
95%	123.4	136.3	119.7	124.9	77.7	120.7	126.3
Max	210.4	300.6	210.3	216.0	125.9	183.2	252.3

Source: The authors, based on CONAMA-Data (MACAM II stations)

are nonetheless way above the threshold, representing a high health risk in the Metropolitan Area of Santiago de Chile. Although it is not yet possible to identify a concentration threshold below which health effects are undetectable, the body of evidence dealing with PM health effects is rapidly expanding and will therefore be discussed more thoroughly in the following. Recent studies demonstrate that particle pollution or particulate matter is related to mortality and morbidity among people regarded as exposed (Pope et al. 1991, 1992, 1995; Wietlisbach et al. 1996; Wichmann et al. 2000; Schwartz 2000, 2001; Brajer and Mead 2004; Boldo et al. 2006; Breitner et al. 2009).

Table 11.2 summarizes the statistics on PM_{10} concentrations for the year 2006 at the seven MACAM II stations. Most of the stations have operated smoothly and rarely had a breakdown. All seven stations, defined as 'urban background stations', show similar characteristics in means of concentration levels and distribution, with the exception of the Las Condes station, which is located in the extreme east of the city and therefore less affected by the adjacent emission sources. Since PM_{10} concentrations are distributed evenly throughout the Metropolitan Area of

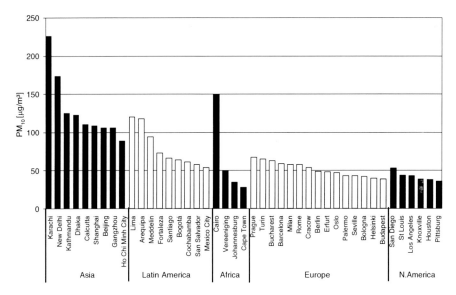

Fig. 11.5 Annual PM_{10} means in selected urban agglomerations worldwide (Source: The authors, based on WHO 2006)

Santiago, the health impact assessment based its evidence on measurements performed by the representative station Parque O'Higgins in the centre of the city.

The World Health Organization (WHO 2006) compared cities across the world based on annual PM_{10} concentrations (Fig. 11.5) and found Asian and Latin America cities to be among the highest polluted regions in terms of PM_{10}. However, concentrations within the Metropolitan Area of Santiago are well below the most polluted global megacities, albeit Santiago remains one of the most polluted cities in Latin America. By comparison, pollution levels in Santiago are significantly higher than the annual means for European and North American cities.

When it comes to identifying local and/or advection (transport) effects for the assessment of the origin of such high PM_{10} concentrations, the basic tool is a pollution rose – a combination of wind direction and concentration measurements. The PM_{10} pollution roses for 2004 show a high degree of similarity in distribution for all stations (Fig. 11.6, right). There is no significant allocable wind direction where higher/lower PM_{10} concentrations occur. For each wind direction, the mean concentrations are of the same order. Only the stations themselves show different levels of absolute concentrations, e.g., the lower concentrations at Las Condes (see also Table 11.2). In the absence of a correlation between PM_{10} concentrations and wind direction, measurement sites are primarily dominated by high urban background concentrations rather than local single street emission sources in close proximity. Although dense street networks also contribute to urban background concentrations, secondary sources are responsible for high PM_{10} concentrations, e.g., build-up processes of secondary organic aerosols (SOA) due to the occurrence

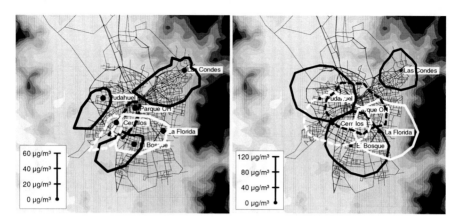

Fig. 11.6 Distribution of O_3 (*left*) and PM_{10} (*right*) concentrations measured in the Metropolitan Area of Santiago in 2004. The lower left scale indicates the concentration benchmark for all stations (Source: CONAMA)

of biogenic and anthropogenic volatile organic compounds (VOC) originating from traffic emissions.

Analysis of the O_3 pollution roses (Fig. 11.6, left), on the other hand, points to a different behaviour. Apart from global background concentrations, here O_3 is a secondary pollutant and depends on primary pollutants such as NO_x and VOCs. As a result of the chemical characteristic of O_3, high concentrations are found in downwind regions of primary emission sources, e.g., urban agglomerations, highways, industrial areas (Suppan and Schädler 2004). Las Condes is one such area in Santiago de Chile and shows the highest O_3 concentrations measured.

11.4 Air Quality Assessment for Santiago De Chile

Based on the methodological approach presented in Sect. 3, this section gives an overview of the results of various measurements and modelling techniques, and serves as a prerequisite for the set-up and analysis of mitigation strategies and scenarios. Traffic emissions in the Metropolitan Area of Santiago were calculated for all streets sections (ca. 15,000) for the present situation (2010), based on the ESTRAUS traffic model and the MODEM traffic emission model. Traffic and fleet composition data was made available by SECTRA. Figure 11.7 summarizes the relative distribution of PM_{10} and NO_x per mileage for all Santiago traffic categories. The figures indicate that the 'Heavy duty vehicles (HDV > 16 t)' category (for further explanation, see Chap. 10) has the highest share of PM_{10} and NO_x per mileage, despite the fact that HDV indicate a rate of less than 4% of the overall mileage. Passenger car mileage (the highest, at more than 50% of all vehicles) has almost no bearing on the overall emission distribution in relation to kilometres

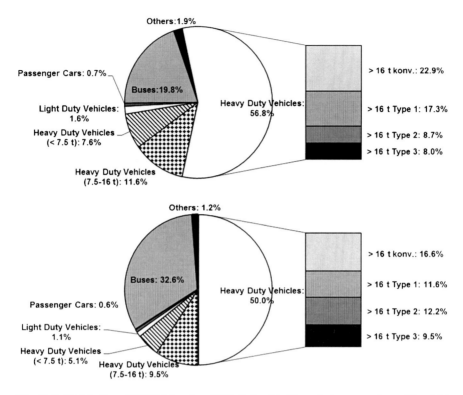

Fig. 11.7 Distribution of PM10 (above) and NO_x (below) traffic emissions per mileage (Nogalski 2010) and traffic category for 2010 (by SECTRA) within the Metropolitan Area of Santiago de Chile

driven. This finding is significant in terms of the associated risk from pollutants induced by traffic emissions. A decrease in passenger car emissions impacts less than reductions in HDV emissions.

Identifying the emission distribution in the individual municipalities of Santiago de Chile is a prerequisite for the investigation of socio-spatial impacts on health. Based on the emission modelling at street level, emission were aggregated at the municipal level. Figure 11.8 shows the distribution of NO_x emissions by municipality. Municipalities in the suburban area, e.g., San Bernardo, Maipu, Quilicura, Huechurba, and in the centre of Santiago show the highest emission rates for NO_x. These areas – apart from the centre – act as a starting point and transition route for people working in the city of Santiago, and in the industrial zones with a high HDV traffic load.

The distribution of NO_x concentrations at the micro scale for traffic emissions are shown in Fig. 11.9 and based on detailed calculations of traffic flow and traffic emissions. The figure represents the annual mean of NO_x in traffic emissions, based on meteorological conditions in 2006. The highest concentrations are dedicated to ring roads and main access roads as well as to the city centre in the municipality of

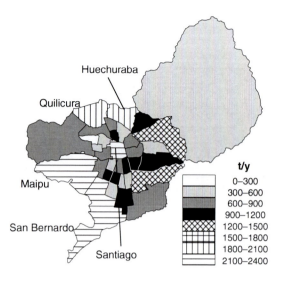

Fig. 11.8 Distribution of NO_x traffic emissions in the municipalities of the Metropolitan Area of Santiago de Chile (tons per year in 2010) (Source: The authors)

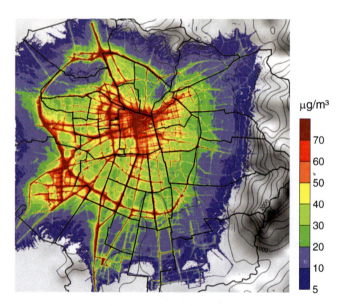

Fig. 11.9 Traffic-related annual mean NO_x concentration levels in µg/m³ within the Metropolitan Area of Santiago (municipality border lines are included in the graphic) based on meteorological conditions in 2006. (Source: The authors)

Santiago. Natural background, domestic heating and industrial sources are not considered.

The micro scale close to the emission sources shows no (or only infinitesimal) evidence of chemical reactions, so that the focus remains on the emissions themselves and their dispersion. Chemical reactions and interactions with other pollutants at the meso scale are important processes, e.g., photochemistry, the formation of secondary aerosols (SOA). In order to consider the principal chemical reactions in the atmosphere, the meso-scale model WRF/chem. was introduced. For mathematical and physical reasons, model calculations at the meso scale are only meaningful down to a resolution of approximately 1 km (depending on the topography and emission inventory available). In combination with the GRAL micro-scale model, however, these calculations become the platform for detailed health impact studies with high spatial and temporal resolution. Figure 11.10 shows, for example, the regional distribution of NO_x for a short episode. Due to the time-consuming simulation, results are shown for 2 weeks in January 2006 only.

Figure 11.10 reflects that the city centre of Santiago is dominated by up to 50 μg/m^3 as a mean value for 2 weeks. The area around the city centre is governed by background concentrations of 5–10 μg/m^3. Contrary to the highly detailed information at the micro scale, the meso scale shows a more homogeneous pollutant distribution, which is of particular significance for the region around the Metropolitan Area of Santiago (background concentration). Also, the secondary chemical reaction processes in the development, e.g., of NO_2 (a component of NO_x), the photochemical processes of O_3 and the development of secondary organic aerosols (SOA) are considered in the fully coupled meteorological/chemical model WRF/chem. The secondary part of the pollutants calculated with WRF/chem., e.g. NO_2 and PM_{10}, must be added to the results of the micro-scale simulations in order to gain a comprehensive picture of pollutant distribution at a smaller scale.

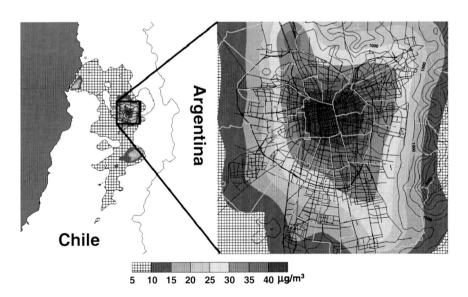

Fig. 11.10 Distribution of NO_x concentrations on the regional scale (*left*) and in the Greater Region of Santiago (*right*) during a 2 week period in January 2006 (Source: The authors)

11.4.1 Health Impact Assessment

Health impact assessment can be carried out by means of monitoring stations and/or results of air quality dispersion modelling. For the analysis presented in this chapter, PM_{10} concentrations for the year 2006 based on data from the Parque O'Higgins monitoring station were initially considered. Since the temporal resolution of health data is limited to daily values, simulation and monitoring data was aggregated from a 1 h resolution to daily data sets.

Table 11.3 gives an overview of deaths caused by cardio-respiratory disease in 2006 in the Metropolitan Area of Santiago de Chile used for this evaluation (Ministerio de Salud, DEIS, personal communication, 2009). Only disease codes with high case numbers were included.

As health effects do not always manifest themselves on day one of high pollution concentrations, consecutive days after the first day of elevated pollutant exposure were also investigated and health effects of PM_{10} up to time lags of 14 days analysed. Evidence of significant risk increases for mortality and morbidity rates were found from the first to the tenth day following elevated exposure levels. Here, time lags and greater exposure depend on the respective disease class. Figures 11.11–11.13 present the risk increase per 10 µg/m^3 PM_{10} increase for the disease groups in Table 11.2 in relevant time intervals as determined by logistic regression analysis.

Table 11.3 Total number of deaths in various cardiovascular and respiratory disease classes in Santiago de Chile, 2006 (ICD-10 codes are an international coding system for disease)

Disease class	Included ICD-10 codes	Case numbers
Hypertensive disease	I10–I15	3,110
Ischemic heart disease	I20–I25	7,943
Other forms of heart disease	I30–I52	3,947
Influenza and Pneumonia	J09–J18	2,984
Chronic lower respiratory disease	J40–J47	2,934

(Source: The authors, based on data from Ministerio de Salud, DEIS, personal communication, 2009)

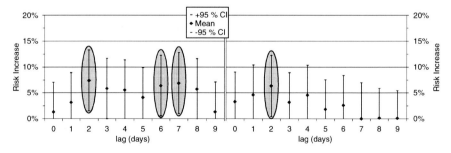

Fig. 11.11 Lagged risk increase per 10 µg/m^3 PM_{10} for hypertensive (*left*) and ischemic heart disease (*right*). Significantly increased values are indicated (CI – 95% confidence intervals) (Source: The authors, based on data from Ministerio de Salud, DEIS, 2009)

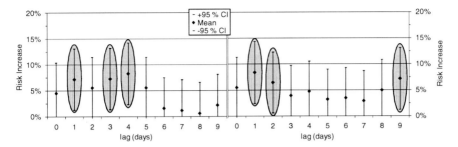

Fig. 11.12 Lagged risk increase per 10 μg/m³ PM$_{10}$ for other forms of heart disease (*left*) and influenza and pneumonia (*right*). Significantly increased values are indicated (CI – 95% confidence intervals). (Source: The authors, based on data from Ministerio de Salud, DEIS, 2009)

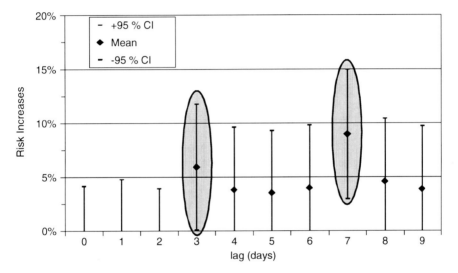

Fig. 11.13 Lagged risk increase per 10 μg/m³ PM$_{10}$ for chronic lower respiratory disease. Significantly increased values are indicated (CI – 95% confidence intervals). (Source: The authors, based on data from Ministerio de Salud, DEIS, 2009)

Figures 11.11–11.13 illustrate that the risk of premature death from either a cardiovascular or respiratory disease is heightened by PM$_{10}$ exposure. The number of deaths in the selected disease groups is dominated by one ICD-10 code. Of the groups in Table 11.3, hypertensive heart disease, myocardial infarction, heart failure, pneumonia and emphysema cause the highest number of deaths. According to Cakmak et al. (2007), risks are higher among the elderly. Different age groups were not considered but the mean risk increase for the whole population of Santiago de Chile was determined. It should be remarked that these investigations cannot reveal causal pathways in the development of the diseases mentioned above, but at least the acute exacerbations of serious and often life-threatening diseases.

Summarizing, the findings demonstrate the critical need for improvement of the air quality in Santiago de Chile with respect to mitigation of premature deaths from cardiovascular and respiratory disease. Exposure to PM_{10} is high, e.g., compared to the target values of the European Commission. The annual mean values for Santiago in 2006 amounted to 74 $\mu g/m^3$ (Table 11.2) in comparison to 20 $\mu g/m^3$, which according to the WHO guidelines (WHO 2006) should not be exceeded. Taking into account that as a rule a daily risk increase of more than 5% was found per 10 $\mu g/m^3$ in Santiago for one single day of high exposure, a major decline in the risk of premature death from the investigated diseases can be expected if PM_{10} levels are reduced.

11.4.1.1 Health Impact Assessment: Outlook

The urgent need to improve air quality in the region of Santiago de Chile and reduce adverse health effects calls for a short outlook in terms of reducing PM_{10} concentrations. Based on the potential to reduce these concentrations, two scenarios showing the characteristic values of PM_{10} concentrations were designed. Although potential developments and the associated PM_{10} threshold values may not reflect real developments, they nonetheless demonstrate the dangers and potentials with respect to air pollution associated health risks. For purposes of comparison, we used a specific value characterizing the current annual average concentration in the Metropolitan Area of Santiago de Chile (see Table 11.1) and target values defined by the European Commission (1999) and the WHO. The starting point for the assessment study is the current air quality state and the annual PM_{10} mean value, which lies at 75 $\mu g/m^3$. Future scenarios were adjusted to European environmental directives (Phase 1 and Phase 2 (also WHO guideline)) with respective annual mean values of 40 and 20 $\mu g/m^3$ (Table 11.4).

Table 11.4 Assignment of PM_{10} concentrations to scenarios

Scenario	Source of reference value	Annual average of PM_{10} concentration
Current state	Current measured annual mean value (2006)	75 $\mu g/m^3$
Phase 1	Directives 1999/30/EC and 96/62/EC, European Commission, Phase 1	40 $\mu g/m^3$
Phase 2	Directives 1999/30/EC and 96/62/EC, European Commission, Phase 2	20 $\mu g/m^3$

Source: The authors

Table 11.5 Maximum daily death risk decreases for selected disease groups. Risk decreases are compared with a business-as-usual scenario, which defines the 100% PM_{10} associated mortality risk

Scenario	I10–I13	I20–I25	I30–I51	J11–J18	J40–J47
Phase 1	−25.0%	−25.3%	−19.8%	−22.7%	−26.8%
Phase 2	−36.3%	−36.8%	−29.3%	−33.3%	−38.7%

Source: The authors

Table 11.5 lists the prognosticated maximum reduction of daily mortality risks by means of these exemplary values. Only unadjusted risk reductions associated with PM_{10} exposure changes are considered. Other risk sources are not included in the assessment but may contribute to the remaining risks.

11.5 Conclusions

A cascade of models and their intertwined linkages is highly useful for detailed health impact studies. In order to carry out an in-depth air quality assessment analysis, in-situ monitoring networks, remote sensing, and coupled micro and meso-scale models beginning with the assessment of traffic emissions should be considered. Nevertheless, due to inhomogeneous data sets, e.g., emission inventories, low availability of monitoring station measurement data, and classified modelling codes, e.g., MODEM, further modelling work will be required should the data become available. Reliable simulations of the current situation are a prerequisite for assessment of the future development of air quality in the Metropolitan Area of Santiago de Chile.

Initial findings of a preliminary assessment of the health impact of PM_{10} concentrations showed a significant increase in the mortality risk for cardio-respiratory disease. On the other hand, the combination of air pollution modelling and epidemiologic analyses establishes the perspective for selective and cost-effective mitigation strategies, identifying pollution sources with the greatest health relevance and targeting mitigation efforts to the maximum benefit of the population of the city. These first evaluations only take PM_{10} into account. In the future, multi-pollutant models will be indispensible to the assessment of the impact on human mortality and morbidity of co-exposure to the main air pollutants PM_{10}, NO_2 and O_3. The findings in the mortality risk assessment demonstrate the urgency of reducing particulate air pollution in the megacity of Santiago de Chile. Additional risk reduction can also be expected from reduced concentrations of other air pollutants, e.g., NO_2, O_3.

Acknowledgments The authors are extremely grateful for assistance from and helpful discussions with the Departamento de Estadísticas e Información de Salud (DEIS) of the Ministerio de Salud (MinSal, Gobierno de Chile). Furthermore we would like to thank Martin Nogalski, Johannes Werhahn, Renate Forkel and Ulrich Uhrner for their support with the modelling work (MODEM, GRAL and WRF/chem.).

References

Alastuey, A., Querol, X., Rodriguez, S., Plana, F., Lopez-Soler, A., Ruiz, C., & Mantilla, E. (2004). Monitoring of atmospheric particulate matter around sources of secondary inorganic aerosol. *Atmospheric Environment, 38*, 4979–4992.

Beelen, R., Hoek, G., Pebesma, E., Vienneau, D., de Hoogh, K., & Briggs, D. J. (2009). Mapping of background air pollution at a fine spatial scale across the European Union. *Science of the Total Environment, 407*, 1852–1867.

Bell, M. L., Dominici, F., Ebisu, K., Zeger, S. L., & Samet, J. M. (2007). Spatial and temporal variation in PM2.5 chemical composition in the United States for health effects studies. *Environmental Health Perspectives, 115*, 989–995.

Berner, A., Galambos, Z., Ctyroky, P., Fruhauf, P., Hitzenberger, R., Gomiscek, B., Hauck, H., Preining, O., & Puxbaum, H. (2004). On the correlation of atmospheric aerosol components of mass size distributions in the larger region of a central European city. *Atmospheric Environment, 38*, 3959–3970.

Boldo, E., Medina, S., LeTertre, A., Hurley, F., Mucke, H. G., Ballester, F., Aguilera, I., & Eilstein, D. (2006). Apheis: Health impact assessment of long-term exposure to PM2.5 in 23 European cities. *European Journal of Epidemiology, 21*(6), 449–458.

Brajer, V., & Mead, R. W. (2004). Valuing air pollution mortality in China's cities. *Urban Studies, 41*(8), 1567–1585.

Breitner, S., Stölzel, M., Cyrys, J., Pitz, M., Wölke, G., Kreyling, W., Küchenhoff, H., Heinrich, J., Wichmann, H.-E., & Peters, A. (2009). Short-term mortality rates during a decade of improved air quality in Erfurt, Germany. *Environmental Health Perspectives, 117*(3), 448–454.

Cakmak, S., Dales, R. E., & Vidal, C. B. (2007). Air pollution and mortality in Chile: Susceptibility among the elderly. *Environmental Health Perspectives, 115*(4), 524–527.

CONAMA (2003). Area Descontaminacion Atmosferica Conama Region Metropolitana de Santiago. www.conama.cl. Accessed 2010.

CONAMA (2004). Reformulación y Actualización Plan de Prevención y Descontaminación Atmosférica para la Región Metropolitana (PPDA - DS.58/2004). www.conama.cl. Accessed 2010.

CONAMA (2006) Informe Seguimiento Plan de Prevencion y de Descontaminacion Para La Region Metropolitana. www.conama.cl. Accessed 2010.

CONAMA (2007). Actualización del Inventario de Emisiones de Contaminantes Atmosféricos en la Región Metropolitana 2005. http://www.conama.cl. Accessed 2010.

Corvalán, R., Osses, M., & Urrutia, C. (2002). Hot emission model for mobile sources: Application to the metropolitan region of the city of Santiago-Chile. *Journal of the Air & Waste management Association, 52*(2), 167–174.

Duzgoren-Aydin, N. S., Wong, C. S. C., Aydin, A., Song, Z., You, M., & Li, X. D. (2006). Heavy metal contamination and distribution in the urban environment of Guangzhou, SE China. *Environmental Geochemistry and Health, 28*, 375–391.

European Commission (1999). Council Directive 1999/30/EC of 22 April 1999 relating to limit values for sulphur dioxide, nitrogen dioxide and oxides. *Official Journal of the European Communities, L 163*, 41–60.

Grell, G. A., Peckham, S. E., Schmitz, R., McKeen, S. A., Frost, G., Skamarock, W. C., & Eder, B. (2005). Fully coupled "online" chemistry within the WRF model. *Atmospheric Environment, 39*(37), 6957–6975.

Gurjar, B. R., & Lelieveld, J. (2005). New directions: Megacities and global change. *Atmospheric Environment, 39*, 391–393.

Gurjar, B. R., Jain, A., Sharma, A., Agarwal, A., Gupta, P., Nagpure, A. S., & Lelieveld, J. (2010). Human health risks in megacities due to air pollution. *Atmospheric Environment, 44*, 4606–4613.

Harrison, R. M., & Jones, A. M. (2005). Multisite study of particle number concentrations in urban air. *Environmental Science Technology, 39*, 6063–6070.

Houthuijs, D., Breugelmans, O., Hoek, G., Vaskovi, E., Mihalikova, E., Pastuszka, J. S., Jirik, V., Sachelarescu, S., Lolova, D., Meliefste, K., Uzunova, E., Marinescu, C., Volf, J., de Leeuw, F., van de Wiel, H., Fletcher, T., Lebret, E., & Brunekreef, B. (2001). PM_{10} and $PM_{2.5}$ concentrations in Central and Eastern Europe: Results from the Cesar study. *Atmospheric Environment, 35*, 2757–2771.

Krzyzanowski, M. (2008). WHO air quality guidelines for Europe. *Journal of Toxicology Environmental Health A, 71*, 47–50.
Krzyzanowski, M., & Cohen, A. (2008). Update of WHO air quality guidelines. *Air Quality, Atmosphere, and Health, 1*, 7–13.
Lianou, M., Chalbot, M. C., Kotronarou, A., Kavouras, I. G., Karakatsani, A., Katsouyanni, K., Puustinnen, A., Hameri, K., Vallius, M., Pekkanen, J., Meddings, C., Harrison, R. M., Thomas, S., Ayres, J. G., Brink, H., Kos, G., Meliefste, K., de Hartog, J. J., & Hoek, G. (2007). Dependence of home outdoor particulate mass and number concentrations on residential and traffic features in urban areas. *Journal of Air Waste Management Association, 57*, 1507–1517.
Molina, M., & Molina, L. (2004). Megacities and atmospheric pollution. *Air & Waste Management Association, 54*, 644–680.
Nogalski, M. (2010). Modellierung und Berechnung der Verkehrsemissionen in Santiago de Chile – Ist-Situation und Prognose für 2030. Diplomarbeit, Garmisch-Partenkirchen: Institut für Meteorologie und Klimaforschung, Atmosphärische Umweltforschung (IMK-IFU) am Karlsruher Institut für Technologie (KIT).
Ostro, B. D., Sanchez, J. M., Aranda, C. & Eskeland, G. A. (1995). *Air pollution and mortality. Results from Santiago, Chile* (Policy Research Working Paper 1453). The World Bank, Development Research Group.
Öttl, D., Sturm, P. J., Pretterhofer, G., Bacher, M., Rodler, J., & Almbauer, R. A. (2003). Lagrangian dispersion modeling of vehicular emissions from a highway in complex terrain. *Journal of the Air and Waste Management Association, 53*, 1233–1240.
Penard-Morand, C., Schillinger, C., Armengaud, A., Debotte, G., Chretien, E., Pellier, S., & Maesano, I. (2006). Assessment of schoolchildren's exposure to traffic-related air pollution in the French six cities study using a dispersion model. *Atmospheric Environment, 40*, 2274–2287.
PNUMA (2003). Perspectivas del Medio Ambiente Urbano: GEO Santiago. Equipo del Instituto de Estudio Urbanos
Pope, C. A., Dockery, D. W., Spengler, J. D., & Raizenne, M. E. (1991). Respiratory health and Pm10 pollution – A daily time-series analysis. *American Review of Respiratory Disease, 144*(3), 668–674.
Pope, C. A., Schwartz, J., & Ransom, M. R. (1992). Daily mortality and PM_{10} pollution in Utah Valley. *Archives of Environmental Health, 47*(3), 211–217.
Pope, C. A., Thun, M. J., Namboodiri, M. M., Dockery, D. W., Evans, J., Speizer, F. E., & Heath, C. W. (1995). Particulate air-pollution as a predictor of mortality in a prospective-study of us adults. *American Journal of Respiratory and Critical Care Medicine, 151*(3), 669–674.
Rappenglück, B., Schmitz, R., Bauerfeind, M., Cereceda-Balic, F., von Baer, D., Jorquera, H., Silva, Y., & Oyola, P. (2005). An urban photochemistry study in Santiago de Chile. *Atmospheric Environment, 39*(16), 2913–2931.
Rutllant, J., & Garreaud, R. (1995). Meteorological air pollution potential for Santiago, Chile: Towards an objective episode forecasting. *Environmental Monitoring and Assessment, 34*, 223–244.
Schmitz, R. (2005). Modelling of air pollution dispersion in Santiago de Chile. *Atmospheric Environment, 39*, 2035–2047.
Schmitz, R., Falvey, M., Clerc, J., Ozimiça, N., & Oporto, L. (2010). Optimización del Modelo Fotoquímico de alta resolución implementado en la fase 2007 y ampliación de su alcance a material particulado respirable y precursores, Final report, CONAMA.
Schwartz, J. (2000). The distributed lag between air pollution and daily deaths. *Epidemiology, 11*(3), 320–326.
Schwartz, J. (2001). Is there harvesting in the association of airborne particles with daily deaths and hospital admissions? *Epidemiology, 12*(1), 55–61.
SECTRA (2001). Metodología para el cálculo de emisiones Vehiculares (MODEM).
Suppan, P. (2010). Assessment of air pollution in the conurbation of Munich – Present and future. *International Journal of Environment and Pollution, 40*(1–3), 149–159.

Suppan, P., & Schädler, G. (2004). The impact of highway emissions on ozone and nitrogen oxide levels during specific meteorological conditions. *Science of the Total Environment, Volumes 334-335*, 215–222.

Tuch, T. M., Herbarth, O., Franck, U., Peters, A., Wehner, B., Wiedensohler, A., & Heintzenberg, J. (2006). Weak correlation of ultrafine aerosol particle concentrations <800 nm between two sites within one city. *Journal of Exposure Science Environmental Epidemiology, 16*, 486–490.

United Nations Environment Program/World Health Organization. (1992). *Urban air pollution in megacities of the world*. United Kingdom: Blackwell.

Wehner, B., & Wiedensohler, A. (2003). Long term measurements of submicrometer urban aerosols: Statistical analysis for correlations with meteorological conditions and trace gases. *Atmospheric Chemistry and Physics, 3*, 867–879.

Wichmann, H. E., Spix, C., Tuch, T., Wolke, G., Peters, A., Heinrich, J., Kreyling, W. G., & Heyder, J. (2000). Daily mortality and fine and ultrafine particles in Erfurt, Germany part I: Role of particle number and particle mass. *Research Report Health Effects Institute, 98*, 5–86.

Wietlisbach, V., Pope, C. A., & Ackermann-Liebrich, U. (1996). Air pollution and daily mortality in three Swiss urban areas. *Sozial-und Präventivmedizin, 41*(2), 107–115.

World Health Organization. (2005a). *WHO air quality guidelines for particulate matter, ozone, nitrogen dioxide and sulfur dioxide (AQGs)*. Bonn Office: WHO.

World Health Organization. (2005b). *Special programme on health and environment. European Centre for Environment and Health*. Bonn Office: WHO.

World Health Organization. (2006). *WHO air quality guidelines – Global update 2005: Particulate matter, ozone, nitrogen dioxide and sulphur dioxide*. Copenhagen: WHO regional office for Europe 2006.

Chapter 12
Risks and Opportunities for Sustainable Management of Water Resources and Services in Santiago de Chile

Helmut Lehn, James McPhee, Joachim Vogdt, Gerhard Schleenstein, Laura Simon, Gerhard Strauch, Cristian Hernàn Godoy Barbieri, Cristobal Gatica, and Yarko Niño

Abstract This chapter analyses the sustainability performance of water resource and water service management in the Metropolitan Region of Santiago de Chile, adopting water resource and water service perspectives. By comparing the targets with the current situation, we address sustainability deficits and identify potential risks to and opportunities for sustainable development. On the basis of this assessment, we summarize some of the most pressing issues that pose risks to sustainability and suggest mitigation alternatives. On the basis of population projections and historical fresh water data, we find that per capita availability could decrease from 767 to 1,100 today to 575–825 m^3 per capita and year by 2030 for a normal water year, shifting Santiago de Chile from a position of water stress to one of water scarcity. This could become critical for semi-rural localities surrounding the urban core, which are currently outside the concession area of the major drinking water utilities. Although sewage treatment has improved considerably in the last 10 years, several reaches of natural streams remain at risk as a result of unregulated liquid emissions and solid waste disposal. Storm water management is still mostly confined to the development of a vast collection and disposal infrastructure, without significant investment in distributed management systems. Hence the risk of flooding in the lower areas of the city remains high, compounding other social problems in the city.

Keywords Drinking water • Integrated water resource management • Irrigation • storm water management • Sustainable use of water • Waste water management • Water resources • Water services

---

H. Lehn (✉) • J. McPhee • J. Vogdt • G. Schleenstein • L. Simon • G. Strauch • C.H.G. Barbieri • C. Gatica • Y. Niño
Karlsruhe Institute of Technology (KIT), Institute for Technology Assessment and Systems Analysis (ITAS), Hermann-von-Helmholtz-Platz 1, 76344 Eggenstein-Leopoldshafen, Germany
e-mail: helmut.lehn@kit.edu

12.1 Introduction

The development of cities, i.e., dense settlements of human beings within a restricted area, requires – among other conditions – the delivery of sufficient quantities of water to the settlements, and the collection and removal of different kinds of waste water from these settlements to avoid the emergence of epidemics. The safe functioning of the water supply and waste water transport technologies were thus prerequisites for the development of cities 8,000 years ago in the ancient civilizations of the Indus valley, Mesopotamia, Greece and Rome.

Referring to the Brundtland Report (WCED 1987), which delivers the widely accepted modern definition of sustainable development, water-related sustainable urban development entails meeting the water-related needs of today's city inhabitants without compromising the ability of future generations to meet their own water related-needs (intergenerational justice) or that of people currently living outside the city to meet theirs (intragenerational justice). Hence sustainable water management must take city water services (present and future) into account and the impact on these services in other urban areas upstream and downstream, as well as on water resources outside the city (e.g., long-lasting ecological functions).

The purpose of this contribution is to define the relevant features for sustainable urban water management in Santiago de Chile and to assess whether and to what extent the performance of both urban water services and water resources management can be regarded as sustainable in the aforementioned sense. Section 12.2 presents the research approach and methods utilized to identify opportunities and risks to sustainability. In Sect. 12.3 we indicate the principal findings of our research, which include a current status analysis of water resources and services, and a distance-to-target assessment. Some of the most pressing issues of water resource management in the Metropolitan Region of Santiago are identified in Sect. 12.4, where we suggest possible courses of action to mitigate risks to sustainability. Finally, Sect. 12.5 presents major conclusions of this study.

12.2 Research Approach

12.2.1 Conceptual Framework: Sustainability Assessment and Risk Concept

To answer the question raised above on the sustainability of the water sector, the Helmholtz Integrative Sustainability Concept (Kopfmüller et al. 2001) was applied and contextualized (see Chap. 4). Following Kopfmüller and Lehn (2006), a distance-to-target analysis was adopted (see Fig. 12.1) and modified to a three-step procedure: (1) target definition, (2) status analysis and (3) distance-to-target evaluation. Within the target definition, appropriate sustainability criteria and

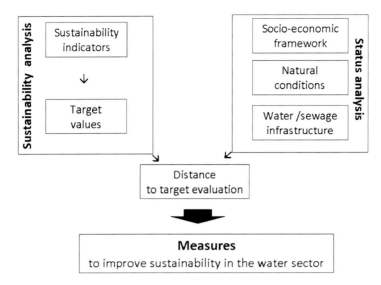

Fig. 12.1 Distance-to-target approach (Source: The authors)

indicators pertaining to the water sector (state and non-state) were developed jointly with regional stakeholders. In the status analysis actual data for these indicators was collected, and a snapshot of the city in terms of sustainability metrics was obtained. The distance-to-target evaluation sets this data in relation to the specifically defined or generally accepted target values of the indicators previously identified. The variance (distance) between the intended target value and the status quo value marks the extent of the sustainability deficit.

The risk dimension of the problem is directly correlated to the distance-to-target analysis, where greater distance to targets is associated with enhanced risk to sustainability. This connection is not yet quantitative in the sense traditionally employed in other risk analysis applications. However, the extension can easily be made once hazard factors and at-risk populations are identified – for example, in the case of local flooding or water shortages that impact on peri-urban localities. Sustainability deficits with a vast temporal or spatial range (including the number of people affected) or with irreversible impacts are regarded as significant risks to the sustainability of water resources and services in the megacity and its surrounding regions.

12.2.2 The 'Water System' of the Metropolitan Region of Santiago de Chile

The description of the 'water system' in the RM Santiago includes a brief geographical and institutional outline. The geographical boundary of the Maipo-Mapocho river watershed largely defines the boundaries of the water system in the metropolitan region. As seen in Fig. 12.2, the catchment areas of these two

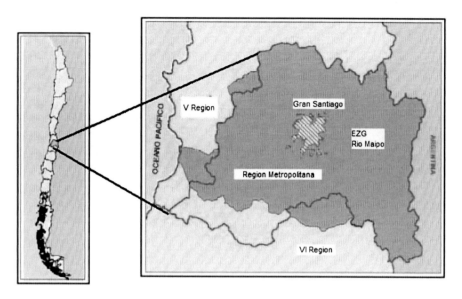

Fig. 12.2 The Santiago Metropolitan Region (within the black bold line) and the catchment area of the Maipo-Mapocho River System (grey) (Source: Bartosch 2007, after DGA 2003)

rivers coincide for the most part with the area of the RM Santiago. Thus, socio-economic data based on the RM Santiago can be related to data on the water resources of the Maipo-Mapocho catchment.

The research perspective in this analysis focuses on a water *resources* and a water *services* perspective. The water resource perspective characterizes ecosystems that provide ground and surface water and the human activities that impact on them. The water services perspective refers to the existing water supply and waste water treatment infrastructure, and its impact on the customers concerned.

The institutional water system is mainly defined by the fact that Chile's water sector is mostly privatized, using a concession model. The Chilean water law and its definition of water user rights is unique. The privatization process of the Chilean water and sanitation sector from 1998 to 2005 represents one of the most far-reaching reforms in Latin America, transferring ownership of assets from the public to the private sector. Water supply and sanitation coverage and quality was to become universal. Investments required to reach the target of treating 100% of Chile's sewage were estimated at approximately US$ 1.5 billion, a sum the Chilean government could hardly afford. Furthermore, service providers should be self-financing with higher tariffs representing real service costs and more efficient performance (Jouravlev and Valenzuela 2007).

The largest water utility in Santiago is the Aguas Andinas Group, which is mainly owned by Aguas Barcelona (AGBAR) with 41.2% of the shares, while the state retains 34.98% through the CORFO agency (CORFO 2010). The Aguas Andinas concession area for drinking water covers most of the urban region, with connections to more than 1,470,000 households. The second biggest utility is the municipal owned SMAPA, which serves 185,473 households in the southwestern commune of Maipú with about 600,000 inhabitants (SISS 2010). As far as waste water collection and treatment are concerned, the concession area of Aguas Andinas includes the commune of Maipú. However, not all inhabitants of the region are connected: rural towns remain outside the concession area, and Aguas Andinas is not obliged to incorporate newly urbanized areas into its network.

The drinking water supply in the concession areas of the two main water suppliers adds up to about 20 m^3/s. Of these, 17 m^3/s (85%) are fed by surface water from the River Maipo, Laguna Negra and River Mapocho; the remaining 3 m^3/s (15%) are supplied from ground water resources primarily extracted by the communal supplier SMAPA in the commune of Maipú (compare Fig. 12.3). Nevertheless, all water utilities and many of the agricultural users own ground-water wells, which are operated when surface resources run dry.

Existing regulatory and planning processes are marked by a pure state-company structure. The Regulation Authority for Sanitation Services (Superintendencia de Servicios Sanitarios (SISS)) allocates concessions, controls the companies, sets tariffs, and defines service and quality standards. It works out investment plans and penalizes non-compliance. Furthermore, it is responsible for consumer protection and represents the customer during the tariff process. Unlike other countries, there is no independent consumer agency in Chile. The Chilean General Water Directorate (Dirección General de Aguas – DGA) is involved in the system to the extent that it grants the water rights required for concession allocation, while the planning ministry MIDEPLAN allocates subsidies and the national environmental ministry MMA (former CONAMA) steers environmental policies.

Water balance challenges and institutional conditions, as well as stream health and the relation between RM Santiago and its hinterland are some of the major water and environmental issues. Water quality can be affected by mining operations in the upper reaches of the Mapocho River; unregulated discharges of liquid effluents and solid waste along the Maipo and Mapocho Rivers threaten the quality of the water at their lower reaches; new hydropower projects (Alto Maipo – AES Gener) have caused conflict among residents of the Upper Maipo Valley; finally, poorly enforced gravel mining operations along the Maipo River pose a threat to other infrastructure facilities, such as bridges and water intakes.

12.2.3 Data and Information Sources

12.2.3.1 Water Budget Investigation

The budget for renewable water resources available, on the one hand, and the anthropogenic need, on the other hand, were calculated via literature research and personal interviews. An extensive bibliographic search was carried out in several organizations related to water distribution. The most important institutions for these investigations are: DGA, Chilean Ministry for Public Works (Ministerio de Obras

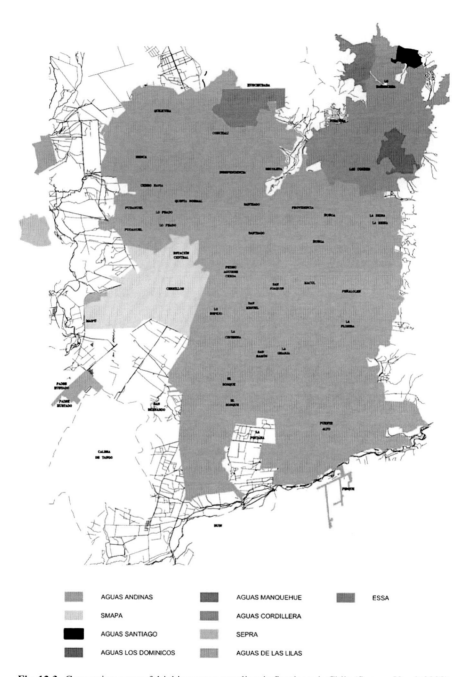

Fig. 12.3 Concession areas of drinking water suppliers in Santiago de Chile (Source: Vogdt 2008)

Publicas – MOP), SISS, MMA at its central and regional branches (SEREMI MMA RM), and the United Nations Economic Commission for Latin America and the Caribbean (ECLAC).

12.2.3.2 Heavy Metal Trace Analysis

For heavy metal analysis (see Sect. 3.2 for results) at each sampling point, a pair of water samples was taken between March 28 and April 10, 2009. They were conserved by nitric acid in suprapure quality and analysed using atomic absorption spectroscopy. Results were obtained by calculating the arithmetic mean of the individual measurements and assessed by comparing them to the Chilean drinking water guidelines (NCh409/1 2005).

12.2.3.3 Modelling of Sediment Management of the Maipo River Basin

A specific activity regarding water and stream bed management was carried out in the form of a Decision Support System (DSS). The DSS involves a simulation model that couples a management model with the Model for Sediment Transport and Morphology, MOSSEM (González 2006; Abarca 2008). The model development considers several modifications, mainly to include gravel extraction from the

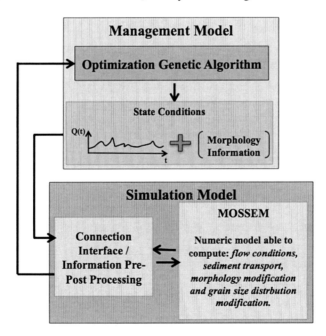

Fig. 12.4 Connectivity implementation between the numerical model (MOSSEM) and the management mode (Source: Melo et al. 2010)

stream bed in the local sediment mass balance. Thus, the modified model can evaluate the dynamic change in morphology as a result of gravel extraction. Based on optimization, the management model is developed, adapting an open source Genetic Algorithm (GA) code. A general scheme of the model is given in Fig. 12.4. It is worth emphasizing that the methodology used in this study for coupling the models can be applied to any morphological model that is based on binary files for inputs/outputs (similar to MOSSEM).

12.3 Results

12.3.1 Target Definition: Development of Sustainability Indicators

In a first step, 64 sustainability indicators for the water sector were developed and discussed with institutional stakeholders. They address the following dimensions (number of indicators related to each dimension in brackets):

- Human life (5)
- Equal opportunities (7)
- Natural resources (17)
- Governance (10)
- Social and cultural resources (12)
- Global stewardship (3)
- Securing society's productive potential (10)

From these 64 indicators, 14 core indicators (shown in Table 12.1) were deduced according to relevance and data availability during a participatory process that included the relevant governmental and non-governmental organizations. Five of these indicators are specifically related to water *resources*. Six indicators address water *services*. Three indicators serve other purposes. The definition of target values proved to be the working step with the highest need for discussion during this process.

12.3.2 Status Analysis

12.3.2.1 Water Resources

Water Supply in the RM Santiago

With the exception of deep ground-water resources and glaciers, the residence time of fresh water resources within a watershed ranges from days (atmosphere, rivers) to years (ground water) (Lehn and Parodi 2009). This means that within the space of

12 Risks and Opportunities for Sustainable Management of Water Resources

Table 12.1 Core sustainability indicators for the water sector

Indicator
Water resources
Relation between natural water offer and human demand, differentiated by aquifer or river basin
Existence of environmental quality norms (secondary norms) related to natural quality of water bodies (expressed in percentage of total number of existing water bodies in the region)
Percentage of length/area of the water bodies suitable for recreation
Degree of norm fulfilment concerning industrial and domestic emissions towards superficial waters
Degree of norm fulfilment concerning industrial and domestic emissions towards ground water
Water services
Percentage of children under the age of five affected by water-related diseases, measured as DALYs according to the WHO definition (sum of years of potential life lost due to premature mortality and the years of productive life lost due to disability), differentiated by commune
Percentage of water samples that fulfil the Chilean drinking water norm, differentiated by supplier
Percentage of the population connected to a safe and hygienic sewage system, differentiated by commune
Percentage of drinking water loss in the distribution system
Cost of domestic water services as percentage of the total income, differentiated by socio-economic groups (quintile)
Tariff rates for the collection of storm water according to the degree of sealed surface of real properties
Others
Water use per capita, differentiated by commune
Participation of NGO/civil society in sanitary sector decisions
Percentage of participation processes where stakeholders are content with the outcome
Source: The authors

one human generation all of the water in the Santiago watershed will have been renewed. Hence fresh water can be considered a renewable resource. According to the former chief economist of the World Bank, Herman Daly (1996), renewable resources should be harvested only at the speed at which they regenerate. Fresh water resources are naturally replenished by precipitation and the inflow of ground and surface water from upstream regions. In the case of the urbanized area of Santiago de Chile this refers to precipitation (primarily in winter) and inflow from the River Maipo and the River Mapocho, both of which originate in the Andes Mountains near the Argentinian border.

Meteorological input is highly variable in the Metropolitan Region. As a result of the ENSO phenomenon with wet (El Nino) and dry (La Nina) years, annual precipitation fluctuates substantially. In Santiago (Quinta Normal meteorological station) the driest years in the period from 1950 to 2004 had values as low as 100 mm/a and as high as 700 mm/a and more (DGA 2007a after Bartosch 2007) – compare Fig. 12.5. The long-term precipitation average in the Central Valley (between the coastal Andes in the west and the high Andes in the east) varies between 261.6 mm in the west of Santiago (Pudahuel) and 347.2 mm in the east at the foot of the Andes (Tobalaba).

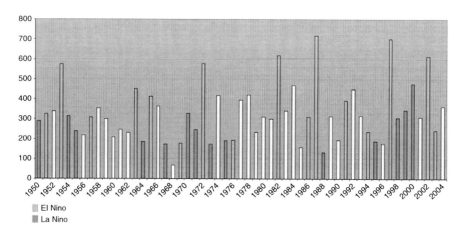

Fig. 12.5 Long-lasting precipitation variations in Santiago – Quinta Normal Station (Source: DGA 2007; Bartosch 2007)

Table 12.2 Absolute and specific amount of renewable fresh water available (supply)

	Available fresh water (km/year)	Specific offer of available fresh water – 6 million people (m/capita year)	Specific offer of available fresh water – 8 million people (m/capita year)
Normal year	4.6–6.6	767–1,100	575–825
Wet year	6.3–7.3	1,050–1,217	788–913
Dry year	2.8–3.2	467–534	350–400

Source: Bartosch 2007

The amount of inflowing water from the Andes can be measured at two gauging stations, El Manzano for the Maipo River and Los Almendros for the Mapocho River. The inflow of the River Maipo varies between 60 and 75 m^3/s in dry and 140 m^3/s in wet years. The corresponding data for the River Mapocho is 2–3 m^3/s in dry and 9–14 m^3/s in wet years (DGA 2007a, after Bartosch 2007).

On the basis of this data we conclude that the supply of renewable fresh water available in the RM Santiago varies between 3 km^3 in dry and 6.8 km^3 in wet years – see Table 12.2. If we consider these values as fixed in time and take into account the population projections for the city of Santiago de Chile, we can compute *specific supply* in terms of m^3 per capita per year for current and future population estimates. The corresponding specific values vary between 467 and 1,217 m^3/capita and year in the case of six million, and between 350 and 913 m^3/capita and year in the case of eight million inhabitants in the region (own estimation for the year 2030). These estimates would transfer Santiago from water stress conditions today to water scarcity conditions in the course of 30 years according to the Falkenmark Index (Falkenmark 1989), not considering hydrologic trends and climate projections. These projections involve the warming and drying of central Chile's climate under most emission scenarios as defined by the IPCC (DGF 2007; Melo et al. 2010).

Water Demand in the RM Santiago

The water demand in the RM Santiago is driven first and foremost by agricultural needs. In the year 2007 this sector accounted for about 74% of the total water demand in the region. The share of drinking water amounted to 17%, while water demand for industrial purposes reached about 9.4% (see Table 12.3). The absolute water demand in the region varies between 3.2 km^3 (wet years) and 4.4 km^3 (dry years), depending on irrigation needs (INE 2002 after Bartosch 2007).

According to Aguas Andinas (2005c), the average specific drinking water consumption per capita in Santiago is 222 l per capita and day. A slight decrease of 6% to 207 l per capita and day is expected by the year 2019 (compare Table 12.4). In contrast, the per capita daily consumption of private households in Berlin in 2006 was 116 l. Specific data on drinking water consumption in various municipal districts of Santiago can be estimated on the basis of development plans for drinking water supplies (Aguas Andinas 2003a, 2005a, b, 2006a, b; Aguas Cordillera 2005; Aguas Los Dominicos 2005; Aguas Santiago Poniente 2006), based on historical and empirical consumption analysis. Waste water quantities are estimated by calculating

Table 12.3 Water usage by different economic sectors in the Metropolitan Region of Santiago in 2007 and the prediction for 2017 and 2032

Usage	2007 km^3/year	2017 km^3/year	2032 km^3/year
Agriculture	2.60	2.54	2.54
Drinking water	0.58	0.74	0.78
Industry	0.33	0.49	0.87
Mining	0.02	0.04	0.06
Total	3.52	3.80	4.25

Source: DGA 2007b

Table 12.4 Per capita drinking water consumption in the concession area of Aguas Andinas S.A for the year 2005 and predictions for the future

Year	Consumption (l/capita day)
2005	222
2006	221
2007	219
2008	218
2009	217
2010	216
2011	215
2012	214
2013	213
2014	212
2015	211
2016	210
2017	209
2018	208
2019	207

Source: Aguas Andinas 2005c

losses arising from, e.g., hosing of gardens, of which there are many in some of the eastern Santiago areas (e.g., Las Condes).

The calculated variation between poor and prosperous areas of the city is striking: estimations of water use variability indicate a per capita consumption of 120 l per capita and day in Bosques de San Luis, for example, and more than 850 l in Santa Rosa del Peral (Vogdt 2008).

Heavy Metals in Surface Waters

In line with Santiago's geological conditions upstream (Frikken et al. 2005), copper mining takes place on a large scale at 'Minera Los Bronces' – more familiar under its previous name 'Minera Disputada de las Condes' – in the Mapocho catchment (Spürk 2010). Smaller mining activities are carried out on the upper reaches of the Maipo River ('Minera Erfurt'). Natural resources and anthropogenic processes suggest the presence of heavy metal in the environment. Due to the high degree of surface water in the drinking water supply (see above), a random test on trace metals was conducted at certain points in the Maipo/Mapocho catchment.[1] Copper concentrations in the Maipo sub-catchment show values below 5 µg/l, similar to the drinking water sample measured for Providencia, which is far below the threshold value of 2,000 µg/l. In the Mapocho sub-catchment, on the other hand, concentrations are higher by a factor of 100–1,000, and occasionally even exceed the drinking water threshold. This indicates the need for a careful analytical survey at regular intervals to guarantee the fulfilment of drinking water norms in the Mapocho supply area (Fig. 12.6).

Arsenic concentrations in water samples taken at the Mapocho, El Volcán and Yeso Rivers in the Maipo sub-catchment range between 1 and 6 µg/l, or approximately 10–60% of the drinking water threshold. Water samples from the upper Maipo River show arsenic concentrations within the range of 15 µg/l. After inflow from the River El Volcán and the River Yeso, these concentrations decrease to a level of 12 µg/l but still exceed the threshold value for drinking water. The fact that concentrations of less than 2 µg/l have been detected in the drinking water in Providencia, which is supplied by water from the Maipo System, and that the water works of La Florida and Las Vizcachas have no purification technology specifically for arsenic leads us to conclude that most arsenic is removed during the drinking water treatment process. The concentration of arsenic in the drinking water sample analysed undercuts the threshold value of 10 µg/l, now called for in the Chilean drinking water guideline.

---

[1] The authors are immensely grateful to Adnan Al-Karghuli and Gerd Schukraft from the Geographical Institute of the Heidelberg University (Germany) for their detailed analysis of heavy metal concentrations.

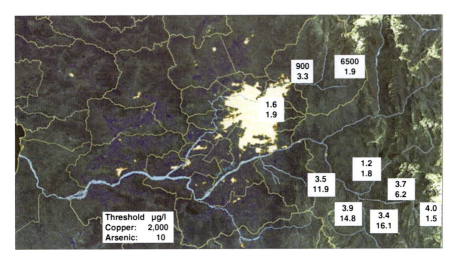

Fig. 12.6 Copper (upper value) and arsenic (lower value) concentrations in the surface waters of the Maipo/Mapocho catchment and drinking water in Providencia (µg/l) (Source: The authors)

12.3.2.2 Water Services

Water services are essential for the smooth running of cities. In terms of sustainability analysis, water services are an indicator of whether the needs of the present urban population are being met. If services are efficient and their performance in compliance with the appropriate indicators, a prerequisite for sustainability has been fulfilled.

Drinking Water Supply

Over 99% of the inhabitants within the urban confines of Santiago have been connected to drinking water supplies since the 1990s (Vogdt 2008). Santiago has therefore reached the drinking water coverage level of developed countries. No sustainability deficit could be detected with regard to current access to drinking water. Aguas Andinas S.A. – along with its subsidiary companies Aguas Cordillera, Aguas Manquehue and Aguas Los Dominicos – is the chief supplier of drinking water in the RM area, using surface water from the River Maipo, Embalse El Yeso, Laguna Negra and the River Mapocho. In the southwestern commune of Maipú, which has about 600,000 inhabitants, the municipal company SMAPA supplies drinking water obtained mainly from ground water. The water supply is steady and water pressure adequate (according to Nch 409/1 2005) (Simon 2009).

Outside the concession areas of the above-mentioned drinking water companies, specific population groups in the more rural periphery of the RM could be at risk of exclusion from basic water services (Durán 2009). These peri-urban settlements rely on self-managed *Rural Drinking Water* supply systems that operate under a cooperative management system with technical assistance from the government. However, technical expertise is frequently a scarce commodity and operations are further complicated by lack of resources. Moreover, these rural systems usually rely on ground-water wells as their sole source of water, making them vulnerable to intense drought, which reduces ground-water levels. The inclusion of these localities in the concession area of the utility company would probably mean improved service quality and reliability, albeit at higher cost. In many instances, however, economic constraints prevent utility companies from expanding their coverage areas. A few of the non-concession localities are adjacent to zones with drinking water supplies provided by a private utility; this stresses the economic origin – rather than geographic nature – of this spatial variability in sanitation coverage status.

According to the SISS (2008a), Santiago drinking water has been fulfilling WHO international quality standards (WHO 2006) since 2005. Recent quality census results show that urban drinking water is 96.1% safe to drink. Water quality, which has increased steadily since the 1990s, is strongly connected to the low child mortality rate in Santiago; the child mortality here is the lowest in South America and comparable with European rates (SISS 2008a). Random water analysis of samples taken in the context of the RHM project revealed low copper concentrations in the drinking water despite elevated concentrations of this heavy metal in the Mapocho River.

Sewage Treatment

By 2006, approximately 90% of the urban population was connected to sewer systems. Waste water collection and treatment is performed by the Aguas Andinas Group (compare Sect. 2.2) for the entire urban area of RM Santiago. The heavy concentration of inhabitants and industrial production in the city results in huge amounts of sewage of a highly complex composition. Total sewage production in Grand Santiago is estimated at 18 m^3/s (equivalent to 1.6 mill. m^3/day or 584 mio. m^3/year). In 2008, 13.2 m^3/s (73.3% of total sewage) were treated in the two large sewage treatment plants, El Trebal and La Farfana. Effluents from these plants and untreated sewage from northern districts such as Colina, Batuco and Lampa are discharged into the Mapocho River. The purification capacity of both plants is comparable to European standards with regard to the decomposition of carbon compounds: the COD is reduced by 92–94% (Berlin: 96.7%), the BOD by 95–97% (Berlin: 99.2%). Nutrient decomposition (nitrogen and phosphorous) is low (25–35%) compared with modern German plants (Berlin: 85–97%) (Lehde 2010).

Current legislation requires disinfection of semi-treated sewage by chlorination in a final treatment step to remove pathogenic micro-organisms. Chlorination of semi-treated sewage carries the risk of synthesizing new chlorinated hydrocarbons in the sewage. Due to inadequate laboratory capacities to analyse these compounds in Chile (Vogdt 2009, personal communication) no information is available on these substances in the watercourses.

To improve the environmental situation within the confines of the city (offensive odour near the river), a large sewer was recently built parallel to the riverbed of the Mapocho River that cuts across the city. Its purpose is to collect sewage from the northern parts of Santiago (5 m^3/s) and conduct it to the southwest where three-fifths of the sewage is released into the river and two-fifths treated in the treatment plants mentioned above (Projecto Mapocho Urbano Limpio). A third purification plant with a capacity of 6.6 m^3/s on the site of the El Trebal plant, called Los Nogales or Mapocho, is in the planning stage. When it begins operations (expected in 2015), Gran Santiago will have a total purification capacity of 19.8 m^3/s, which exceeds the amount of waste water currently collected.

The treatment of sewage sludge has a negative impact (offensive odour) on the neighbourhood of the Maipú commune. For this reason a large share of the sludge is now dumped at the 'Lomas Los Colorados' landfill approx. 100 km north of Santiago. In the long run sewage sludge from altogether three purification plants will be treated in the new sludge treatment plant 'El Rutal' located close to the 'Lomas Los Colorados' site. Sewage sludge production is expected to reach an annual 139,000 tons (dry matter), which is equivalent to roughly 14,500 truckloads of 25 t. If the sludge is dried beforehand, it would yield a dry matter proportion of 38%.

Rural and semi-rural settlements co-exist on the periphery of Santiago, outside the Aguas Andinas Group concession areas. Here, more decentralized waste water treatment systems are in place (septic tanks and compact treatment plants), which are typically beset by problems related to odour, maintenance and operation of the plant, and especially sludge disposal (Inter-American Development Bank 2007). According to a recent SISS study (IASA 2007) nationally reported issues on odours emanating from sewage treatment plants did not refer to the size of the plant. However, most problems associated with offensive odours did relate either to the sewage system or sludge treatment.

Finally, practical/field experience described in a number of studies (e.g., IASA 2005) also indicates that treatment could be improved in terms of comfort and the security of the population.

12.3.3 Distance-to-Target Analysis

12.3.3.1 Water Resources

Quantitative Aspects

As far as the sustainable use of water resources in the RM Santiago is concerned, the most important deficit is the relationship between available fresh water and water usage. A comparison of the data in Tables 12.2 and 12.3 clearly indicates that current and future water demands cannot be met in dry years. The water demand of

3.5 km^3 in 2007 accounts for 53–76% of the total fresh water available in a normal year. The estimated demand of 4.3 km^3 in 2032 represents even 66–92% of the freshwater offer available in a normal year. The water demand in dry years now already exeeds the amount of renewable fresh water.

In her global overview of water stress in river catchments, Döll (2008) assessed the Maipo/Mapocho catchment to be under very high water stress. The ratio between water extraction and renewable water resources (precipitation minus evapotranspiration average for 1961–1990) is higher than 0.8 – the worst classification, and comparable to North Africa or the Middle East. See Fig. 12.7 below.

Falkenmark classifies a region's water status in terms of per capita availability. A fresh water availability of more than 1,700 m^3/capita year is not a cause for concern. Availability ranging between 1,000 and 1,700 m^3/capita year, on the other hand, is defined as water stress conditions. A figure below 1,000 m^3/capita year implies water scarcity, while regions with values under 500 m^3/capita year face severe water scarcity (Falkenmark 1989). In line with the data in Table 12.2, the RM Santiago currently oscillates between water stress and water scarcity. If the statistical prognosis on population growth materializes (INE, CEPAL, no date) by 2050, the situation will deteriorate to oscillation between water scarcity and severe water scarcity levels.

In accordance with these findings, the ground-water table at 'Estación Consejo Nacional de Menores' in the central sector of Santiago fell from 14 to 26 m below the surface in the period from 1971 to 2001, (DGA/MOP 2005). The persistent

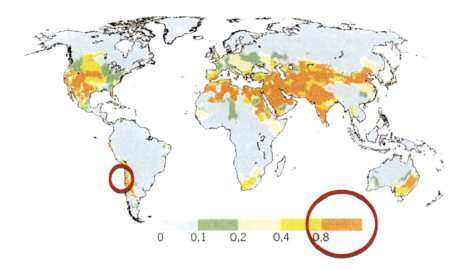

Fig. 12.7 Water stress in river catchments in the year 2000 calculated as the ratio between water extraction and renewable water resources (precipitation minus evapotranspiration 1961–1990) (Source: The authors, based on Döll 2008)

decrease in the ground-water table over decades is a clear indicator that water use exceeds the rate of water renewal. These facts render the situation unsustainable when Herman Daly's above-mentioned criteria are taken into account.

Following a study by Aguas Andinas in 2003, about 94 mill. m^3 (about <20%) of ground water is used by the water suppliers Aguas Andinas, Aguas Cordilleras and Aguas Manquehue, whereas 555 mill. m^3 (about >80%) is surface water from rivers, lagoons and dams (Aguas Andinas 2003b). Since all surface water resources have already been allocated (Simon 2009), the sustainable exploitation of the ground-water resources in the Santiago Valley aquifer is crucial to the Santiago water supply.

Munoz et al. (2003) have evaluated ground-water availability for the upper Santiago Valley aquifer. Recharge by non-point infiltration is sourced from precipitation, river water infiltration, and agricultural and urban irrigation – depending on the degree of urbanization. The latter is a significant measure for decisions on the sustainable exploitation of ground water in the urban region of Santiago. In accordance with assumptions about the efficiency of the water distribution network, seepage recharge contributes to 0.08–0.15 $Ls^{-1}km^{-1}$ (Munoz et al. 2003). The scenario calculations by Munoz et al. allowed for estimation of the reliability of ground-water use under different hydrologic conditions and showed a deficit at the level of maximum demand. The data made it likewise possible to calculate the risk of being unable to cover ground-water demand. The results indicate that simultaneous extraction of ground water by all users would satisfy only 67% of the water rights granted in the basin. Such considerations do not include a potential shift in the water regime in the RM Santiago as a result of global climate change.

Water Quality Aspects

As shown in Sect. 3.2, the negative impact of sewage from the megacity Santiago on downstream areas has continuously been reduced and performance is improving. Now that the third sewage treatment plant has gone into operation, status and target values of norms for treated sewage emissions should no longer differ greatly to treatment levels in threshold countries. The most important open questions in this field concern the effect of chlorinating the residual carbon compounds in semi-treated sewage and the possible new synthesis of chlorinated hydrocarbons and the related environmental and (eco-)toxic effects.

Concerning urban water supply in the future, climate projections indicate that the region will experience an increase of mean temperature ranging from 2°C to 4°C and a reduction in seasonal precipitation from 10% to 25% by the end of the twenty-first century (DGF-CONAMA 2007). Under these conditions surface water will become a scarce resource (Melo et al. 2010). The growing demand for water resources must be met by the sustainable use of ground and surface water.

The growth of Santiago has brought with it an increase in the extent of impervious surface, altering the hydrological behaviour of urban areas. The expansion of the city has impacted on the water cycle in the urban region in the form of sealing recharge areas in the surroundings of the River Maipo and the River Mapocho, the two main rivers. Additionally, there is an enhanced risk of diffuse contamination from leaky sewers.

The impact of urban recharge into urban aquifers can be evaluated with environmental isotopes. Iriarte et al. (2006) show high concentrations of nitrates and sulfate for the alluvial aquifer of the Mapocho River Basin – below the old city of Santiago. As shown from evidence gathered by N-/S isotopes from ground water and surface water, leaky pipes in the sewage collection system were partly responsible for the recharge into the urban ground water (Iriarte et al. 2006).

Our own isotope studies demonstrate that the water in the River Maipo and the River Mapocho springs from the high altitude source region. However, the water quality of both rivers is affected by typical anthropogenic indicators in the surroundings of Santiago such as boron, sodium, chloride and sulphate.

The fulfilment of emission norms (Table 11.1, indicator Nos. 4 and 5) lies at about 90% for industrial direct discharge and >97% for the discharge of municipal waste water (with and without treatment!) (SISS 2008a, b, 2009, 2010). Secondary environmental quality norms (indicator 2) are proposed for the whole catchment area. Appropriateness for human recreation (indicator 3) should be assessed specifically in terms of substances and the related water courses. As seen for the arsenic and copper content measured in the Maipo/Mapocho catchment (Fig. 12.6), a general assessment of the water quality cannot be given. With respect to the improvement of sewage treatment in Gran Santiago, water quality can be expected to improve once the water authorities have been strengthened.

12.3.3.2 Water Services

A comparison between the current situation and the sustainability target shows little or no variation in the water service core indicators 7, 8, 9 and 11. This means that drinking water supply and sewage connection rates, and the attendant hygienic standard attained are satisfactory and affordable. According to the ministry of planning (MIDEPLAN) and the national sanitation authority, Superintendencia de Servicios Sanitarios (SISS), the cost of water services should not exceed 3% of the household income (MIDEPLAN n./a.). As a result of economies of scale, the cost in Santiago varies between 1% and 2.5%, the lowest national value (Simon 2009).

Specific water consumption (indicator 10) in the wealthier parts of the city is assessed as unsustainable if it exceeds 230 l per capita and day. This value was put forward after several discussions with local stakeholders and calculated by doubling the according value for Berlin, and is due to irrigation of private gardens and the semi-arid climate of Santiago. Personal observations produced evidence of unsatisfactory irrigation management: the hosing of gardens frequently continues although water has already begun to flow down the streets and footpaths. Furthermore, it tends to take place in the early afternoon when evaporation prevents a high percentage of irrigation water from reaching the roots of the garden plants.

According to the SISS, both water suppliers, Aguas Andinas and SMAPA, have high losses of about 30–40% in the drinking water supply system, which is an indicator for weak infrastructure (indicator 12). These losses are assessed as unsustainable, since a value of 10% is estimated as state of the art (Howard 2002). In 2007, for example, drinking water losses in the pipe network of the state of Baden-Württemberg (Germany) were approximately 12% (Heitzmann 2009). Moreover, water transport distances in this German state are much longer (200 km and more) than those of Gran Santiago. From a systemic perspective, on the other hand, water losses in the supply network recharge ground water resources and become available to other users, occasionally even to Aguas Andinas itself. Hence on a basin-wide scale, water efficiency is probably much higher, albeit difficult to estimate. An approximation of the optimal percentage of distribution losses in the system would require detailed analysis.

At present 72% of domestic sewage in Santiago is treated (El Mercurio 2010). When construction of the large sewer parallel to the Mapocho River ('Mapocho urbano limpio') is complete and the third sewage treatment plant goes into operation, between 80% and 100% of the city's sewage will be treated (El Mercurio 2010). Santiago will then be the Latin American city with the highest degree of treated wastewater. This measure will not only improve the quality of life in the city (offensive odour) but, more importantly, allow for irrigation of 130,000 ha of agricultural land downstream from Santiago, where crops will benefit from the radical reduction in the risk of exposure to fecal micro-organisms resulting from urban waste water (Bernales 2008).

Storm water management is another water service with a performance below par. As economics play a major role in Chile, an economic rather than a technical indicator was chosen to assess storm water management. Indicator 6 suggests fixing tariff rates for storm water collection according to the percentage of sealed real estate. This would provide the necessary financial resources for construction and maintenance of a storm water treatment system and at the same time motivation to avoid or minimize soil surface sealing.

12.4 Water Sector Challenges and Mitigation Alternatives

After several workshops and meetings with local stakeholders associated with the water sector, we identified several challenges facing Santiago and the Metropolitan Region. Beyond appropriate coverage of drinking water and sewage services at the basin scale, more local issues require special attention, since they typically affect economically disadvantaged subsets of the population.

12.4.1 Storm Water Management

Subsequent to the privatization of the water sector in the 1990s, various authorities (e.g., DOH, municipalities, SERVIU) were involved in the management of storm water. With the exception of some inner-urban municipalities that operate a combined sewage system (e.g., Santiago central and Providencia), waste water sewers exist only in other parts of the city. Storm water is discharged by streets and often causes flooding in the lower – mostly poorer – areas of Santiago. Government agencies and the private water/sewage utility Aguas Andinas pursue different strategies to improve the situation. New storm water sewers are in the planning phase or under construction by DOH, and include centralized approaches to storm water management, e.g., temporary flooding of parks. The Aguas Andinas group, the only private water service supplier expected to run the storm water treatment system, has plans for an alternative. The idea is to transport storm water with the sewage, thereby converting the system into a combined sewage system that would serve the entire city.

Decentralized storm water management facilities (e.g., infiltration systems, green roofs, retention systems) are nonexistent. If these technologies were to be legally sanctioned as flood prevention installations (with the relevant financial support), they would spread more rapidly.

12.4.2 Gravel Mining and Sediment Management at the Maipo River Basin

Gravel mining is a common practice in Chile, based on the demand for this type of material for construction. The driving force behind the extraction of aggregates from rivers are: (1) easy access to the source, (2) high quality material, and (3) a wide range of material sizes, reducing the cost of processing the material (Kondolf 2002). In central Chile, sources of fluvial sediments are concentrated in the Andean mountain region, from where they are transported downstream to the river basins.

At the same time, many rivers are subject to extraction of large amounts of water for a number of uses, consumptive and non-consumptive, such as irrigation, drinking water supplies and hydropower generation. As a result of the close relationship between the sediment transport capacity and the flow discharge of the rivers, it is not uncommon for the hydrodynamic behaviour and morphologic development of the latter to be controlled by external activities. These are planned without taking into account the inherent dynamics of these systems in terms of sediment transport processes and the consequent degradation/aggradation of the stream bed.

The issue of gravel extraction is therefore a visible indication of the relationship between the city and its hinterland. Easy access to aggregates is a function of the river, but their extraction is both an operational and an ecological consideration that must be addressed by the city. The economic and ecosystem services provided by the river to the city and its inhabitants becomes seriously endangered when the integrity of its course and riparian habitats is compromised.

12.4.3 Stream Protection

Although significant investment have been made to prevent raw sewage from entering the water courses (e.g., Mapocho Urbano Limpio, sewage treatment plants), many reaches of the Mapocho and Maipo Rivers are continually at risk due to unregulated solid and liquid waste discharges. In the case of solid waste, the stumbling block comes in the form of illegal garbage collectors and informal recyclers who dump unwanted residue along the river banks. The Regional Public Health representative has ranked this situation a major public health concern (personal communication), since contact waters leach contaminants from the garbage and carry them along the river and through the irrigation channel network. In terms of liquid sewage, unregulated discharges originate mostly from industrial activities in the northern part of the watershed. Mellado (2008) and Cox (2007) identified the drawbacks affecting the Batuco lagoon, a wetland located 30 km northwest of downtown Santiago. Here the difficulty is the lack of SISS capacity to enforce environmental regulations. Indeed, most data collected by SISS on liquid residue discharge is self-reported by the dischargers themselves, and the number of personnel available to SISS to enforce compliance is limited.

12.4.4 Water Supply in Semi-rural Areas

According to the regulatory framework, the water utility (in this case Aguas Andinas) retains the right to decide on the areas to be incorporated into its concession zone. Because this decision is based on economics i.e., the balance of connection costs against water sale revenues, less densely populated areas adjacent to the urban zone are often not connected to the utility network. This is true for many localities in the northern and southern periphery of Santiago. They satisfy their water needs with self-managed, cooperative supply systems (Sistemas de Agua Potable Rural, APR, or rural drinking water systems) that generally rely for water on a single ground-water well. Although APRs receive technical support from the state via a special unit of DOH, regional public representatives expressed concern about the reliability of these systems, claiming that as a rule the quality of the service was inferior to what could be achieved if the localities were connected to the main utility network. All of this poses a risk to sustainability, since the combination of population growth and ground-water depletion could render a number of APRs unviable, particularly under drought conditions.

12.4.5 Mitigation Alternatives

In the following we present actions that could alleviate some of the issues identified above. These alternatives are presented here as first-order technical options and not as definitive answers to the water problems identified in Santiago de Chile. An intervention of any kind would naturally require the appropriate technical, economic and environmental assessment before being adopted.

12.4.5.1 Reducing the Flow Speed and Keeping Water Longer in the Catchment

As a result of the vast difference in altitude between the Andes Mountains and the Pacific Ocean, more than 5,000 m for a distance of approx. 100 km, the flow velocity of streams and rivers is high. This may have been an advantage in the past because the hydraulic energy transported water pollution rapidly to the sea. Under today's increasingly unbalanced water situation, this becomes a disadvantage because a surfeit of water is 'lost' to the sea in wet periods.

- The classic option for reducing flow velocity is the construction of dams. This has already been carried out for the upper Maipo catchment area and the Yeso River (Embalse de Yeso). Another option is to build several small dams in the Central Valley where the Maipo River flows into the south of Santiago. The upstream dams could be used to catch sediments, while those further downstream could contribute to water-related recreation close to the city. Both types could help to recharge the ground-water body beneath Santiago by infiltration of surface water.
- With reforestation more water could be stored in the soils of the catchment's mountain area. Since the transpiration rate of trees is much higher than that of agricultural plants, water losses to the atmosphere might increase, making a thorough assessment of reforestation strategies advisable before realizing this option.
- There are other ways of artificially recharging ground-water resources apart from the infiltration of surface water, e.g., the infiltration of storm water in the built-up area of the RM Santiago. Indeed, this connects to option *a* above (dams), as small dams located along the Mapocho River bed could be used to infiltrate storm water to the aquifers during wet periods.

12.4.5.2 Making new Water Resources Available for Anthropogenic Purposes

1. Rainwater harvesting in built-up areas could make use of water quantities that would otherwise gush away to no avail, or worse, cause flooding. Rain can be collected decentrally at the individual real estate level or more centrally along the watercourses and in parks. Since rainfall events in the RM Santiago are concentrated over a few days in winter, the cost effectiveness of the necessary infrastructure (construction and maintenance) is low. It is therefore recommended that the advantage of preventing flood damage be included in the calculation. This alternative also relates to water service issues, as indicated below.

2. The southern part of Chile is rich in water. Technically, as demonstrated in many other countries (e.g., Germany, Libya, Australia), the transport of water in long water pipes or canals is not rocket science. Supplying water to the RM Santiago via long-distance water transport systems is more likely to be an economic or legal rather than a technical issue, given the existing water rights to the prospective source rivers. On the other hand, long-distance water supplies would increase the dependency of the city on distant areas, and carry the risk of externalizing water-related problems to other regions and neglecting the city's local resources.
3. Domestic and industrial water usage does not always imply consumption. Although polluted or heated after use, it may be still available. Household sewage contains carbon compounds, nitrogen and phosphorous, and can be seen as a precious resource for agriculture. Thus more attention should be given to treating urban sewage in such a way that the water and its nutrients can be diverted to agriculture without the risk of microbiologic or toxic poisoning of agricultural soils and products.
4. Water transfers are possible within the current regulatory framework, i.e., in the form of water right transactions carried out on the private market. It is assumed that water transfers from the irrigation to the municipal sector will occur in the future. The price of water in these transactions may affect water tariffs, although to what extent has not yet been calculated.

12.4.5.3 Integrated Institutional Arrangement for Storm Water Management

The appropriate management of storm water could impact on both water resources and water services. From the services point of view, an integrated approach has the potential to reduce the negative outcome of extreme events, e.g., in terms of infrastructure damages, loss of productive time, inconvenience to inhabitants. A number of institutions currently have jurisdiction over this topic, usually through the planning and funding of drainage infrastructure (Municipalities, Housing Ministry-Serviu, Public Works Ministry-DOH). From the resources standpoint, integrated storm water management typically entails a decentralized approach, which in turn includes enhanced infiltration of storm water into the soil. The latter has the double effect of decreasing peak flows (thus reducing the need for large infrastructure) and enhancing the recharge of subsurface sources.

12.4.5.4 Increased Enforcement Capabilities for Water-Related Institutions (SISS, Seremi Salud, DOH)

Many of the quality problems affecting streams in the Metropolitan Region derive from unregulated activities (liquid and solid discharges, excessive gravel mining). Institutional capacities to enforce regulations are limited; increased accountability through enhanced enforcement has the potential to mitigate these problems. A similar effect can be achieved by self-regulation. However, this requires a cultural shift that may take years, intensive environmental education notwithstanding.

12.4.5.5 Incorporation of Stream Uses to Evaluation Processes

Due to the current institutional framework, water-related decision-making is confined to water right holders through a free-market system. Hence, only economic uses are regarded in this process (in-stream flow requirements do not apply in the Maipo-Mapocho Rivers, since all water rights were assigned prior to the passing of the Water Codex modification in 2005). In the Maipo-Mapocho basin, water right holders are mostly farmers or water utility and hydropower companies. As a result water management tends to represent the interests of these stakeholders. Opening the decision-making process to civic participation and giving more weight to non-economic water uses (such as environmental conservation and restoration), increases the potential to enhance environmental water services and make them available to a larger share of the population in the city and its hinterland.

12.5 Conclusions

In this chapter we presented opportunities for and risks to the sustainability of water resources and water services in Santiago de Chile and its surrounding watershed. We adopted a sustainability indicator/distance-to-target conceptual framework to categorize and rank the most relevant aspects of the topic, based on interviews, workshops and meetings with local public and private stakeholders.

From a comprehensive literature review, field measurements and modelling, we characterized the current status of Santiago de Chile in terms of the sustainability indicators. As far as water resources are concerned, attention should be given to the most pressing sustainability deficit: the quantitative relationship between the available fresh water offer and water demand. With respect to anthropogenic water demand, the per capita availability today of 500–1,200 m^3/capita year marks oscillation between water stress and water scarcity. Sheltering eight million inhabitants in the future, the Metropolitan Region of Santiago may find itself in a more dire situation if added resources are not allocated to municipal and industrial supply. Then, by providing a reduced water quantity of a mere 350–800 m^3 per capita and year, the RM Santiago would proceed from being a region with water scarcity to one with severe water scarcity.

Today more than 80% of available water resources have been extracted. In other words, water reserves are at a dangerously low level. Population growth and mounting economic activity speak for an unbalance between the natural water supply and anthropogenic demand. Even without considering the impact of global climate change, the situation may well deteriorate radically in the future.

Table 12.5 Options for measures to overcome water stress and water scarcity in the RM Santiago

Reduce flow speed – keep water longer in the catchment	
– Dams	
– Afforestation	
– Artificial ground-water recharge	
Make new water resources available	
– Rainwater harvesting	
– Long-distance water transport	
– Use more virtual water	
– Regard waste water as a resource	
Improve water efficiency	
– Agriculture: more crop per drop	
– Industry: from throughput to water cycling	
– Households: use water in cascades: drinking – > toilet flushing – > irrigation	
Increase water sufficiency	
– Propagate 150 l per person	
Capacity development	

Source: The authors

The findings of this water sector analysis have led to recommendation of the following options (Table 12.5).

From the point of view of water services, aggregated coverage of the drinking water supply and sewage collection is almost complete. Sewage treatment has increased dramatically in the past 10 years due to investments made in the context of water utility privatization. In this sense, Santiago boasts one of the best situations in the region. Yet problems exist on a local scale. These are related primarily to service deficiencies in semi-urban areas, which are located in the immediate vicinity of Santiago but not connected to the chief distribution networks. Likewise, water quality deficits and stream degradation pose a risk to sustainability at some reaches of the Maipo and Mapocho river networks.

References

Abarca, D. (2008). Adaptation and implementation of a mathematical numerical model for the analysis of riverbed evolution. Undergraduate Thesis, Department of Civil Engineering, Universidad de Chile.
Aguas Andinas, S. A. (2003a). *Plan de Desarrollo, Rinconada de Maipú*. Santiago de Chile.
Aguas Andinas, S. A. (2003b). *Memoria Annual Aguas Andinas*. Santiago de Chile.
Aguas Andinas, S. A. (2005a). *Plan de Desarrollo Sistema Gran Santiago*. Santiago de Chile.
Aguas Andinas, S. A. (2005b). *Plan de Desarrollo, Ampliación de Concesión Vizcachas III*. Santiago de Chile.
Aguas Andinas, S. A. (2005c). *Actualizacion Plan de desarrollo. Sistema Gran Santiago*. Santiago de Chile
Aguas Andinas, S. A. (2006a). *Plan de Desarrollo – Ampliación de Concesión de Servicios Sanitarios Las Casas de Quilicura III*. Santiago de Chile.

Aguas Andinas, S. A. (2006b). *Plan de Desarrollo – Ampliación de Concesión El Trebol, Comuna de Padre Hurtado*. Santiago de Chile

Aguas Cordillera, S. A. (2005). *Actualización del Plan de Desarrollo – Aguas Cordillera*, Santiago de Chile.

Aguas Los Dominicos, S.A. (2005). *Actualización Plan de Desarrollo Sistema Agua Los Dominicos 2005–2009.* Santiago de Chile.

Aguas Santiago Poniente (2006). *Actualización Plan de Desarrollo Loteo Industrial Lo Prado.* Santiago de Chile.

Bartosch, A. (2007). *Die Wasserversorgung in einer Metropolregion in Lateinamerika.* Das Beispiel Santiago de Chile. Diploma thesis, Jena.

Bernales, C. (2008). Proyecto Mapocho Urbano Limpio: En vías del 100% de las aguas saneadas en 2011. *Technología Construccio, Proyecto Mapocho Urbano Limpio: En vías del 100% de las aguas saneadas en 2011*, 35(4), 4–8.

Corporación de Fomento de la Producción – CORFO (2010). *Corporacion de Fomento de la Produccion-Matriz*. Santiago de Chile.

Cox Oettinger, C. (2007). *Metodología de diseño de una red de monitoreo de recursos hídricos para humedales: aplicación en la Laguna de Batuco.* Undergraduate Thesis, Universidad de Chile. Santiago de Chile.

Daly, H. (1996). *Beyond growth. The economics of sustainable development.* Boston: Beacon Press.

Departamento de Geofísica, Universidad de Chile – DGF (2007). *Estudio de la variabilidad climática en Chile para el siglo XXI.* Technical Report, prepared for Comisión Nacional de Medioambiente – CONAMA.

Dirección General de Aguas – DGA (2003). *Evaluacion de los Recursos Hidricos superficiales en la Cuenca del Rio Maipo.* Santiago de Chile.

Dirección General de Aguas – DGA (2007a). *Bases para la Formulacion de un plan director para la gestion de los recursos hidricos.* Cuenca del Rio Maipo. Etapa I. Diagnostico. Informe Final. Santiago de Chile.

Dirección General de Aguas – DGA (2007b). *Estimaciones de Demanda de Agua y Proyecciones Futuras. Zona II. Regiones V a XII y Región Metropolitana.* DGA Publikation N°123, Santiago de Chile.

Dirección General de Aguas/Ministerio de Obras Publicas – DGA/MOP (Ed.), (2005). *Informe Tecnico N.166 de aprovechamiento comun de declaracion área de restricción Santiago de Chile.*

Döll, P. (2008). Wasser weltweit – Wie groß sind die globalen Süßwasserressourcen, und wie nutzt sie der Mensch? *Forschung Frankfurt, 3*(2008), 54–59.

Durán, G. (2009). Water governance and new urban poverty in Santiago de Chile. In: *Risk Habitat Megacity – A Helmholtz Research Initiative*, Ph.D. progress reports I (pp. 111–151).

El Mercurio (2010). Río Mapocho Estará 100% Libre de Aguas Servidas en 2012. http://www.estrategia.cl/detalle_noticia.php?cod=28645. Accessed 16 Oct 2010

Falkenmark, M. (1989). The massive water scarcity now threatening Africa – why isn't it being addressed? *Ambio, 18,* 112–118.

Frikken, P., Cook, D., et al. (2005). Mineralogical and isotopic zonation in the Sur-Sur Tourmaline Breccia, Río Blanco-Los Bronces Cu-Mo Deposit, Chile: Implications for ore genesis. *Economic Geology, 100,* 935–961.

González, J. (2006). Reservoir sedimentation considering the effect of turbidity currents. Development of a mathematical numerical model. Master of science thesis, Department of Civil Engineering, Universidad de Chile.

Heitzmann, D. (2009). Wassergewinnung für die öffentliche Trinkwasserversorgung in Baden-Württemberg. Statistisches Monatsheft Baden-Württemberg, 9, 31–35. http://statistik-portal.de/Veroeffentl/Monatshefte/PDF/Beitrag09_09_06.pdf. Accessed 13 May 2010.

Howard, K. W. F. (2002). Urban groundwater issues – An introduction. In K. W. F. Howard & R. G. Israfilov (Eds.), *Current problems of hydrogeology in urban areas, urban agglomerates and industrial centers* (NATO Science Series IV. Earth and Environmental Sciences, Vol. 8, pp. 1–15). New York: Springer.

12 Risks and Opportunities for Sustainable Management of Water Resources

IASA – Ingeneria Alamana. (2005). *Reposicion Planta de Tratamiento de Aguas Servidas Los Pellines – Comuna de Llanquihue.* Santiago de Chile: IASA.
IASA – Ingeneria Alamana. (2007). *Propuesta metodológica para el establecimiento de indicadores de calidad de servicio basados en Paneles de Olores.* Santiago de Chile: IASA.
Instituto Nacional de Estadisticas – INE (Ed.), (2002). *Estadísticas del Medio Ambiente 1996–2000.* Santiago de Chile.
Instituto Nacional de Estadisticas/Cepal – INE/Cepal (Eds.), (n/a.). Chile: *Proyecciones y Estimaciones de Poblacion. Total pais periodo de informacion: 1950–2050.* Santiago de Chile.
Inter-American Development Bank (2007). *Análisis Ambiental Programa de Saneamiento Rural (CH-L1025).* Santiago, April 2007.
Iriarte, S., Aravena, R., & Rudolph, D. (2006). The use of multiple isotope tracers to evaluate the impact of urban recharge in an alluvial aquifer located underneath the city of Santiago, Chile. *Geophysical Research Abstracts, 8*, 10264.
Jouravlev, A., & Valenzuela, S. (2007). *Servicios urbanos de agua potable y alcantarillado en Chile: factores determinantes del desempeno.* Santiago de Chile.
Kondolf, G. M. (2002). Channel response to increased and decreased bedload supply from land-use change: Contrasts between two catchments. *Geomorphology, 45*(1–2), 35–51.
Kopfmüller, J., & Lehn, H. (2006). Nachhaltige Entwicklung in Megacities: Die HGF-Forschungsinitiative "Risk Habitat Megacity". In J. Kopfmüller (Ed.), *Ein Konzept auf dem Prüfstand – Das integrative Nachhaltigkeitskonzept in der Forschungspraxis* (pp. 269–282). Berlin: Edition Sigma.
Kopfmüller, et al., (Ed.) (2001). *Nachhaltige Entwicklung integrativ betrachtet. Konstitutive Elemente, Regeln, Indikatoren.* Berlin.
Lehde, M. (2010). *Klärschlammbilanzierung für Santiago de Chile unter Berücksichtigung der sich verändernden Abwassersituation.* Aachen.
Lehn, H., & Parodi, O. (2009). Wasser – elementare und strategische Ressource des 21. Jahrhunderts. I. Eine Bestandsaufnahme. *Umweltwiss Schadst Forsch, 21*, 272–281.
Mellado Tigre, CA. (2008). *Caracterización hídrica y gestión ambiental del humedal de Batuco.* Masters Thesis, Universidad de Chile, Santiago, Chile.
Melo, O., Vargas, X., Vicuna, S., Meza, F., & McPhee, J. (2010). Climate change economic impacts on supply of water for the m & i sector in the Metropolitan Region of Chile. 2010 Watershed Management Conference: "Innovations in Watershed Management Under Land Use and Climate Change", August 23–27.
Ministerio de Planificación Nacional y Política Económica – MIDEPLAN (n./a.). *El subsidio al pago del consume de agua potable y servicios de alcantarillado de aguas servidas.* Santiago de Chile.
Munoz, J. F., Fernandez, B., & Escauriaza, C. (2003). Evaluation of groundwater availability and sustainable extraction rate for the Upper Santiago Valley Aquifer, Chile. *Hydrogeology Journal, 11*, 687–700.
Norma chilena Oficial – NCh409/1 (2005).www.dinta.cl/docs/NCh409_1_2005.pdf. Accessed 16 May.2011.
Simon, L. (2009). *Die "Nachhaltigkeitsperformance" der Wasser- und Sanitärversorgung in Santiago de Chile. Eine politisch-geographische Untersuchung.* Diploma thesis, Münster.
Superintendencia de los Servicios Sanitarios – SISS (Ed.). (2010). *Informe de gestión 2009.* Santiago de Chile.
Spürk, S. (2010). *Die Kupfergewinnung in "Los Bronces" und ihre Einwirkungen auf die Umwelt mit besonderer Berücksichtigung des Umweltkompartimentes Wasser.* Aachen: Studienarbeit.
Superintendencia de los Servicios Sanitarios – SISS (Ed.), (2008a). *Informe de gestión 2007.* Santiago de Chile.
Superintendencia de los Servicios Sanitarios – SISS (Ed.), (2008b). www.siss.cl/articles-7505_calidad_tratamiento_agua.xls.

Superintendencia de los Servicios Sanitarios – SISS (Ed.), (2009). *Informe de gestión 2008*. Santiago de Chile.

United Nations World Commission for Environment and Development – WCED. (1987). *Our common future*. Oxford: Oxford University Press.

Vogdt, J. (2008). Entsorgung von Abwasser und Abfall in Gran Santiago de Chile – Zwischenbericht N° 1 für das Themenfeld "Wasser – Ressourcen und Dienstleistungen." Santiago de Chile.

World Health Organization – WHO (2006). Guidelines for drinking water quality. http://www.who.int/water_sanitation_health/dwq/fulltext.pdf.

Chapter 13
Municipal Solid Waste Management in Santiago de Chile: Challenges and Perspectives towards Sustainability

Klaus-Rainer Bräutigam, Tahnee Gonzalez, Marcel Szanto, Helmut Seifert, and Joachim Vogdt

Abstract This chapter gives an overview of Municipal Solid Waste (MSW) management in the Metropolitan Region of Santiago de Chile (RMS). On the basis of this data, MSW management in RMS is assessed with selected sustainability indicators and their corresponding target values. In addition, the most urgent problems are identified. To evaluate options for the reduction of greenhouse gas emissions from landfills, the results of model calculations are presented, taking into account the segregated collection of the organic fraction of MSW in RMS.

Keywords Greenhouse gas emissions • Landfill • Separate collection • Sustainability analysis • Waste management

13.1 Introduction

The production of goods and their use and disposal is invariably associated with the generation of waste. It is well known that a rising standard of living increases the amount of waste produced per capita. The Metropolitan Region of Santiago de Chile (RMS) has experienced a huge growth in population in recent years and a rise in the standard of living. As a result the amount of municipal solid waste (MSW) produced has grown from roughly 1.8 Mio tons in 1995 to about 3.0 Mio tons in 2007. Production per capita increased from 0.8 to 1.2 kg for the same period. To avoid negative impacts on health and the environment, waste needs to be handled

---

K.-R. Bräutigam (✉) • T. Gonzalez • M. Szanto • H. Seifert • J. Vogdt
Karlsruhe Institute of Technology (KIT), Institute for Technology Assessment and Systems Analysis (ITAS), Hermann-von-Helmholtz-Platz 1, 76344 Eggenstein-Leopoldshafen, Germany
e-mail: klaus-rainer.braeutigam@kit.edu

and disposed of appropriately. High population density and a good infrastructure allow for an effective waste collection system (almost 100% of households have access to MSW collection services). Two new sanitary landfills went into operation in 2002 (the first began operations in 1996), allowing for the proper disposal of the total volume of MSW – excluding the fraction that is recycled, primarily by the informal sector.

Decomposition of the organic fraction of MSW produces leachate and the landfill gas (mainly CH_4 and CO_2,) that contributes to global warming, local air pollution, odour and nuisance, and increases the risk of fire, explosions and the potential exposure of workers to toxic emissions. Landfill gas emissions and their associated impacts are mitigated by collecting gas from the landfill to be burned or used as a source of fuel; however, only about 50% of the generated landfill gas can be collected. The capturing and flaring of landfill gas is financed by Clean Development Mechanism (CDM) projects, and in some instances constitutes an extra income for the companies operating the landfills.

Bearing in mind the negative impacts of disposing untreated waste in landfills, the amount of waste to be disposed – particularly the organic fraction – should be reduced as much as possible.

The west European solid waste management system is highly advanced compared with that of Chile. Numerous improvements have been made in recent years to reduce the amount of waste production, e.g., with waste reduction incentives, separate collections and the recycling of different waste streams. The amount of waste finally disposed of in Germany, the European country with one of the highest recovery rates (in terms of material and energy), has dropped considerably since 2005, when the final disposal of untreated waste has banned. Instead, mechanical, biological and/or thermal treatment measures are carried out prior to final disposal. Comparisons with other European waste management systems offer a glimpse at possible perspectives for the development of MSW management in Santiago de Chile.

The following research questions arise and are the focus of the next paragraphs:

1. How sustainable is the current MSW situation in RMS?
2. What is the role of the informal sector in the current waste management system?
3. What technical options for waste management could improve the sustainability of the system?

The basis for the evaluation of waste management in RMS is a detailed analysis of the current system as described in Sect. 13.2. The methodology used to identify sustainability deficits and risks is discussed in Sect. 13.3. Section 13.4 presents sustainability indicators, deficits and risks with regard to the current waste management system. Section 13.5 gives an example of how to improve the system by reducing landfill gas emissions from sanitary landfills. Finally, key findings and their consequences are summarized in Sect. 13.6.

13.2 Municipal Solid Waste Management in Santiago de Chile: The Current Situation

13.2.1 Waste Generation

According to official statistics (Gobierno de Chile 2009) the total amount of waste (all types) produced in RMS was estimated at about 5.75 Mio tons for the year 2006, while municipal solid waste (MSW) accounted for roughly 2.65 Mio tons in the same year (see Table 13.1).

Rapid economic growth in Chile over the last 10 years brought about several benefits for most of its inhabitants, including a rise in income levels across all population groups. On the other hand, the rising Gross Domestic Product (GDP) shows a strong correlation with the amount of MSW produced per capita (Blight and Mbande 1996; Vehlow 2008), which is likewise apparent for the specific production of MSW in the RMS. The MSW per capita generated in Santiago de Chile between 1995 and 2007 is depicted in Fig. 13.1 in terms of the GDP of the Region for the same period. It clearly indicates that economic development had an impact on the amount of MSW produced. The first dot in Fig. 13.1 below

Table 13.1 Quantity of different types of waste in RMS in 2006

Type of waste	Total amount (Mg/a)	Percentage (%)
Municipal solid waste	2,645,966	46
Industrial waste	920,336	16
Construction waste	2,128,277	37
Hospital waste	57,521	1
Total	5,752,100	100

Source: Gobierno de Chile 2009

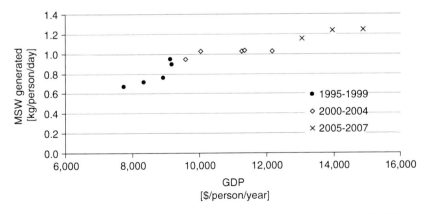

Fig. 13.1 Correlation between MSW deposited in landfills and economic growth in RMS (Source: The authors, based on Szantó 2006; Chilean Central Bank 2003)

corresponds to the year 1995, when the GDP in Santiago amounted to approximately 7,700 dollars per person and year, and the average amount of waste generated to 0.67 kg per person and day. When the GDP rose to approximately 14,800 dollars per person in 2007, the amount of waste increased to 1.24 kg. In short, an annual increase of 7.7% in the GDP of Santiago de Chile led to an increase of 7.1% in the amount of MSW produced (see Fig. 13.1).

The amount of waste produced per capita has increased in recent years from about 0.8 kg/(person*day) in 1995 to about 1.2 kg/(person*day) in 2007. In addition, rising income levels correlate with a change in consumption patterns, which in turn influences the composition of MSW. The most noticeable changes occurred in the organic fraction, which declined from 68% in 1990 to 50% in 2007 (Szantó 2006). Hence the share of other fractions increased, e.g., paper and cardboard from 15% to 18% and plastics from 6% to 10%. Parallel to an increase in the number of people living in the city, the total amount of MSW produced in RMS rose from about 1.8 Mio tons in 1995 to about 2.5 Mill. tons in 2007 (see Table 13.2). These – steadily growing – quantities of MSW must be managed properly if negative impacts on health and the environment are to be avoided. This presents a challenge to stakeholders involved in waste management, whose task it is to develop appropriate strategies for waste reduction and waste treatment within the existing legal, social and financial framework.

13.2.2 Waste Processing, Treatment and Final Disposal

Figure 13.2 shows the current waste mass flow in RMS. As mentioned above the total MSW generation amounted to almost three million tons in 2007, most of

Table 13.2 Quantities of municipal solid waste in RMS in the years 1995–2007 and calculated recycling rate

Year	Landfill (Mg/a)	Recycling (Mg/a)	Total (Mg/a)	Recycling rate (%)
1995	1,789,599	2,891	1,792,490	0.16
1996	1,870,868	5,020	1,875,888	0.27
1997	1,952,137	18,466	1,970,603	0.94
1998	2,156,446	73,787	2,230,233	3.31
1999	2,270,311	94,048	2,364,359	3.98
2000	2,405,433	139,928	2,545,361	5.50
2001	2,331,173	195,973	2,527,146	7.75
2002	2,273,897	205,932	2,479,829	8.30
2003	2,336,474	229,368	2,565,842	8.94
2004	2,373,228	234,319	2,607,547	8.99
2005	2,449,831	347,809	2,797,640	12.43
2006	2,502,000	–	–	–
2007	2,524,000	409,000	2,933,000	13.94

Source: Szantó 2006; Health Ministry Seremi Salud 2008

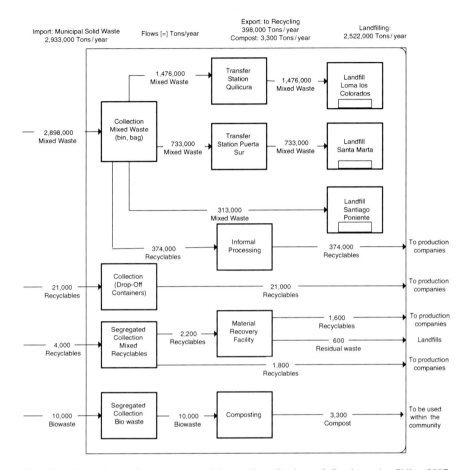

Fig. 13.2 Mass flow of waste in the Metropolitan Region of Santiago de Chile, 2007. Stock flows, emissions and residue fluxes have been omitted to simplify the diagram (Source: The authors)

which (2.8 Mio tons) was deposited in bags or small containers on the streets of Santiago de Chile. This is represented by "collection of mixed waste" in Fig. 13.2. From this point there are four flow alternatives for mixed waste. If the waste is collected by publicly organized collection systems, it is disposed of at one of the three landfills in RMS. About 1.5 million tons were disposed of in Lomas Los Colorados via the transfer station Quilicura, about 0.7 million tons in Santa Marta via the transfer station Puerta Sur and about 0.3 million tons were directly disposed of in Santiago Poniente. The fourth flow alternative is informal collection and separation by waste workers, prior to the formal collection. Roughly 375,000 t of waste, largely paper, cardboard and metals, were separated and collected by the informal sector in 2007. A recycling rate of almost 14% was achieved purely as a result of the high contribution from the informal sector.

As evident in Fig. 13.2, there are three main alternatives for publicly organized recycling: drop-off systems, and the segregated collection of inorganics and biowaste. Their contribution to recycling, however, is negligible. Drop-off systems amounted in 2007 to about 21,000 t (less than 1%). The segregated collection of biowaste amounts to about 10,000 t per year, from which roughly 3,300 t of compost is produced, while the segregated collection of recyclable materials amounts to approximately 4,000 t per year, half of which can be sold as secondary raw materials.

MSW quantities for the years 1995–2007 are shown in Table 13.2. The amounts recycled and those disposed of are shown separately. The recycling rate is also indicated (Szantó 2006). The latter was almost zero in 1995, over 5% in 2000, reaching almost 14% in 2007. It should be noted, however, that recycling data is merely a rough estimation, since most material for recycling is collected by the informal sector. The data is based on questionnaires distributed to middlemen and industries that buy the materials.

Recycling activities contribute to a reduction in the total quantity of waste to be disposed of in sanitary landfills. In addition recyclable materials are used for the production of several goods, which contributes to saving renewable and non-renewable primary materials. Hence a high recycling rate is desirable. In order to improve these rates, it is vital to analyse the entire recycling system of RMS, including the informal sector. The latter is comprised of a vulnerable group of people confronted with numerous problems, including inadequate or lack of labour legislation, lack of social security, and limited bargaining power to increase the price of materials they sell (Flodman Becker 2004).

Formal recycling activities have been introduced in some municipalities and include recycling programmes to cover composting, segregated collection and drop-off systems for certain residues. Activities include education and campaigns to enhance awareness of the environmental benefits of recycling practices among the residents of these communities.

The schemes differ for the most part on how waste is collected and divided into 'differentiated collection' and 'drop-off systems'. Participation in differentiated collection is voluntary. Citizens are encouraged to separate their recyclable materials, which are then collected from their homes by collection trucks. Differentiated collection systems exist in La Pintana (biowaste), in Ñuñoa, La Florida (inorganic waste), and María Pinto (biowaste and inorganic waste).

The drop-off systems are organized by production companies and charity foundations. The scheme consists of containers located in public places, where citizens deliver their recyclable materials. One example is the Clean Point in the municipality of Vitacura, where containers have been set up in an area of 500 m for the recycling of paper, cardboard, plastics, ferrous and non-ferrous metals, yard waste, batteries and old medicines.

For the year 2007, collection from drop-off containers, a less expensive option than separate collection systems, amounted to about 21,000 t of recyclable materials, whereas the differentiated collection of mixed recyclables produced roughly 4,000 t. Approximately 10,000 t of biowaste was collected separately. This data is based on literature studies, surveys in RMS communes with recycling programmes, interviews with people from the informal sector and own estimations (CONAMA 2005a, b).

The amount of recycling material collected by containers associated with charity foundations is about five times higher than the recycling material emanating from differentiated collection. This could be attributed to the following:

- Recycling containers have been set up throughout RMS, whereas differentiated collection takes place on a small scale in fewer than ten municipalities.
- People prefer to collaborate in recycling for social rather than environmental reasons, possibly because of recycling advertisements associated with charity institutions.

The informal sector is a key component of the economy and the labour market in many low and middle-income countries, and instrumental in the creation of employment, production and income (Hussmanns and Farhad 1999). Collecting, sorting, trading and recycling disposed materials provides income to thousands of people. As a rule, these people work parallel to the formal waste management system. They work on their own, however, and are not contracted by the municipalities or any other entities associated with the waste area; they do not pay tax and are excluded from social welfare and insurance schemes (Wilson et al. 2006).

Inofficial estimates indicate that between 4,000 and 15,000 people work as primary collectors in Santiago de Chile (Astorga 2008; MNRCH 2009). They collect valuable materials from the streets of residential and commercial zones, use tricycles as transport and working tools. They separate and classify the materials, improving their monetary value in the process, and sell them to middlemen who deliver them to production companies as secondary raw materials.

Figure 13.3 shows the total recycling rate for different materials, as well as the contribution from formal and informal systems. The contribution from the formal sector (drop off) is almost negligible. The exception is glass, since quite a number of containers for glass collection have been in operation in RMS for more than 10 years.

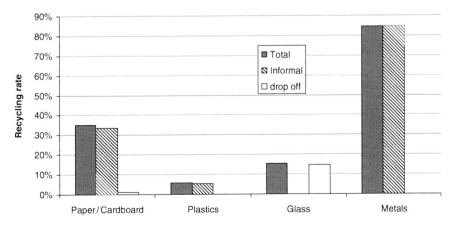

Fig. 13.3 Recycling rates for different materials – formal and informal contributions, 2007 (Source: The authors)

13.2.2.1 Final Disposal of MSW

Waste management in RMS is primarily based on final disposal. RMS utilizes three landfills: one began operations in 1996, the other two in 2002. All of them are equipped with a bottom liner and a collection system for leachate. When the organic fraction of the waste decomposes, it produces landfill gas. Some of the gas is captured and flared (in some cases, the energetic use of landfill gas is planned). Capturing and flaring landfill gas is financed by Clean Development Mechanisms (CDM) projects and constitutes extra income for the operating companies of the three landfills in Santiago de Chile.

Due to the steady increase in waste and the limited capacity of each landfill, gaining information on the actual quantity of waste hitherto disposed of in the landfills and on their remaining lifespans is vital. This helps to plan for a new landfill or for expansion of an existing landfill in due course. The total quantity of MSW disposed of is calculated by weighing the waste at the gate to the landfill. This overall volume of MSW also includes (unknown) quantities of waste from markets, restaurants, small businesses and street cleaning.

Lomas Los Colorados landfill is both the largest and the oldest of the three landfills in operation and is situated in Til Til county, 60 km north of Santiago. It began to receive waste from the transfer station in Quilicura in 1996, initially by truck and since 2003, by train (in both cases using silos with external compaction). The current landfill input rate is approximately 150,000 t of waste per month. Although the current contract ends in 2011, it will be renewed automatically for 16 years if none of the parties cancel the contract earlier. The landfill, operated by Kiasa-Demarco S.A (KDM), an international company specialized in solid waste management, is anticipated to reach capacity around 2045. Landfill gas, for the most part carbon dioxide (CO_2) and methane (CH_4), is collected and flared (CDM-Executive 2006). According to KDM, energetic utilization of landfill gas began at the end of 2009, generating electricity with a capacity of 2 MW (KDM 2009).

Santa Marta landfill is the second largest of the three disposal sites in operation. It was opened in 2002 and will continue landfilling until at least 2022, when the contract ends (although landfill capacity allows for extended operation time). The current waste disposal rate is approximately 80,000 t per month. The total amount of waste disposed of at Santa Marta Landfill by the end of 2008 was about 4.7 million tons. Roughly 80% of the waste disposed is received at the transfer station (direct discharge, no compaction) and subsequently transported to the landfill in special containers. Landfill gas is flared by five flare stacks, each with a capacity of 1,000 m^3/h (CDM-Executive 2006; Wens 2008; CSM 2009).

Santiago Poniente landfill is the smallest of the three landfills. It started operations in October 2002 and shows a current filling rate of approximately 45,000 t per month. The landfill is expected to reach capacity around 2025. Waste is transported directly to Santiago Poniente (without a transfer station). Proactiva-Chile, the operating company, was recently given environmental authorization for the capture of landfill gas, treatment and utilization in the gas system.

13.2.3 Summary of Identified Problems

Analysis of the current state of waste management in RMS indicates that almost all MSW is collected areawide. The main bulk is disposed of at sanitary landfills, and accompanied by landfill gas emission and the production of leachate over a long period of time as a result of decomposition of the organic fraction of MSW. Landfill gas emissions contribute to global warming, while leachates are a source of groundwater contamination. Additional problems are the growing quantity of MSW and low recycling rates. Most recycling is carried out by people from the informal sector who work independently and are heavily disadvantaged both socially and economically.

13.3 Methodology

13.3.1 Risk Assessment

In order to evaluate the current situation of MSW management in RMS under sustainability aspects and to carry out a risk analysis, the *distance to target concept* is applied (see Chaps. 1 and 4).

The following steps are undertaken:

1. Selection of sustainability indicators. Criteria for the selection of these indicators include:

    (a) Validity, i.e., indicators should reflect how the sustainability of waste management is affected by changes in indicator values;
    (b) Data availability;
    (c) Possibilities to define quantitative goals, and
    (d) Indicators should be easy to understand, even by people or working groups not associated with the field.

2. Determination of target values for the indicators. The selection of target values is based on a hierarchical approach that considers:

    (a) Political targets at the national level (these objectives may have been defined by political parties, governmental or non-governmental institutions);
    (b) Scientific, political or societal debates on such targets;
    (c) Practice in comparable countries, and
    (d) Values set and achieved at international level.

3. Identification of sustainability deficits by comparing current indicator values with determined target objective.

4. Identification of risks in two ways:

   (a) Indicators with the greatest distance to target are considered core risks to the attainment of sustainable development, and
   (b) Sustainability deficits are defined as risks if (a) they are irreversible, (b) they show a particular spatial range (e.g., in terms of numbers of people affected) and/or (c) they show a particular temporal range.

13.3.2 Model Calculations to Reduce Greenhouse Gas Emissions from Sanitary Landfills

As will be shown in Sect. 13.4, landfill gas emission is one of the most severe sustainability deficits of waste management in RMS. The volume of biowaste disposed of in landfills and landfill gas collection efficiency influence the amount of landfill gas emitted. Model calculations were performed to estimate the amount of landfill gas emitted and to assess the impact of waste composition and the capturing and flaring of the gas on landfill gas emission. Various shares of segregated biowaste collection were assumed, ranging from 0% to 90%. Furthermore, three landfill gas efficiencies were tested: 10%, 30% and 65% of gas capture.

Landfill gas emission was calculated with the aid of a simple formula, i.e., that of Tabarasan and Rettenberger (Tabasaran and Rettenberger 1987). Section 13.5 describes the formula and the model assumptions in detail.

13.4 Risk Assessment

This section evaluates MSW management in the Metropolitan Region of Santiago de Chile, using the distance-to-target approach.

13.4.1 Selection of Sustainability Indicators for Waste Management

On the basis of an extensive review of the relevant literature on general and region-specific sustainability indicators, a set of more than 100 indicators was established and assigned to appropriate sustainability rules, in accordance with the Helmholtz Integrative Sustainability Concept.[1] Based on the current situation and the problems identified, and taking into consideration the criteria for the selection of relevant indicators given in Sect. 13.3.1, the following indicators were chosen for further evaluation of MSW management in RMS:

---

[1] For more detailed information on this concept, see Kopfmüller et al. (2001).

13.4.1.1 Specific production of MSW [kg/(person*day)]

The amount of waste generated per capita in RMS has almost doubled in recent decades and is essentially a result of Chile's economic development during this period.

The Chilean legal framework is wanting in terms of political measures to prevent waste or strategies to minimize it. Without incentives, it seems logical that specific waste production will increase further.

International trends were analysed in order to define a target value. They indicate that an upper level of specific waste production is reached when a certain degree of economic wealth has been achieved. The upper level corresponds to about 2 kg/(person*day) (IMF 2010; OECD 2006; PAHO 2005). In comparison, the amount of MSW production in Germany, a country with a host of regulations and incentives for waste reduction, was at 1.57 kg/(person*day), lower than might have been expected in light of its economic prosperity (Statistisches 2009). In Berlin the value came to 1.4 kg/(person*day) (Senatsverwaltung für Gesundheit, Umwelt und Verbraucherschutz Berlin 2009). Both figures refer to the year 2007.

For this reason future development in MSW generation in RMS should be decoupled from economic development and, based on international comparisons, a maximum target value of 1.6 kg/(person*day) is suggested for 2030 for RMS.

13.4.1.2 Waste Fraction Recovered as Material or Energy [kg of Waste Recovered (Material or Energy)/Total Waste Arising]

With a recycling rate of 14%, recycling in RMS is well placed compared with other Latin American countries, albeit its recycling potential has not been fully exhausted. A recycling strategy for RMS was developed at the national level by CONAMA (2005c), the goal of which was to achieve a recycling rate of 20% in 2006, a value that has not been achieved so far. In 2009, a new target was set at 25%, to be fulfilled by 2020.

Recovery targets in RMS should not only address recycling of inorganic fractions, but also the biogenic fraction, since it corresponds to almost 50% of the total MSW. The informal sector should also be involved in the solid waste recycling system. Based on political targets already set for recycling and alternative energy uses in Chile, a total recovery of 36%, including composting and energetic recovery, is suggested for 2030.

13.4.1.3 Amount of Mixed Waste Pre-treated to Reduce Organic Carbon Content in Relation to Total Mixed Waste [%]

Currently, MSW is sent to landfills without pre-treatment and there are no political targets for pre-treatment of waste in Chile. In the European Union, on the other hand, the landfill guideline 1993/31/EG instructs member states to gradually reduce the organic fraction of MSW brought to landfills. In Germany, final disposal at landfills is now limited to waste with an organic content of not more than 3%. Additionally, special limit values for organic waste that has undergone mechanical-biological treatment were introduced. Since the deadline, the amount of landfilled municipal waste has dropped to 1%.

In order to reduce the negative impact of waste disposal at landfills, pre-treatment of mixed waste in RMS should increase to 100% of collected mixed waste. Due to the comparatively little experience in Latin America with these issues, however, the suggested target value to be met in RMS by 2030 is 50%.

13.4.1.4 Emission of Greenhouse Gases Caused by Waste Management Activities [kg/(person*year)]

The most critical greenhouse gas related to MSW in Santiago is methane, produced in the course of anaerobic biodegradation of waste disposed at landfills. Own calculations result in an emission value of approximately 143 kg CO_2 equivalents/(person*year) for RMS. For Germany, the corresponding values are 50 kg CO_2 equivalents/(person*year) for 1990 and about 13 kg CO_2 equivalents/(person*year) for 2005 (Butz 2009). The collection and flaring of landfill gas is one method of reducing these emissions and, financed by CDM projects, has been carried out to a certain extent on landfills in RMS. Furthermore, separate collection of biodegradable waste and aerobic or anaerobic treatment of this fraction would reduce landfill gas emissions. The separate collection and composting of 50% of food and garden waste in RMS, for example, would reduce landfill gas emissions (measured as CO_2 equivalents) by about 30% (Bräutigam et al. 2009). Therefore a target value of 71 kg CO_2 equivalents/(person*year) by 2030 for RMS (50% of the actual value) is suggested (see also Sect. 13.5 for model calculations).

13.4.1.5 Income Level of Informal Workers in Relation to Individual Household Income [%]

A further sustainability requirement is that all members of society be in a position to secure their livelihood with work that they have chosen. In the specific case of RMS, the informal primary collectors emerge as a vulnerable group. The correlation between their income level and the average individual household income is used to operationalize the sustainability goal of ensuring an independent livelihood.

The earnings of informal primary collectors are difficult to determine. First of all, there are variations in the price of materials, as well as in the amount collected. Not all waste pickers work the same number of hours or days, or work in the same areas with the same collection capacity. Hence their income is variable.

Several publications have analysed that scavenger cooperatives improved the living conditions of waste pickers and increased their earnings. Moreover, the incorporation of scavengers into formal systems can save the city money and, at the same time, provide those concerned with a steady income. The target value for the income of informal collectors is set at 100% of the average individual household income. It is assumed that this value is sufficient to guarantee basic needs, improve living standards and reduce poverty.

13.4.1.6 Cost of MSW Management in Relation to GDP [%]

Investments in MSW management services should be affordable. Similar levels of investment in countries with disparate economic histories implies that less developed countries invest a much larger fraction of their GDP in MSW management.

The selected indicator to measure the affordability of a particular waste management strategy in a particular region corresponds to the fraction of GDP spent on MSW management.

On the whole, the fraction of GDP spent on MSW management lies between 0.2% and 0.5%. An increase in the current value of 0.22–0.30% in 2030 is therefore suggested for RMS.

Table 13.3 shows a summary of selected sustainability indicators for MSW with their current and target values. These values serve as a basis for the ensuing risk analysis.

13.4.2 Risk Analysis

The objective of the distance-to-target approach is to identify the most urgent concerns and prioritize them for action. As observed in Table 13.3, one of the most critical issues (largest distance to target) corresponds to the "amount of pre-treated waste that is sent to adequate landfills in relation to total waste arising". Hence this indicator has been chosen to exemplify the risk analysis and the required measures for improvement Sect. (13.5).

Table 13.3 Sustainability indicators for MSW management

Indicator	Current value	Target value (2030)
Specific waste arising [kg/person/day]	1.2	Max. 1.6
Amount of mixed waste pre-treated to reduce organic carbon content in relation to total mixed waste [%]	0	50
Greenhouse gases emitted during waste management [kg CO_{2eq} /person/year]	143	71
Waste fraction recovered as material or energy [%]	14	36
Cost of MSW in relation to GDP [%]	0.22	0.30
Income level of informal workers in relation to individual household income [%]	76	100

Source: Seidl 2008; González et al. 2009

As already mentioned in Sect. 13.4.1, most of the MSW collected – with the exception of waste collected separately (formally and informally) for recycling – is transported without pre-treatment to one of the three existing sanitary landfills, resulting in an actual value of 0 for the above-mentioned indicator. This can be attributed to:

- Lack of legislation in Chile on waste pre-treatment;
- The preference of Chilean institutions responsible for financing waste management schemes for projects related to closure of old dumping sites, improvement of technical landfill standards, rather than those dealing with alternative waste treatment technologies, and
- The low cost of landfilling, which makes implementation of alternative waste management technologies unattractive, e.g., composting or mechanical-biological pre-treatment.

Because MSW is not pre-treated in Santiago, vast quantities of waste are taken directly to the landfill and produce the following negative impacts:

- High MSW transport costs.
- Installation and operation of transfer stations means land consumption, higher costs, health hazards to employees due to contaminants, and exposure of employees and residents to noise.
- Reduced landfill lifespan makes it necessary to search for new sites. The build-up of new landfills leads to additional land consumption and an increase in waste management costs. Should the new landfill not begin operations as planned, the appropriate collection and disposal of waste could become a problem.
- Generation of methane and carbon dioxide in the landfill due to decomposition of the organic fraction of MSW. The formation of methane can cause fires and explosions. Methane must therefore be captured and flared under supervision. This again is associated with costs. Even when systems for capturing landfill gas are introduced, there is no method of capturing the gas to a satisfactory degree. Furthermore, landfill gas emission impacts on global climate.
- Development of leachate containing contaminants. Leachate must be captured and treated. Landfill leaks carry the risk of ground-water contamination. If ground water is used for irrigation or the production of drinking water, the impact on people's health will be negative.
- Due to the long time span of methane development, landfills must be monitored after closure. Methane and carbon dioxide continue to develop (impact on global climate) and the risk of ground-water contamination increases (growing probability of landfill leaks). In addition, landfill monitoring after closure implies higher costs.

To determine whether the above-mentioned negative impacts of waste disposal pose a risk to the achievement of sustainable development in the city, it must first be assessed whether the impacts are irreversible and/or have a specific spatial / temporal range (see Sect. 13.3.1).

13.4.2.1 Irreversibility

- Landfills can be revegetated. The process is highly complex, however, and can only be completed long after landfill closure.
- Exposure to highly toxic chemicals, e.g., cyanides, mercury and polychlorinated biphenyls, if released untreated can lead to chronic disease, cancer, infections or death.
- Since reversing the contamination of ground water is such a complex process, it must be considered irreversible.
- The impact of CH_4 and CO_2 emissions on climate is irreversible.

13.4.2.2 Spatial Range

- CH_4 and CO_2 emissions impact on global climate, i.e., on all world regions, and affect a vast number of people.
- Land consumption as a result of landfill construction is a regional problem.
- The impact of improper waste management on health is both a local and a regional problem.
- Ground-water contamination is a regional problem that affects a large section of the population living in MRS.

13.4.2.3 Temporal Range

- Ground-water contamination and the impact of CH_4 and CO_2 emissions on global climate is a long-term concern.
- Infections and disease caused by inadequate waste management can be long term.

Consequently, CH_4 and CO_2 emissions and ground-water contamination caused by the disposal of untreated solid waste constitute a risk to the achievement of sustainable development in RMS.

13.5 Alternatives to Improve MSW Management in Santiago de Chile: Separate Collection of Biowaste

In RMS, virtually all MSW is collected, transported and disposed of in approved sanitary landfills. The decomposition of the organic fraction of MSW produces landfill gas, thereby contributing to global warming, local air pollution, odour and nuisance. The potential of landfill gas to migrate and infiltrate zones outside the boundaries of the landfill puts the surrounding population and built-up area at risk and can lead to fires, explosions and the exposure of workers to toxic emissions (see Sect. 13.4: risk analysis).

Landfill gas emissions and their associated impacts can be mitigated if gas is collected from the landfill, burned in flares or used as a fuel source for energy generation. This has already been carried out in some landfills and is financed by Clean Development Mechanism (CDM) projects.

On the other hand, minimizing the amount of organic material disposed of at landfills – either by separate collection of this fraction or mechanical-biological pre-treatment of mixed waste – would not only reduce the production of landfill gas more effectively, but in addition mitigate most of the more notorious environmental and social impacts of current waste management practices (odours, presence of sanitary vectors). Furthermore, mitigating landfill gas emissions can also be encouraged economically by the generation of CO_2 certificates in the frame of Clean Development Mechanisms (CDM).

In order to demonstrate the benefits of cutbacks in landfill gas emissions by reducing the amount of organic material disposed of there, some model calculations will be presented.

13.5.1 Model Assumptions

A set of data and boundary conditions is defined in order to set up and evaluate different alternatives for waste management in RMS that could improve the sustainability of the system.

13.5.1.1 Time Horizon

Since one of the three landfills in RMS – the Santa Marta landfill, which began operating in 2002 – is to close after 2022, model calculations take the time span 2001–2022 into account.

13.5.1.2 Population Development

The last census in Chile, which was conducted by the Instituto Nacional de Estadisticas (INE) in 2002, was taken as the basis for population development for the period 2001–2022 (INE 2009). The population growth rate was taken from the same source, starting with 5.8 million people in 2001 and culminating in a total population of 7 million in 2022.

13.5.1.3 Waste Arising

Municipal solid waste that arrives at one of the three landfills in Santiago de Chile is weighed at the entrance gate. Data on the total amount of MSW deposited in RMS in recent years is available (Secretaria Regional Ministerial de Salud, Region

Metropolitana 2009). It needs to be modified for model calculations, since the municipal solid waste from households is mixed with waste from public areas not accounted for separately. Since these amounts do not correlate with the number of people living in RMS, they cannot be considered.

To estimate the share of waste from public areas, the disposal at the *RS Santiago Poniente* landfill was observed on site over a period of 2 weeks, during which time large quantities of waste accumulated from cleaning public areas could be identified from the type of delivery vehicle. The amounts were calculated and extrapolated to the whole of RMS and are estimated at an annual 356,000 t (Wens 2008).

With the above-mentioned data, specific arisings of municipal solid waste (kg/(person*day)) sent to landfills were calculated for the years 2001–2007. For the years 2008–2022 a constant value of 1 kg/(person*day) was chosen, resulting in total waste arising for the model calculations of about 1.8 million tons in 2001, about 2.2 million tons in 2007 and about 2.5 million tons in 2022 with a total for the time span 2001–2022 of about 50 million tons.

13.5.1.4 Waste Composition

Data on waste composition was extracted from a study carried out by the Universidad Catolica de Valparaiso in 2006 (UCV 2006). Waste composition was assumed to be constant over the time period under consideration. Separate collection was taken into account in the model calculations for food (37.41% share) and garden waste (12.08% share) only.

13.5.1.5 Waste Characteristics

Data on the water and carbon content of waste was sourced in the literature (Bayrisches Landesamt für Umweltschutz 2003; Rolland and Scheibengraf 2003) due to absence of this information in (UCV 2006). Table 13.4 shows the waste characteristics used for model calculations: the share of different waste fractions, their water content and their organic carbon content. The last column gives the calculated values for the organic carbon content in 1 kg of MSW, resulting from the specified waste fraction; in other words the contribution of food waste (residuos de alimentos) to the organic carbon content of 1 kg of MSW is 65.1 g, that of garden waste (residuos de jardín y poda) 19.79 g. In total, in the case shown in Table 13.4, the organic carbon content of 1 kg of MSW amounts to 156.78 g. This value will change when different fractions of organic waste are separated and no longer disposed of at landfills.

13.5.1.6 Composition of Landfill Gas

The share of CH_4 in landfill gas is 55% by volume, while the share of CO_2 is 45% by volume.

Table 13.4 Waste characteristics used for model calculations

	Fraction[a] (%)	Water content[b] (%)	c_{org}[c] (g/kg dry matter)	c_{org} calculated (g/kg of total waste)
Organic matter (food residues)	37.4	62	458	65.10
Yard waste	12.1	62	431	19.79
Paper	15.1	22	377	44.41
Cardboard	2.8	22	397	8.74
Plastic	13.6	18	0	0.00
Tetrapack	0.6	19	276	1.26
Diapers and hygienic articles	6.9	63	389	9.95
Rubber	<0.1	7	0	0.00
Leather	<0.1	7	0	0.00
Glass	4.7	1	0	0.00
Metals	1.8	2	0	0.00
Wood	0.2	14	0	0.00
Textiles	2.0	15	314	5.39
Dust, ashes	1.2	28	105	0.93
Batteries	<0.1	1	0	0.00
Bones	0.2	2	204	0.49
Fruit pit	0.3	14	0	0.00
Ceramics	0.6	2	0	0.00
Other	0.1	28	204	0.21
MSW	0.4	28	204	0.51
Total	100.00			156.78

[a] Wens 2008, [b] CSM 2009, [c] Tabasaran and Rettenberger 1987

13.5.2 Compilation of Waste Management Alternatives

Model calculations were performed for the following waste management alternatives:

- The total amount of waste is deposited on landfills (approx. 50 million tons).
- Due to the absence of segregated biowaste collection and the long time period required for its implementation, it is assumed that different fractions of organic waste (food waste, garden waste) are separated and composted, and the rest is brought to landfills.
- Due to variation in gas collection efficiency, it is assumed that different shares of landfill gas (10%, 30% and 65%) are captured and flared.

13.5.3 Assumptions for the Calculation of Greenhouse Gas Emissions

Calculations were performed for total emissions of greenhouse gases (CH_4 and CO_2) for the total amount of waste deposited, as well as for the total amount of waste composted in the time span 2001–2022.

Total greenhouse gas emissions were calculated according to the formula developed by Tabarasan & Rettenberger (Tabasaran and Rettenberger 1987).

$$G_e = 1.868 * c_{org} * (0.014 * T + 0.28)$$
G_e = total amount of gas produced in Nm^3/t of waste
c_{org} = organic carbon content in kg/t of waste
T = temperature in the body of the landfill in $°C$

The following data was used for the calculations:

- Temperature in the landfill: 39°C;
- CH_4 emissions from composting 0.65 kg/(ton of organic substance) (Cuhls et al. 2008);
- CO_2 emissions from composting 200 kg/(ton of organic substance) (Cuhls et al. 2008);
- With corg = 156.78 g/kg (see Table 13.4) the decomposition of 1 t of waste, specified in Table 13.4, produces a total of approx. 240 Nm^3 of landfill gas or 132 Nm^3 of CH_4. In other words, the mass specific Nm^3 generation amounts to 95 kg CH_4/t MSW or 1,995 kg CO_2 equivalents/t MSW.

No CH_4 is emitted during composting with an optimal supply of oxygen. However, optimal oxygen supplies cannot be assured in all places at all times, so that small amounts of CH_4 will be emitted during composting. Data on emissions from various composting plants were compiled and evaluated in Cuhls et al. (2008). The values chosen for the calculations in this study are taken from Cuhls et al. (2008).

Secondary emissions such as traffic exhaust fumes, emissions resulting from landfill and composting plants operations, or from energy required for separate collection or mechanical-biological treatment, and emission reductions due to the application of compost rather than nitrogen fertilizers are not considered. According to the Kyoto Protocol (UN 1998), CO_2 emissions caused by the decomposition of organic material are not taken into account, since this CO_2 has already been taken up from the atmosphere during the period of growth of the organic material.

Separating different amounts of organic waste from the total waste volume (10–90% of food waste and garden waste) alters the organic carbon content of each kg of waste brought to the landfills. Table 13.5 shows the corresponding calculated values.

Table 13.5 Calculated values for the organic carbon content of disposed waste

Percentage of separated organic waste (food waste and garden waste)	c_{org} calculated (g/kg of total waste)
0	156.78
10	155.34
20	153.89
30	152.45
40	151.00
50	149.55
60	148.11
70	146.66
80	145.22
90	143.77

13.5.4 Results of Model Calculations

Figure 13.4 shows the model calculation results. Normally, less than 50% of the landfill gas produced can be captured. This is due to lack of cover in the first few years of landfill operations, leaks in the cover system and inherent technical faults.

If 65% of landfill gas is captured and flared (an optimistic assumption), CO_2 equivalents can be reduced from 35 million tons (no separate collection of organic waste) to 17.9 million tons, e.g., emissions halve when 90% of the organic fraction (food and garden waste) is composted.

If 30% of landfill gas is captured and flared, CO_2 equivalents can be reduced from 70 million tons (no separate collection of organic waste) to 36 million tons when 90% of the organic fraction (food and garden waste) is composted.

If 10% of landfill gas is captured and flared, CO_2 equivalents can be reduced from 90 million tons (no separate collection of organic waste) to 46 million tons when 90% of the organic fraction (food and garden waste) is composted.

It is worth noting that energetic utilization of landfill gas has not yet been considered.

Collection and flaring and/or utilizing landfill gas would improve the local environment by reducing the quantity of noxious air pollution arising from the landfill, which in turn reduces the amount of nuisance caused by odours and the health risks associated with emissions. In short, it would have a positive effect on people's health and their local environment. In addition, proper collection and destruction of flammable landfill gas would reduce the risk of explosions in and around the landfill. A landfill gas collection system would minimize the potential

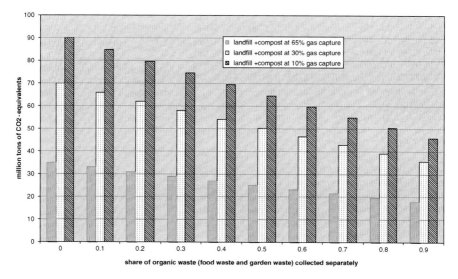

Fig. 13.4 Calculated emissions of CO_2 equivalents for "RMS" for shares of organic waste collected separately and capture rates of landfill gas (Source: The authors)

for landfill gas migration, whereby gas infiltrates zones outside landfill boundaries, posing a threat to the surrounding population and their building structures.

Regarding the organic fraction, the results underline the importance of this fraction for the reduction of landfill gas emissions. Apart from reducing the latter, separate treatment of the organic fraction would produce economic advantages. Given that disposal at landfills is state of the art in Chile today, the separate collection and treatment of biodegradable waste fractions would be an opportunity to introduce emission certificates within the framework of CDM projects.

13.6 Conclusions

In order to evaluate municipal solid waste management in RMS under sustainability aspects and to identify the associated risks, the current situation was analysed. This includes waste arisings, waste composition, waste collection, waste treatment and waste deposition. The results indicate that economic progress in Chile and the growth in population in RMS have led to increased MSW generation in RMS. Municipal solid waste management depends chiefly on final disposal in sanitary landfills and on the improvement of these disposal sites. Apart from recycling activities, which are for the most part carried out by the informal sector, almost no waste treatment activities have taken place.

An assessment of the system by selected sustainability indicators points to several deficits. The fact that nearly all MSW is sent to landfills without pre-treatment constitutes the largest risk, since it leads to long-term landfill gas emission.

In order to demonstrate risk minimization, model calculations that considered the segregated collection of the organic waste fraction of MSW were made. Findings show that segregated collection of at least 50% of the organic fraction would reduce landfill gas emissions by approximately 30% in the analysed time span of 20 years.

The segregated collection of biowaste is one method of diverting MSW from landfills and thus of decreasing the associated negative impacts. Its application in Santiago de Chile could be undertaken at the municipal level, as is the case in La Pintana, one of the poorest municipalities of Santiago. However, the economic feasibility of such strategies should be assessed and additional support for the application of these alternatives provided by a political and legal framework.

References

Astorga, A. (2008). Personal interview with Andrés Astorga, national consultant for the scrap and metal business, May 2008, Santiago de Chile.
Bayerisches Landesamt für Umweltschutz (Ed.), (2003). *Zusammensetzung und Schadstoffgehalt von Siedlungsabfällen.*
Blight, G. E., & Mbande, C. M. (1996). Some problems of waste management in developing countries. *Journal of Solid Waste Technology and Management, 23*(1), 19–27.

Bräutigam, K. R., Gonzalez, T., Seifert, H., Vogdt, J., & Wens, B. (2009). Landfill gas emissions from landfills in Santiago de Chile – strategies to reduce impact on local environment as well as on global climate. Proceedings of the II Simposio Iberoamericano de Ingeniería de Residuos Barranquilla, 24 y 25 de septiembre de 2009.

Butz, W. (2009). Deponien und Klimaschutz. In G. Rettenberger, & R. Stegmann (Eds.), *Trierer Berichte zur Abfallwirtschaft, Band 18, Stilllegung und Nachsorge von Deponien 2009 – Schwerpunkt Deponiegas* (pp. 13–21).

CDM-Executive Board (2006). Clean Development Mechanism Project Design Document Form (CDM-PDD) Version 02 – in effect as of: 1 July 2004 – Santa Marta Landfill Gas (LFG) Capture Project – Version 06 – 05/12/06, http://cdm.unfccc.int/UserManagement/FileStorage/L26258L8W1D3SUXE7YYKF6Y2OV4EQH. Accessed 2 July 2010.

CSM – Consorcio Santa Marta (2009). Site visit and verbal information from the company.

Cuhls, C., Mähl, B., Berkau, S., & Clemens, J. (2008). *Ermittlung der Emissionssituation bei der Verwertung von Bioabfällen.* Ingenieurgesellschaft für Wissenstransfer mbH – gewittra – Studie im Auftrag des Umweltbundesamtes, Förderkennzeichen 206 33 326, Berlin: Umweltbundesamt.

Chilean Central Bank (Banco Central de Chile) (2003). *Statistic Synthesis of Chile 1998–2002, 2003–2007* (Síntesis estadísticas de Chile). Department of Study Division. Santiago

CONAMA – Comisión Nacional de Medio Ambiente (2005a). *Estrategia de Reciclaje de Residuos Sólidos Domiciliarios de la Región Metropolitana.* Area de Gestión de Residuos y Materiales Peligrosos. Santiago de Chile.

CONAMA – Comisión Nacional de Medio Ambiente (2005b). *Metropolitana de Santiago. "Sistemas de Reciclaje, Estudio de Casos en la Región Metropolitana".* Santiago de Chile.

CONAMA – Comisión Nacional de Medio Ambiente (2005c). *Estrategia de Reciclaje de Residuos Sólidos Domiciliarios en la Región Metropolitana* (Recycling Strategy for Domiciliary Solid Waste in the Metropolitan Region). Area de Gestión de Residuos y Materiales Peligrosos.

Flodman Becker, K (2004). The informal economy. Sida, Department for Infrastructure and Economic Co-operation.

Gobierno de Chile (2009). Secretaria Regional Ministerial de Salud Region Metropolitan (Seremisalud RM), http://www.asrm.cl/paginasSegundoNivel/NivelTecnico.aspx?param1=197&param2=125/196/197&param3=−1. Accessed 2 July 2010.

González, T., Bräutigam, K.-R., & Seifert, H. (2009). Sustainability of the waste management situation in the Metropolitan Region of Santiago de Chile. Presentation at the international conference "megacities: Risk, vulnerability and sustainable development". Leipzig.

Health Ministry, Seremi Salud (2008). *Solid waste deposited in landfills.*

Hussmanns, R., & Farhad, M. (1999). *Statistical definition of the informal sector, international standards and national practices.* Geneva: International Labour Office (ILO)/Bureau of Statistics.

INE – Instituto Nacional de Estadísticas. (2009). http://www.ine.cl/canales/chile_estadistico/demografia_y_vitales/proyecciones/MenPrincOK.xls. Accessed 2 July 2010.

IMF – International Monetary Fund (2010). Washington: USA. World economic outlook database. Available at: http://www.imf.org/external/data.htm. Accessed 2 July 2010.

KDM – Keyneborn – Demarco (2009). Private communication.

Kopfmüller, J., Brandl, V., Jörissen, J., Paetau, M., Banse, G., Coenen, R., & Grunwald, A. (2001). *Nachhaltige Entwicklung integrativ betrachtet. Konstitutive Elemente, Regeln, Indikatoren.* Berlin: Sigma.

Movimiento Nacional de Recicladores de Chile (MNRCH) (2009). Personal contact with Alvaro Alaniz, Director from "Fundación Avina".

OECD – Organization for Economic Cooperation and Development (2006). Environmental data. compendium 2006–2007, waste, environmental performance and information division. Working group on environmental information and outlooks. http://www.oecd.org/dataoecd/22/58/41878186.pdf. Accessed 2 July 2010.

PAHO – Pan-American Health Organization. (2005). *Report on the regional evaluation of municipal solid waste management services in Latin America and the Caribbean.* Washington, DC: PAHO.

Rolland, C., & Scheibengraf, M. (2003). *Biologisch abbaubarer Kohlenstoff im Restmüll.* Wien: Umweltbundesamt. Berichte BE-236.

Secretaría Regional Ministerial de Salud, Región Metropolitana (2009). http://www.asrm.cl/paginasSegundoNivel/NivelTecnico.aspx?141. Accessed 2 July 2010.

Senatsverwaltung für Gesundheit, Umwelt und Verbraucherschutz Berlin (2009). Abteilung Umweltpolitk – Siedlungsabfallwirtschaft in Berlin.

Seidl, N. (2008). *Selection and analysis of sustainability indicators for waste management in Santiago de Chile* (Auswahl und Analyse von Nachhaltigkeitsindikatoren für den Bereich Abfall für Santiago de Chile). Diploma thesis. Karlsruhe: Forschungszentrum Karlsruhe.

Statistisches Bundesamt (2009). – Umwelt – Abfallbilanz. http://www.destatis.de/jetspeed/portal/cms/Sites/destatis/Internet/DE/Navigation/Statistiken/Umwelt/UmweltstatistischeErhebungen/Abfallwirtschaft/Tabellen.psml. Accessed 2 July 2010.

Szantó, M. (2006). *Domiciliary solid waste management in Santiago de Chile (Manejo de residuos sólidos domiciliarios en Santiago de Chile).* Valparaíso: Pontifica Universidad Católica de Valparaíso, Grupo de Residuos Sólidos.

Tabasaran, O., & Rettenberger, G. (1987). Grundlagen zur Planung von Entgasungsanlagen. In: Hösel, Schenkel, Schurer (Eds.), *Müll-Handbuch.* Bd. 3. Berlin: E. Schmidt.

UCV –. Universidad Catolica de Valparaiso (Ed.) (2006). Estudio Caracterización de Residuos Sólidos Domiciliarios en la Región Metropolitana.

UN – United Nations (Ed.), (1998). *Kyoto protocol to the United Nations framework convention on climate change.*

Vehlow, J. (2008). Waste – an often overlooked source in bioenergy promoting programmes. Proceedings of the 5th international conference on "combustion, incineration/pyrolisis and emission control". Chiang Mai.

Wens, B. (2008). *Einsparungspotenzial von Deponiegasemissionen durch die getrennte Erfassung organischer häuslicher Abfälle in der Region Metropolitana in Chile.* student research project. Rheinisch-Westfälische Technische Hochschule Aachen, Fachgruppe für Rohstoffe und Entsorgungstechnik.

Wilson, D., Velis, C., & Cheeseman, C. (2006). Role of informal sector recycling in waste management in developing countries. *Habitat International, 30*, 797–808.

Part IV
Synthesis and Perspectives

Chapter 14
How Sustainable is Santiago?

Jürgen Kopfmüller, Jonathan R. Barton, and Alejandra Salas

Abstract The objective of this chapter is to measure the performance of the Santiago Metropolitan Region and to demonstrate how the Helmholtz Integrative Sustainability Concept and a set of indicators can serve as a tool to support decision-making by public, private and civil society actors for sustainable development. The chapter combines results for selected headline indicators with those of sustainability performance in the various fields presented in more detail in Chaps. 6–13 of this volume. The exercise of setting target values as necessary reference lines to identify existing strengths and weaknesses is clearly an incentive to goal-oriented policy and planning. The analysis reveals positive trends for some of the indicators, which deserve continued support, but also tremendous challenges in others bearing negative trends. The chapter concludes with a synthesis of the sustainability challenges ahead. This includes reflections on the conceptual and methodological dimensions of this exercise, and suitable institutional responses.

Keywords Santiago de Chile • Sustainability indicators • Sustainability performance • Sustainable development • Urban and regional planning

14.1 Introduction

A sustainable development approach to urban and regional policy and planning requires the effective integration of different variables and dimensions present in the city region, and the tracking of these over time against targets. This chapter seeks to bring together the different thematic issues noted in the previous chapters of this volume and to present them in a synthetic, manageable way. In order to make

---

J. Kopfmüller (✉) • J.R. Barton • A. Salas
Karlsruhe Institute of Technology (KIT), Institute for Technology Assessment and Systems Analysis (ITAS), Hermann-von-Helmholtz-Platz 1, 76344 Eggenstein-Leopoldshafen, Germany
e-mail: juergen.kopfmueller@kit.edu

statements about the development of a city or region, it is vital that basic performance information be available to decision-makers.

By giving an overview of regional performance through the selection of key variables that can be traced over time and provide headline or general sustainability indicators, it is possible to contextualize the more detailed data that is provided for certain sectors but often not linked to others. This is commonly referred to as the 'silo' approach to information storage. By looking at several variables and their trajectories over time, and fixing specific targets according to normative criteria, narratives of city region development as opposed to specific projects (e.g., the Sanitation Master Plan or Transantiago) or territorial spaces (e.g., municipalities or peri-urban localities) can be constructed. It is also possible to establish the synergetic impact of policies, plans and investments as city region development is generated over time. While it is imperative that specific features of urban and regional life are resolved with targeted investments, e.g., in housing and infrastructure, it is also vital that the city region is not understood as a series of fragments. A holistic, integrative perspective on the city region is both feasible and necessary, and calls for difficult decisions on determining what the 'vital signs' of sustainable development actually are.

This chapter seeks to develop a short list of indicators for the Santiago Metropolitan Region in order to measure the overall performance of the city region. It acts as the glue that binds more sectoral views on metropolitan challenges, capacities and responses. It also serves as an overarching contribution to understanding metropolitan performance according to sustainability criteria, established via the Helmholtz Integrative Sustainability Concept (see Chap. 4), in a particular case. The outcome of this type of analysis serves to inform decision-makers by identifying sustainability challenges and prioritizing adequate responses. An overview of existing indicators and their application in the Santiago Metropolitan Region is given in Sect. 14.2. Whereas Sect. 14.3 refers to overarching headline indicators for Santiago, Sect. 14.4 provides a synthesis of indicators that emerge from specific policy fields. Section 14.5 is a brief reflection on indicator design and application.

14.2 The Design and Application of Indicators in the Santiago Metropolitan Region

Chapter 40 of Agenda 21 clearly states the importance of being able to monitor and evaluate policy and planning processes that seek to generate more sustainable development. Article 40.1 makes the point that "In sustainable development, everyone is a user and provider of information considered in the broad sense. That includes data, information, appropriately packaged experience and knowledge." To operationalize this, the objectives outlined in Article 40.5 point to the value of "timely, reliable and usable information" across different political and planning scales, as well as the need "to make relevant information accessible in the form and at the time required to facilitate its use." Despite the apparent logic of the statements, the challenge has rarely been met in these processes, although progress has been made in the sustainability sciences in this regard.

The statement that best sums up this situation can be found in Article 40.17: "There already exists a wealth of data and information that could be used for the management of sustainable development. Finding the appropriate information at the required time and at the relevant scale of aggregation is a difficult task."

As with any planning or management process, the need to assess performance over time is a sine qua non. There is little evidence of this in most urban contexts, however, nor has it been debated much in public arenas. Consequently, little is known about how the metropolitan system operates and whether interventions, from programmes to projects, can steer it towards more sustainable outcomes. Piecemeal interventions are therefore supported by piecemeal information, which often complicate discussions on how urban development is evolving. We have only scant information about system performance, about its operational elements and their synergies, or about how individual and societal welfare can be maximized, while at the same time critical natural capital is maintained.

In their development of systemic sustainability analysis, Bell and Morse (2001) identify the selection and application of indicators as crucial to moving from an initial position to one that has advanced and become more sustainable. As well as pointing out the urgency of context-specific understanding (diminishing the 'one size fits all' approach to urban indicator sets) and the role of stakeholders in identifying these, they use the abstraction of 'amoeba', and the need to respond to 'good amoeba' and 'bad amoeba'. If indicator sets are presented as rose charts or similar formats and reveal changes over time in absolute or indexed formats, they can appear as non-uniform masses of amoeba that mutate with each data update. To move in the direction of 'good amoeba' calls for an ex-ante conceptualization of desirable indicator trends and a shift towards this particular outcome.

Three further elements worthy of consideration are provided by Pinter et al. (2005), who suggest, firstly, that a small set of indicators has greater relevance for decision-makers than for technical specialists in thematic arenas. This idea is also supported by Rogers et al. (2008), who refer to the 'seven, plus or minus two' law of social psychologist George Miller (1956), stating that individuals can manage between five and nine independent facts of a linear problem at any given moment. Although computers can manage a surfeit of data, the generated output reverts to this law at some point, as decision-makers have to apply it to a specific issue or planning process. This is probably why the concept of headline indicators has emerged so strongly within the sustainability indicator experience. Despite the complexity of the sustainability of socio-ecological systems, a small set of key indicators must still be made available to decision-makers, regardless of the wealth of data behind them. Secondly, Pinter et al. (2005) highlight the significance of indicators being related to policy targets and, thirdly, that socio-ecological indicators are compatible with macro-economic indicators and the budgeting process. This third point is as critical as the first two in terms of these indicators becoming mainstream within public financial administration.

The elements mentioned above are not easy to remedy. The challenge of creating viable indicator sets that are both used and useful in decision-making processes remains significant. Nevertheless, the rising interest in the development of aggregate indices (Barton et al. 2007; Schushny and Soto 2009), the recognized opportunities of headline indicators, the emergence of goal-oriented indicators, and the push towards performance measurement in models of contemporary public policy assessment all point to the search for suitable remedies.

Indicator sets that have been generated since the Rio Conference reveal that sustainable development indicator sets co-exist with three other principal indicator set types: public performance planning indicators; sectoral management and planning indicators; and 'state of the environment' planning indicators (Gudmondsson 2003). This typology is also evident in the Chilean context. Although it is difficult to identify to date what might be termed sustainable development indicator sets at national, regional or local levels, these were considered by the National Environment Commission (CONAMA) during the late 1990s, with several authors contributing to the process (Blanco et al. 1997; Quiroga 2001). However, they were never implemented. Neither the national Sustainable Development Council nor the national environmental policy for sustainable development (both 1998) established sustainable development indicators (see Barton and Reyes 2008). The environment information system (SINIA), on the other hand, is an outcome of this post-Rio (1992), post-basic environmental law (1994) process, albeit relatively weak in terms of data provision.

On the environmental side, there are the reports published by the National Statistical Institute (INE) and the irregular 'state of the environment' reports compiled by the University of Chile in collaboration with CONAMA and the United Nations Environment Programme (UNEP). In terms of socio-economic data, this is more widespread with longer time series but has little relevance to changes in environmental quality and resource use. The challenges of improving environmental data provision and of linking socio-economic with bio-physical transformations remain crucial.

At the multilateral level, the incorporation of environmental data into the CEPAL statistical yearbook marks the growing recognition of environmental data as accompaniment to social (e.g., the Human Development Index published by the United Nations Development Programme UNDP) and economic data managed by diverse authorities (e.g., the National Statistical Institute-INE, Central Bank, Finance Ministry and Economy Ministry in the Chilean case). In terms of public performance planning and sectoral planning and management, numerous instruments have been generated and although indicators do exist, they are often specific to projects or programmes and biased towards programme budgetary management.

Much of the latter development in Chile is due to modernization of the government programmes introduced under the leadership of President Eduardo Frei in the late 1990s, i.e., tighter administration of public spending and systems in terms of budgetary allocation, spending and objectives. This has led to a centralization of power in the budgetary directorate of the Treasury (DIPRES), which to date does not require government programmes and investments to meet sustainable development objectives (as outlined in the 1998 policy, for example). However, there may be an opportunity to move in this direction in the near future, since legislation that created the Ministry of the Environment, the Environmental Evaluation Service and the Environment Superintendency contemplates the establishment of a transversal Sustainable Development Council of Ministers to ensure that sustainable development resonates across the ministries and is not only pursued by one.

Instruments such as the *Ficha CAS* (defining which households benefit from state subsidies), the CASEN survey (a panel-based household survey for baseline socio-economic and consumption data) and the SECTRA origin-destination transport survey are examples of data sets currently used. Whether they provide indicators as such or are linked to drivers of development and other data sets is a moot point, since indicators should have clear parameters as well as target values and timetables. It can therefore be stated without hesitation that although considerable information in the Chilean national and regional context is available, it has rarely been converted into or used to analyse indicators. Perhaps the most familiar indicator set that followed a public performance planning logic is the Ambient Air Quality Monitoring (MACAM) network in Santiago, which produced an air quality index based on illness costs (ICAP) and subsequent control measures for transport and industry according to endangerment levels (alert, pre-emergency, emergency). What is clearly lacking is a sustainable development indicator set that operates on a national and regional scale, bridges sectoral and territorial issues in an integrative manner, and provides headline indicators following the philosophy of the 'seven, plus or minus two' law that for policy purposes is crucial to decision-makers across the public and private sectors and civil society.

Despite the current absence of a single set of sustainability indicators for the metropolitan region, existing initiatives have clearly been driving in this direction in recent years, and it is thus likely that it will become a sine qua non of regional development planning in the near future. It is with this context in mind that the following sections outline results for some headline indicators.

14.3 Sustainability Performance Based on General Indicators: Strengths and Weaknesses

The analysis of the sustainable development performance of the Santiago Metropolitan Region is based on a key set of 12 headline indicators. It complements the indicator sets used to determine sustainability performance in the policy fields presented in Chaps. 6–13. An original set of 120 indicators was reduced to 18 and based on data series and scale availability (i.e., municipal, regional and national samples), for example, and data quality. Of these 18 indicators, six were finally eliminated for reasons explained below. Target values based on recent trends, proposals by supranational organizations, international comparisons, and existing needs and capacities were developed for the remaining 12 indicators. Being normative, they are neither correct nor incorrect. Instead, they represent performance goals generated via an expert group process. Stronger participatory processes can help to ensure the legitimacy of outcomes with regard to targets and timetables.

Some examples of the six indicators omitted are as follows. The indicator 'rate of unionization' is useful in terms of participation and potential governance issues. However, although employed in European contexts, it is particularly problematic in the Chilean case. During the 17-year dictatorship in Chile, unions were banned and most prominent unionists were 'disappeared' or went into exile. The union movement failed to return to pre-1973 levels of relevance and membership following the transition to democracy in 1990. In fact, unions remain a marginal actor, despite the sectoral importance of the copper workers, for example. Nevertheless, the push towards sub-contracting and other precarious contracts (to reduce permanent staff and increase flexible exploitation of employees) has severely impacted on the labour movement and its mobilization capacity. As a consequence, and given the low levels of union membership in the region (and the fact that membership is mostly national rather than regional), this indicator was deleted.

Another example is 'public debt per capita', which is not measured regionally. Furthermore, municipalities, as administrators of education and local health budgets, are subject to control by the national auditing office and not permitted to draw on capital markets. As such, only the national government holds public debt. Thus, it is difficult to establish the situation at regional level, despite the potential use of such an indicator to determine the long-term stability of the local economic system and public finance.

'Biodiversity' and 'Protected areas' are also problematic. Although the region is a global hotspot of Mediterranean biodiversity according to the Millennium Ecosystem Assessment (Reid et al. 2005), available information is scant. Failure to value biodiversity and protected areas for ecological services and other benefits leads such issues to be viewed as marginal in many planning mechanisms. Finally, 'Educational spending', which is the responsibility of the municipalities, varies considerably as a percentage of total spending (between 37% and 42%), making it a difficult measure to analyse and assess. Nevertheless, increased spending on this item compared to others should be a measure of investment in public education, albeit the high level of spending on private education should not be forgotten. In view of the need to improve the quality of education and the equality of access to it, the focus on public education is justified.

The final 12 indicators are presented as a metropolitan headline sustainability indicator set suitable for the description of transversal development outside more specific policy fields. As with all sets, these indicators have their strengths and weaknesses. Given the lack of regional data on carbon emissions, for example, and the potent role of carbon in diverse local and global environmental impacts, national emission data is adopted, with the awareness of the obvious approximation problems of doing so. A further example of complexity is the use of Internet connections per household that reveals a great deal about access to information, which is central to education and participation in decision-making. Bearing in mind that recent technologies are currently shifting Internet users from cable to mobile connections, this indicator, although useful to show trends to date, will have to be redefined accordingly in the near future.

Concerting a conceptual framework of indicators into practice and ultimately into decision-making processes is clearly a complex undertaking and involves numerous concessions along the way. These practical steps of grounding a somewhat abstract scientific concept in management processes entails responding to data availability restrictions and a range of inadequacies, such as the absence of time series or changing measurement frequencies, spatial scales or methodologies over time. Consequently, the final list is neither consistent with the initial proposition nor fully developed in terms of systemic interaction complexities (Gallopin 2003). Nonetheless, it is viable in its current form in terms of its advantage to decision-making processes, which is its ultimate goal.

The 12 indicators that ultimately remained on the list are presented in Table 14.1. They fulfil the basic ambitions of the Helmholtz Integrative Sustainability Concept outlined in Chap. 4, drawing on the socio-economic and socio-ecological dimensions of regional development. Admittedly, the precise mechanisms by which different variables influence each other are not entirely clear and require further cross-impact analysis (e.g., Renn et al. 2007) and research. On the other hand, the variables do serve their purpose of suggesting the various balancing forces at work in the region. Other methodologies that operate from basic data trends and generate more complex indices, such as the Genuine Progress Indicator, have been applied in the region to cover the second half of the 1990s to the early 2000s, and complement the indicators sketched above (effectively building on these second-generation lists to produce a third-generation indicator or index) (Barton et al. 2007).

Initial results of the indicator analysis, mainly based on data time series and distance-to-target considerations (as far as target values exist), reveal that there are several positive trends in the Santiago Metropolitan region, including declining poverty levels, the HDI index (based only on data for 1994 and 2003) and a decrease in infant mortality, all of which implies an improved sustainability performance in terms of these criteria. Positive, primarily socio-economic, indicators must be contextualized with other, more negative trends, particularly in the field of environmental quality, e.g., CO_2 emissions per capita, but also crime levels. There are improvements in many of the indicators, although much discussion revolves

Table 14.1 General sustainability indicators

Infant mortality rate	Gini coefficient
Young people (14-17y) not attending school	Percentage of overcrowded houses
CO_2 emissions per capita	Human Development Index
Violent crime rate	Poverty rate
Green area per capita	Percentage of Internet connected households
Degree of male/female wage equality	Unemployment rate

Source: The authors

around the rapidity of change and the distribution of the corresponding effects, very much along the lines of the problematic income distribution indicator measured by the Gini coefficient, which has changed little in recent decades. These indicators and their significance are discussed in the following paragraphs.

In terms of the conceptual basis of the analysis, variables are clearly aligned with the key topics described by the substantial sustainability rules. 'Securing human existence', for instance, is closely associated with infant mortality, the Gini coefficient, wage inequality, poverty and unemployment rates, and overcrowding. 'Maintaining society's productive potential' refers for the most part to CO_2 emissions, green space and educational attainment. Finally, 'preserving society's options for development and action' specifies Internet access and violent crime rates. In terms of instrumental rules, there is repetition of the above-mentioned variables, indicating that these are not exclusive to particular rules but transversal, e.g., wage equality, educational spending, the Gini coefficient and unemployment rates.

In terms of rules and indicators relevant to 'securing human existence', it is fair to say that Chile has made great strides since its return to democracy in 1990. In particular, poverty rates have fallen significantly (see Fig. 14.1). Poverty is the principal factor that precludes people's capacity to meet their basic needs autonomously and lead a life of human dignity – the basic sustainability goal at the individual level. Here, according to the definition of the Planning and Cooperation Ministry (MIDEPLAN), a person is defined as "living below the poverty line" if he/she is forced to live on a monthly income that is less than twice the value of the basic commodity basket (excluding public transfers).

This decline in the poverty rate also implies improvements in related indicators, such as infant mortality (here defined as the risk of babies dying before the age of 1), which halved between 1990 and 2007 (see Fig. 14.2).

These two variables reveal the close relationship between certain indicators, as they reflect different facets of the same phenomenon, more aligned to the UNEP indicator approach that focuses on so-called Driving force-Pressure-State-Impact-Response

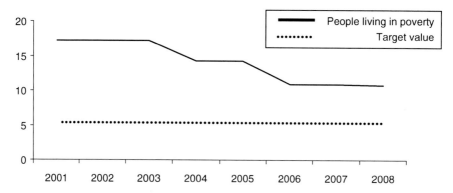

Fig. 14.1 Percentage of people living in poverty (Source: National System of Municipal Information 2010)

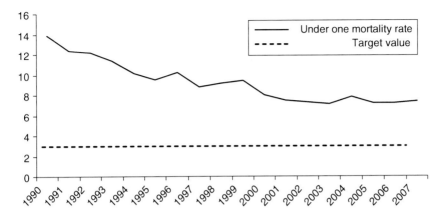

Fig. 14.2 Infant mortality rate (per 1,000 live births) (Source: Ministry of Health 2008)

processes. There is little doubt that strong economic growth rates at national and metropolitan levels, in the Chilean case driven directly and indirectly by copper and non-traditional agricultural products (NTAX), have played a key role in the declining poverty rate. However, social policies should not be overlooked in terms of the attempt to target specific groups for redistribution of benefits. The *Ficha CAS* – a household application form for public benefits – was used as a tool by MIDEPLAN to identify the poorest families in society and facilitate welfare payment transfers during this period (see Clert and Wodon 2001). As with many socio-economic indicators, it should be recognized that the democratic period witnessed a medium-term recovery process following the economic crisis of 1982–1983 which seriously affected Chile due to its heavy export orientation (as it had also done in 1929–1932) and its dependency on copper export revenues.

This positive trend notwithstanding, a consideration of the target values for these two indicators – based on supranational organizations such as the WHO and the Millennium Development Goals perspective – makes it clear that a further 50% reduction is imperative if a standard comparable to OECD countries is to be achieved. This should be the claim and the ambition of the principal Metropolitan Region of a country like Chile.

Rates for household overcrowding – an example of the failure to meet basic needs – also declined over the same period (see Fig. 14.3). This is the case in particular with respect to 'medium overcrowding', i.e., three people sleeping in one room. 'Critical overcrowding', i.e., more than three people sleeping in one room, has also dropped, but to a lesser degree. To a large extent, this trend is due to construction of a vast number of social housing units dating from the early 1990s (marking the highest number of units ever built). Aligned to OECD rather than other Latin American countries, target values should be set at a level of approximately 50% below current values, revealing the need for corresponding programmes and (public) investments.

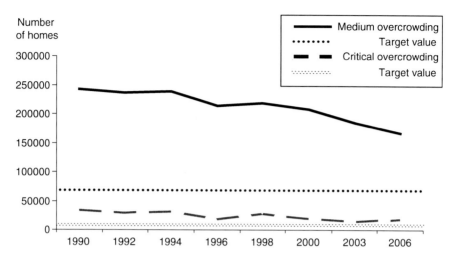

Fig. 14.3 Household overcrowding (Source: Ministry of Planning 2006)

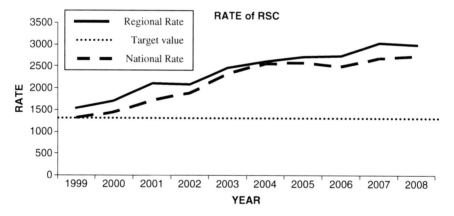

Fig. 14.4 Reported serious crime rate (per 100,000 people) (Source: Ministry of the Interior 2009)

Against these positive trends, concerns remain about rising violent crime rates, for instance (see Fig. 14.4). This aspect is all the more acute, since personal security and the perception thereof are critical to human well-being and quality of life. The indicator used here is 'Reported serious crime cases per 100,000 inhabitants per year', where 'serious crime' includes robbery with violence or intimidation (surprise or force), vehicle theft, theft, injury (mild or serious), manslaughter and rape. Delinquency and crime, in particular violent crime, and the fear of crime remain the key preoccupations of urban inhabitants across social groups. Additionally, delinquency is in many instances associated with drug use and commercialization. While Santiago compares favourably with other cities in

Latin America (see Chap. 2 of this volume), the trend for this indicator has taken a negative turn since the 1990s – not unlike other metropolitan and national scales. Concerns about the ongoing increase of violent crime are currently the principal challenge to public decision-makers. Gated communities and private security companies in many areas of the city are a visible response to this process, while in poorer communities local gangs are active in destabilizing community cohesion and security.

In terms of tolerance, solidarity or adequate conflict solution mechanisms, these factors tend to endanger or even erode social cohesion, a precondition for sustainable development. Target values based on pre-2000 levels, justified by their orientation towards the level of OECD countries, point to the extensive political and societal measures demanded.

Income (in)equality is one of the few areas with a systematic measurement of the distributional issues particularly relevant to sustainable development. Unequal income distribution generates inequality of opportunity for a dignified human life. By analogy, wage equality, e.g. with respect to gender, is a precondition for gender-related equality of opportunity and should be pursued in sustainability strategies. Here, income distribution is captured by two indicators: the Gini coefficient and wage (in)equality by gender (see Figs. 14.5 and 14.6). The Gini coefficient is the real distribution of total population income (Lorenz curve) compared with a perfectly equal distribution. It can range from 0 (meaning complete equality) to 1 (complete inequality). Wage (in)equality is defined here as female average income as a percentage of male average income. This is not a representation of equal wage for equal work, since occupation profiles for each gender are in fact different.

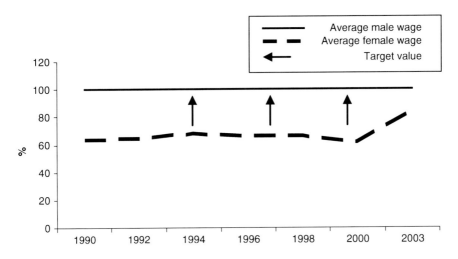

Fig. 14.5 Wage inequality by gender (Source: Ministry of Planning 2004)

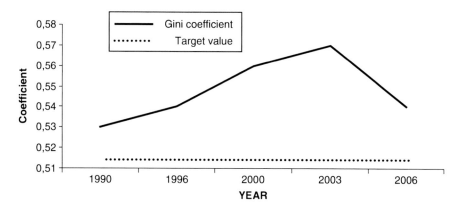

Fig. 14.6 Gini coefficient (Source: Ministry of Planning 2008)

Both indicators show an absence of significant progress for more than 20 years. In 2007, a commission (the Mellor Commission on Work and Equality) was established to identify ways of moving the equality agenda forward. However, very little planning is evident for this complex undertaking. Consequently, the Santiago Metropolitan Region suffers from the Chilean national syndrome of being among the 10% worst countries in the world for income distribution, constantly bearing Gini coefficient values above 0.5 (see UNDP 2009). An orientation in terms of the average for European countries would imply a Gini coefficient target value of about 0.3. The minimum goal should be below the 0.5 line, which is regarded – for instance, by the World Bank – as a threshold value for high inequality (see the World Bank's annual World Development Reports). With respect to the wage indicator, Santiago is no exception to the widespread global phenomenon of unequal pay for equal work along gender lines. Here, the target value should be 100%, conceding that a more informed view would require the consideration of occupational qualities and profiles. Given the persistent 'machismo' in Chilean society, the structural and cultural obstacles to moving forward are obvious. Nevertheless, the situation alters with each generation, as female educational and employment aspirations are realized and family roles change. In this sense the election of the first female president in 2006 was symbolic.

With regard to 'maintaining society's productive potential', progress is less clear than in the previous rule set. CO_2 emissions per capita, a key indicator in terms of the logic of decoupling societal development from fossil fuel use and its deleterious effects on climate change, continue to rise. This is intensified by the heavy increase in private vehicle traffic, the difficulties in the energy carrier mix and energy security, primarily as a result of shortages in gas deliveries from Argentina, one impact of which is a surge in the use of diesel (for further detail, see Chap. 9). The chief concern of this indicator is to raise data on the Metropolitan

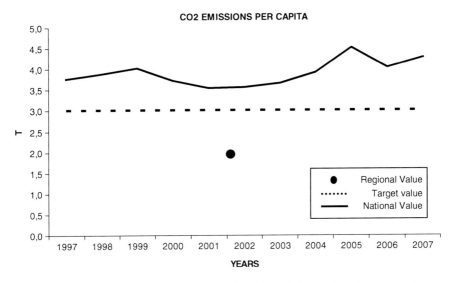

Fig. 14.7 CO_2 emissions per capita (Source: National data: United Nations (sf); Regional data: Barton et al. 2007)

scale. The calculation in Barton et al. (2007) for the year 2002 is based on work carried out in collaboration with the National Energy Commission, but not on a regular basis. Interestingly, in 2002 Santiago performed about 40% better than the national average, although both remained at a low level, at least in OECD terms (see Fig. 14.7).

Despite these data shortcomings, the high relevance of this indicator make its use essential to at least give an impression of the current situation. Discussing a target value here is not an easy task, since it preempts questions such as "What economic development path should countries like Chile take?", "What level of resource use efficiency is feasible?", and – above all – "What right do industrialized countries, as the historically predominant CO_2 polluters, have to demand emission limitations from developing countries?". Against this background, 3 t/capita could be a reasonable target value for Chile and Santiago in 2030, allowing in the meantime for some increase above this limit.

Another indicator deals with green space, which is a key issue for human well-being and heads concerns for urban improvements in the 2008 survey of urban life by the Ministry of Housing and Urbanism (MINVU). Green space softens the 'urban fabric', provides areas of recreation, and contributes to diverse environmental services. In the definition here, only public green space is considered. Not considered are: airfields, agricultural areas, aggregates, park avenues, soccer and other sports complexes, cemeteries, hills, buildings, hospitals, ecological reserves, protected areas, streams, military establishments, roundabouts and universities. Since 2000, the figure for this indicator has remained at approximately 3 m^2 per capita (see Fig. 14.8), which is well below the WHO guideline of 9 m^2 per capita.

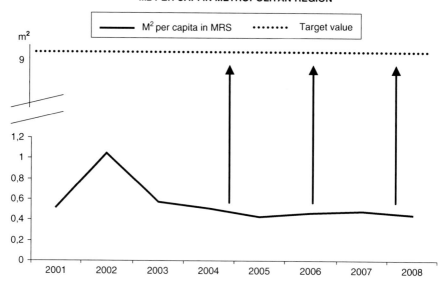

Fig. 14.8 Public green area per capita (Source: National System of Municipal Information 2009)

The distribution of green space is highly variable, favouring higher income groups in the east of the Metropolitan Area. This environmental justice issue has been highlighted by the work of Romero and Vásquez (2005). Since green areas are important not only for leisure and recreation (as identified in the MINVU survey), but also for temperature regulation ('heat island' effect), storm water infiltration and other ecological services, their absence remains a significant social justice challenge. As such, efforts will be necessary both to meet the WHO standard and to reduce the inequitable spatial distribution of green space to an acceptable level.

Internet connectivity (or the possibility of using it) has already and will in the future gradually become a precondition for suitable access to information. This is vital to sustainability with respect to objectives such as equality of opportunity, participation in societal decision-making, and improvement of human capital. The indicator here is defined as the percentage of households connected by cable. The last decade saw an increase in this percentage from 25 to almost 50% (see Fig. 14.9). Given the wide social divisions left unexamined by regional data, however, acquisition of data on Internet access remains a problem. Internet access is growing, but its unequal distribution among socio-economic groups – a phenomenon found in many other countries – leads to an exacerbation of equality issues in this regard. A target value of 80% by 2030 seems reasonable, assuming that in 2030 wireless technologies will have replaced cable connections.

Unemployment is the chief reason why people are unable to meet the needs required to live autonomously. In a society with few public transfers for the unemployed, this leads to a heightening of informal activity. Informal activity does not cover pensions and health care costs, is precarious, and can lead to

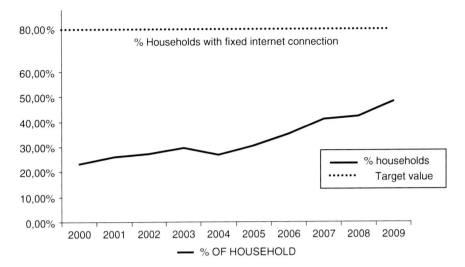

Fig. 14.9 Percentage of households with Internet access (Source: Undersecretariat of Telecommunications 2010)

increased risk and vulnerability in terms of family well-being over time. Here, the unemployment rate – defined as the percentage of unemployed inhabitants aged fifteen and over – is used as an indicator. From a level of around 15% in the mid-1980s, statistics have varied between 7 and 11% in the last 15 years (see Fig. 14.10), a trend that can be seen as stable and at a level comparable to that of OECD countries. Thus, a target value for 2030 of 4% should be realistic, including a possible 6% by 2015 (by comparison: economics conventionally considers 3% full employment). Describing and analysing the topic of unemployment remains complex, given current data collection procedures and the difficulty of distinguishing between informal sector and self-employment. This lack of clarity on formality and informality complicates the overall picture. Furthermore, the number of informal workers has declined over time and the stability of certain kinds of contract work is still open to question.

In terms of creating and improving capacities for future generations, both basic and higher education are crucial to realizing equal opportunities at the individual level and increasing capacities for societal decision-making. Consequently, the international sustainable development agenda is now focusing on this issue with the UNESCO Decade of Education for Sustainable Development (2005–2014). From the large variety of existing indicators dealing with the topic of education, the percentage of the population aged 14–17 not attending school was selected. It covers both the area of basic education and the precondition for higher education. Over the last 15 years, the performance of the Metropolitan Region, similar to the national level, has improved. Both levels currently share the same percentage (see Fig. 14.11). However, the growing number of average school years per capita is hampered in particular by relatively low attainment levels in international ranking

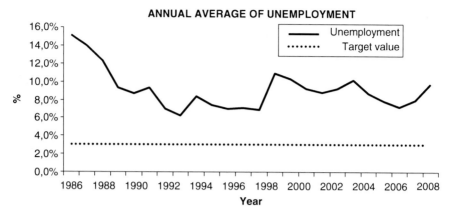

Fig. 14.10 Unemployment rate (Source: National Statistical Institute 2009)

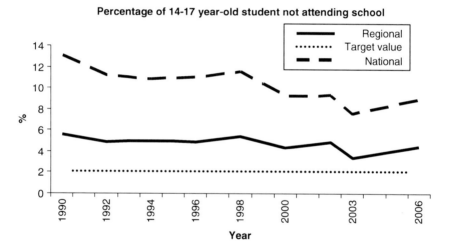

Fig. 14.11 Population aged 14–17 not attending school (Source: Ministry of Planning 2006)

exercises. An increase in the number of young people entering higher education may reduce some of these problems.

14.4 A Synthesis of Sustainability Performance Across Thematic Fields

In addition to performance analysis, the policy fields examined in Chaps. 6–13 will be scrutinized for their representation of the typical issues of metropolitan development. The results described in detail in these chapters demonstrate the key role of the Santiago Metropolitan Region as, for example, a consumer of resources and a producer of waste and pollution. As a habitat of diverse risks and with strong evidence of the critical distributional aspects of mega-urban development processes, the use of indicators to reveal possible linkages and interdependencies between thematic fields is vital.

With respect to *land use*, the expansion dynamics of recent decades, for the most part housing and infrastructure driven by population and economic growth, have impacted heavily on the environment, both in the urban centre and on the fringe. Additionally, land-use changes facilitated or even driven by Metropolitan planning instruments, in particular the increase in built-up areas and the loss of agricultural land, often lead to a shift from natural hazard to risk as a result of increased settlement in flood and landslide-risk areas (see Chap. 7). Against a background of risk in the form of flooding, landslides and earthquakes, the principal deficits are highlighted, e.g., the lack, in planning institutions, of a systematic understanding of risk within the complex system of the region. There is also evidence of a disregard for vital aspects of vulnerability and a focus on technical adaptation measures, while at the same time neglecting long-term planning and risk prevention approaches. Consequently, an overall planning and management instrument, integrating different sectoral approaches and environmental targets more clearly into decision-making processes, is noted as an urgent requirement.

Over the past two decades, the field of *transportation* has primarily – due to changing life styles and the land-use changes mentioned above – been characterized by a modal split, expressed in a greater use of private rather than public transport, and by a substantial increase in urban highway infrastructure. The primary risks associated with transportation are high congestion levels, unreliability of and limited access to public transport, health impairments resulting from air pollution, and deaths or injuries due to traffic accidents. The expert and decision-maker based proposal of a package of political measures to tackle these problems includes investments in higher bus frequencies and new metro lines, a congestion charge in central areas of Santiago, and a demand-side separation of time. However, implementation would require the relevant political will to alter previous approaches in the attempt to meet certain set goals.

The performance in the *energy sector* primarily focuses on the national level and is characterized – similar to many industrialized countries – by a rapid growth of energy demand and a high dependency on fossil fuels, in the case of natural gas, which until recently was imported principally from Argentina. At the same time, Chile – and thus the Metropolitan Region to a certain extent – holds a large potential for renewable energy resources. So far this has only been used for electricity production (hydropower-based to 50%). While the current performance for some indicators is positive, e.g., increased access of households to electricity supplies and low system interruption, other indicators, e.g., energy intensity or the share of renewables in total energy production, remain distant from reasonable sustainability targets. Despite the fact that Chile has already taken steps towards a more sustainable energy system, substantial political support will be necessary to deepen this process, particularly if projected economic growth and population development are considered. The focus here is on improving energy efficiency and increasing the share of renewables, both of which require perseverance along the existing path, with further price-related measures and standard-oriented regulation.

The *water system* is characterized institutionally by the fact that Chile is the only country worldwide with a wholly privatized water sector. With respect to water services, i.e., to access or connection to drinking water, and to sewage infrastructure and hygienic standards, the current performance of the Metropolitan Region can be seen as positive, keeping in mind international comparisons and national targets. A major deficit is urban storm water management, pointing to the need to install new storm water sewers. Nevertheless, the range of water consumption per capita varies dramatically between households and, considering watershed capacities, exceeds sustainable levels for many consumers. Given that approximately 75% of fresh water is used in agriculture beyond the Metropolitan Area, the core challenge will be the shift in water resource management within the catchment area. Possible approaches include maintaining water for longer in the catchment, e.g., by dams, exploiting new water resources, or increasing water use efficiency. This challenge will gain momentum once projections regarding reduced water availability in the future due to climate change are factored into the equation.

A look at the performance of the *waste management* sector, including generation, composition, collection, treatment and disposal of waste, shows that the continuous increase in municipal solid waste volumes is the key issue. Solid waste management in the Metropolitan Region consists predominantly of final disposal in sanitary landfills, whereas recycling levels are low and essentially based on informal worker activities. There are few alternative treatment activities in place. The most urgent problem is that almost none of the waste sent to the landfills is pre-treated, leading, for instance, to long-term landfill gas emissions. Separate collection and treatment or use of a major share of the organic fraction of solid waste seems a suitable measure to substantially reduce these emissions.

Health impairments caused by air pollutants pose major problems and risks to people living in the Metropolitan Region, as in most urban agglomerations, depending among other factors on individual vulnerability criteria such as age, health condition and location. In Santiago, these problems are associated with transportation, and industrial and household emissions, due for the most part to energy use. Besides an appropriate selection of indicators, the availability and assessment of suitable data for the relevant mobile, stationary and area emission sources pose the toughest challenge. So far, the analysis of combined air pollution and epidemiological models indicates a significant increase, albeit socio-spatially differentiated, in the mortality risk for cardio-respiratory diseases as a result of current and projected PM_{10} concentrations. They also point to the political and technical measures required to reduce this threat, concentrating on the transportation and household sectors.

Finally there is the field of *socio-spatial differentiation* processes, which have a long tradition in Santiago and frequently end in segregation as a phenomenon of social inequality. Caused by drivers such as demography, immigration and intra-urban migration, land markets and property prices, on the one hand, and public welfare policies such as subsidies and housing, and land-use planning instruments, on the other, socio-spatial differentiation processes have led to socio-economic concentration and diversification, as well as social inclusion and exclusion tendencies in the peripheral municipalities. At the same time, the city centre is characterized by a gentrification trend, mainly brought about by the immigration of young people, accompanied by the growing attractiveness of more peripheral areas for higher income group settlement. The challenge here is to design housing, land and other policy and planning frameworks in such a way that they will balance social diversification and proximity, and create a social environment that produces greater social cohesion and less segregation and exclusion in order to tackle some of the problems mentioned above in Sect. 14.3.

14.5 Reflections on Sustainability Indicator Design and Application

Against this background of indicator-based sustainability analyses carried out for the Santiago Metropolitan Region and as a conclusion to this chapter, several analytical and strategic remarks are presented.

The precondition for sustainability analysis is absolute clarity about the conceptual and methodological approach, and about the basic objective of the work and its recipients. This includes answering questions, for instance, associated with the structure and thematic focus of the indicator set used, or with the form and intensity of stakeholder involvement in the process. In the work presented here, decisions were made to apply a limited and operable set of indicators, focusing on selection criteria such as the availability of suitable data.

The findings show a mix of sustainability performance strengths and weaknesses for Santiago. This is not, in itself, surprising. It does, however, contribute to the debate on the kinds of indicators to be included in an appropriate metropolitan sustainability indicator set. One important argument emphasizes the need to focus on indicators that depict and reveal existing deficits, pointing to a need for action. Clarity about the actors who should be taking these actions is likewise vital. The work presented here aimed to provide input into the scientific debate on mega-urban development and orientation for political and societal decision-makers. By presenting both positive and negative urban tendencies, the complexity of urban systems becomes apparent, as do the difficulties facing the politicians and other actors involved. Rather than trade-offs between the two tendencies, however, the goal should be to seek 'win-win' outcomes as effectively as possible, whereby improved welfare outcomes are realized with a reduced impact on the resource base and environmental quality. The incorporation of a broad range of actors in the project from the outset sought to enhance societal reflection and ensure the legitimization of project findings as far as possible.

The indicator analyses also reveal that a variety of processes are at work within the socio-ecological system that is the Metropolitan Region. Improvements in criteria of a more social nature are not matched with protection of ecological resources, although this is mostly due to a lack of cross-sectoral thinking and planning aimed at more 'win-win' approaches and solutions. Furthermore, there are structural constraints in terms of how to move beyond a situation with significant levels of social injustice, e.g., access to quality housing, health, education, green space. While the basic needs agenda has been engaged with effectively, such as poverty levels and infant mortality, the principal challenges to be faced lie in this transition to an industrialized, middle-class society that can still generate more equitable development and a greater degree of environmental quality and ecological service provision in the process.

With respect to methodology, three crucial issues or challenges for future analysis with indicators should be highlighted: firstly, it would be more revealing to differentiate the data for a number of indicators even further, e.g., by socio-economic criteria, gender or spatial scales. This has not been possible to date due to time and resource constraints but could provide valuable information about distribution patterns and help to design political measures that are more reflexive in this respect. Secondly, differing spatial scales of indicator analysis were used by the various working groups, focusing on the so-called Metropolitan Area of 34 municipalities, on the enlarged area of 38 municipalities, on a so-called Greater Metropolitan Area of 39 municipalities, or on the entire Metropolitan Region of 52 municipalities. Although all decisions were based on comprehensible motivation and justification, the outcome was an inevitable limit on the comparability of results and a degree of confusion, at least of observers external to the project. Here the objective of further research should be to use common spatial scales where possible. Thirdly, in many cases interdependencies between indicators in terms of cause-impact relations were identified but not analysed further. Hence, systematic so-called cross-impact analysis is called for if such links are to be detected and information is to be provided for the design of measures that fully consider them.

All of the above points to the need for more data and more data quality. However, a common difficulty for most of the thematic fields was the lack of suitable data for the Metropolitan Region or smaller scales. Here, continuous efforts at data collection, interpretation and access are required of official statistical institutions, as well as of political decision-makers, assumed to be one of the principal user groups.

Santiago de Chile, not unlike other cases, has generic topics or challenges and those that are more specific. It is precisely the definition of these challenges, their prioritization, the evaluation of the current 'performance' (or diagnosis) of urban development, and a look at the future through scenario constructions that offer a way forward in the praxis of sustainable development. The identification of water, energy, segregation and the existing complex, multi-level and multi-actor governance regime as core priorities in assessing future options, and an understanding of the synergies between them that transcend mainstream sectoral 'solutions' is the essence of what sustainable development analysis can offer to urban planning, urban policy and decision-making at multiple scales.

Rather than a binary division between sustainable and unsustainable city regions, the task is to identify degrees of sustainability and of sustainability potential, and to strengthen sustainable development processes accordingly in each case. This requires a framework that suggests transparency in conceptual and analytical terms as a minimum condition. As a paradigm for urban development, the renaissance of sustainability thinking presents multidisciplinary challenges, however, to urban managers, urban representatives, and civil society organizations and firms in terms of complexity, of inter- and intragenerational justice, and not least of increased participation. There is no easy solution to managing or responding to these dimensions. Instead, complexity is accepted as inherent to city regions as a consequence of overcoming limited, short-term reductionist 'solutions' that have characterized city planning in recent decades.

Referring to the aforementioned political consultancy perspective of the analyses, most of the indicators outlined above were included on a list prepared for the Regional Government (GORE) as input into the decision process of selecting indicators for a metropolitan sustainable development strategy. In this sense, the findings of the indicator analyses, including target values and appropriate target performance comparisons, form the basis for a comprehensive overview of Santiago Metropolitan Region sustainability performance trends, the aim of which is to stabilize or enhance positive trends and to mitigate those that are negative. This will be carried out in the future in close collaboration with the GORE planning division, as the lead agency for the current designing of a new Regional Development Strategy, the flagship instrument for strategic planning.

References

Barton, J., & Reyes, F. (2008). Una década de gobernanza para el desarrollo sustentable: evaluando el impacto de la política ambiental para el desarrollo sustentable (1998). In V. Duran et al. (Eds.), *Desarrollo Sustentable: gobernanza y derecho*. Santiago de Chile: Universidad de Chile.

Barton, J., Jordán, R., León, S., & Solis, O. (2007). *¿Cuán sustentable es la Región Metropolitana de Santiago?* Metodologías de evaluación de la sustentabilidad. Santiago de Chile: Naciones Unidas.

Bell, S., & Morse, S. (2001). Breaking through the glass ceiling: Who really cares about sustainability indicators? *Local Environment, 6*(3), 291–309.

Blanco, H., Wautiez, F., Llavero, A., & Riveros, C. (1997). Indicadores regionales de desarrollo sustentable en Chile: ¿Hasta qué punto son útiles y necesarios? *EURE, 27*(81), 85–95.

Clert, C., & Wodon, Q. (2001). The targeting of government programs in Chile. In E. Gacitúa-Marió, & Q. Wodon (Eds.), *Measurement and meaning. Combining quantitative and qualitative methods for the analysis of poverty and social exclusion in Latin America.* (pp. 43–68). World Bank Technical Paper No. 518, Washington, DC.

Gallopin, G. (2003). *A systems approach to sustainability and sustainable development*. Santiago de Chile: Naciones Unidas.

Gudmondsson, H. (2003). The policy use of environmental indicators – Learning from evaluation research. *The Journal of Transdisciplinary Environmental Studies, 2*(2), 1–12.

Miller, G. (1956). The magical number seven, plus or minus two: Some limits on our capacity for processing information. *Psychological Review, 63*, 81–97.
Ministry of Health Department of Health Statistics and Information (2008). Mortalidad por grupos programáticos. http://deis.minsal.cl/deis/ev/index.asp. Accessed May 2010.
Ministry of the Interior (2009). Informes anuales de estadísticas delictuales http://www.seguridadciudadana.gob.cl/ano_2002.html. Accessed May 2010.
Ministry of Planning (2008). *Región Metropolitana de Santiago - Evolution of income inequality 1990–2006*. CASEN Survey. Santiago
Ministry of Planning (2006). Encuesta CASEN: educación. http://www.mideplan.cl/casen/modulos/educacion/1990/educacion1990cuadro17.xls. Accessed May 2010.
Ministry of Planning (2006). Encuesta CASEN: vivienda. http://www.mideplan.cl/casen/Estadisticas/modulos/vivienda/2006/vivienda2006cuadro18.xls. Accessed May 2010.
Ministry of Planning (2004). Encuesta CASEN: ingresos. http://www.mideplan.cl/casen/Estadisticas/modulos/ingresos/1990/ingresos1990cuadro4.xls. Accessed May 2010.
National Statistical Institute (2009). Serie empalmada población de 15 años y mas por situación en la fuerza de trabajo. http://www.ine.cl/canales/chile_estadistico/mercado_del_trabajo/empleo/empalmadas/f98/fuerza13.XLS. Accessed May 2010.
National System of Municipal Information (2009). Selección de búsqueda. http://www.sinim.gov.cl/indicadores/busq_serie.php. Accessed May 2010.
Pintér, L., Hardi, P., & Bartelmus, P. (2005). *Sustainable development indicators: Proposals for a way forward*. New York: United Nations.
Quiroga, R. (2001). *Indicadores de sostenibilidad ambiental y de desarrollo sostenible: estado del arte y perspectivas*. Santiago de Chile: Naciones Unidas.
Reid, W., et al. (2005). *Millennium ecosystem assessment*. Washington, DC: Island Press.
Renn, O., Deuschle, J., Jäger, A., & Weimer-Jehle, W. (2007). *Leitbild Nachhaltigkeit. Eine Normativ-Funktionale Konzeption und ihre Umsetzung*. Wiesbaden: VS Verlag für Sozialwissenschaften.
Rogers, P., Jalal, K., & Boyd, J. (2008). *Introduction to sustainable development*. London: Earthscan.
Romero, H., & Vásquez, A. (2005). Evaluación ambiental del proceso de urbanización de las cuencas del piedemonte andino de Santiago de Chile. *EURE, 31*(94), 97–118.
Schushny, A., & Soto, H. (2009). *Guía metodológica: diseño de indicadores compuestos de desarrollo sostenible*. Santiago de Chile: Naciones Unidas.
Undersecretariat of Telecommunications (2010). Estadísticas de Servicio de Acceso a Internet. http://www.subtel.cl/prontus_subtel/site/artic/20080509/asocfile/20080509130640/1_series_conexiones_internet_dic_2009_v3.xls#'9.Co_Rg_D'!A1. Accessed 5 Nov 2010.
UNDP – United Nations Human Development Programme. (2009). *Human development report 2009. Overcoming barriers: Human mobility and development*. New York: Palgrave MacMillan.
United Nations (n.d.). Millennium Development Goals Chile. http://www.indexmundi.com/chile/carbon-dioxide-emissions-(co2),-metric-tons-of-co2-per-capita-(cdiac).html#G, Accessed May 2010.

Chapter 15
Dealing with Risks: A Governance Perspective on Santiago de Chile

Corinna Hölzl, Henning Nuissl, Carolin Höhnke, Michael Lukas, and Claudia Rodriguez Seeger

Abstract Governance is required to manage risks that occur in the course of urban development, but it can also become a source of risk. For this reason, sustainable urban development presupposes knowledge on governance-driven risks. In gathering such knowledge, this chapter combines empirical material, research results and conceptual considerations from different sources: firstly, the findings we obtained from an empirical study of stakeholders in Santiago; secondly, observations we made on how the governance matters of decentralization, privatization, participation and informality (outlined in Chap. 5) actually fall into place in Santiago; thirdly, our reflections on the key tasks in various urban policy fields (cf. Chaps. 6–13). In sum, as crucial sources of – governance-driven – risks, we can identify the extensive power of the private sector and clientelism, over-centralization and coordination deficits, the predominance of technocratic and neoliberal thinking, and low civil society engagement. Finally, we come up with recommendations on how to deal with these risks. They include an increase in private sector regulations, facilitation of multi-level governance strategies, empowerment of civil society and transformation of the political culture in Chile.

Keywords Governance-driven risks • Stakeholders • Policy recommendations • Multi-level governance • Political culture

C. Hölzl (✉) • H. Nuissl • C. Höhnke • M. Lukas • C. Rodriguez Seeger
Department of Geography, Humboldt-Universität zu Berlin, Berlin, Germany
e-mail: corinna.hoelzl@geo.hu-berlin.de

15.1 Introduction

Governance is expected to manage or remove the risks that emerge in megacities. On the other hand, it can itself become a source of risk. This occurs when governance fails to realize the challenges that lie ahead and to develop strategies to provide appropriate solutions; when it falls short of implementing these strategies, aggravates existing risks or worse, creates new dangers. The present chapter intends to formulate some recommendations on the implementation of policy measures aimed at sustainable urban development and the elimination of risks provoked by existing governance structures.

The many problems and risks in urban policy fields, e.g., air pollution, social exclusion and urban sprawl, have been revealed and discussed in detail in previous chapters. The possibility of natural disasters, such as earthquakes, poses likewise significant risks to the Chilean capital, risks that are augmented by the powerful dynamics of urbanization. Against this backdrop and with a view to sustainable development (see Chaps. 4 and 14), a set of policy objectives for areas in Santiago with a particularly strong need for action were identified: (1) social integration and cohesion; (2) controlled densification of urban areas; (3) increased attractiveness of public transport and non-motorized transportation; (4) mitigation of risks emerging from natural hazards; (5) reliable and affordable (long-term) provision of drinking water and energy, and (6) promotion of renewable resources.

It became clear in Chap. 5 that the task of addressing the governance matters of decentralization, privatization, participation and informality is no less important than determining the challenges and risks of urban policy and setting sustainable development goals. Based on an analysis of urban policy actors, relationships, institutions, and actual decision-making, as well as of key governance matters and governance-related aspects in various urban policy fields, this chapter seeks to identify the principal governance-driven risks to urban development.

We have endeavoured to link our research, scholarly discussion and consultation of the relevant literature with the experience and views of those affected on site. In the next section we present the findings of our empirical research on how stakeholders from various actor groups in Santiago view current governance arrangements and processes in their city (Sect. 15.2). Reflecting on stakeholder observations and analyses, we discuss what we consider to be the main governance-driven risks in Santiago (Sect. 15.3). This is followed by recommendations that integrate stakeholder proposals and build on existing opportunities in Santiago (Sect. 15.4). The chapter concludes with a brief summary (Sect. 15.5).

15.2 Urban Governance in Santiago: A Stakeholder Perspective

This section provides an overview of how urban development stakeholders in Santiago perceive and assess a set of governance aspects in the Metropolitan Region of Santiago de Chile. The insights detailed in this chapter are based on an empirical study that employed three methods of exchange with the stakeholders:

- An online survey on governance challenges in Santiago and stakeholder expectations for the future, which was conducted in early 2009 (January to March). Of the 170 stakeholders invited to take part in the online survey (conducted via e-mail), 49 responded. At 28%, this response rate was comparatively high (cf. Dillman 2007). Respondents were political decision-makers, scholars and stakeholders from the private and non-profit sectors.
- A focused round table discussion with five stakeholders was held in April 2009 (some of whom were participants in the survey); two stakeholders came from the public sector (regional government), two from the private sector (planning agencies) and one from civil society (NGO representative).
- Debates on governance issues and risks were held within the framework of six stakeholder workshops in November 2009 (related to the research presented in Chaps. 6–13). Approximately 15 external stakeholders representing the above-mentioned sectors participated in each of these full-day workshops.

The intention of this empirical research was to identify key issues and typical stakeholder perceptions, and incorporate their perspective into our recommendations, rather than to gain representative results on the attitude of the actors involved. In this sense, we too hoped to become familiar with the priorities and assumptions of the stakeholders, since it is they who will shape the discourse of future urban development and be vital to any attempt to change urban policy in Santiago. Taking up the Delphi study principle (cf. Häder and Häder 2000), these consecutive methodological steps (albeit not a Delphi study in the strict sense) made it possible to capture stakeholder reflections on analysis findings (predominantly the online survey) and break them down to the level of selected empirical fields. Stakeholder statements in this section are arranged according to the categories described in the theoretical framework introduced in Chap. 5, i.e., the stakeholder analysis of actors, relationships, institutions and decision-making processes. A final section will address their recommendations.

15.2.1 Important Actors and Their Influence on Governance in Santiago

Urban development stakeholders in Santiago seem to have largely converging ideas on the most influential urban governance actors in Santiago. They perceive the private sector, followed by central state entities, as having a strong influence on urban governance. The vast majority of stakeholders (more than 70%), however, regarded the influence of private actors on urban governance as too strong. In addition, specific public and private actors, such as real estate developers, the Chilean Chamber of Construction and infrastructure firms, as well as the Ministry of Public Works (MOP) and the Ministry of Housing and Urbanism (MINVU), were deemed to have exceptional power. Since the MOP is responsible for the country's infrastructure, it has even more control over developments in Chilean cities than the MINVU, whose chief task is to provide low-cost housing (cf. also Ducci 2004). This opinion was echoed by participants of the stakeholder workshops (see Fig. 15.1).

The importance of other actors in the governance process seems limited. According to the stakeholders, whether municipalities play a decisive role in urban affairs depends largely on the respective policy sector and specific issue. Municipalities are authorized to issue building permits, for example, and are thus seen as influential in the context of local real estate development projects. In general, however, the local level was considered wanting in terms of competence and resources. Although stakeholders acknowledged that the regional government (GORE) was currently active in trying to shape regional development, its influence was nonetheless seen as minor. Due to another revision of the existing Regional Development Strategy (EDR) and the envisaged drafting of a Regional Urban Development Plan (PROT), some interviewees expected an increase in the relevance of GORE (see Chap. 14). Others were less optimistic, since PROT will not be binding as a planning tool (see also Schiappacasse and Müller 2004). With respect to civil society actors, stakeholders saw their influence as negligible or virtually nonexistent, although some did anticipate change.

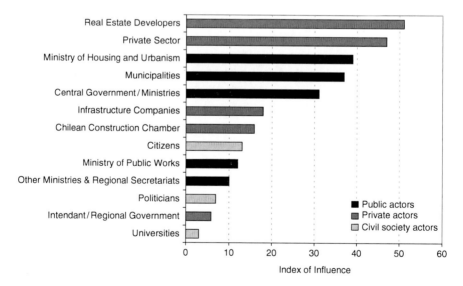

Fig. 15.1 Actors with a major influence on governance in Santiago. $N = 50$ *(weighted additive index based on stakeholder rankings of the most influential actors, whereby a value of 3 denotes the most important, 2 the second most important and 1 the third most important actor)* (Source: Hölzl and Nuissl 2010)

15.2.2 The Nature of Actor Relationships

A common viewpoint among stakeholders with regard to the relationships of the different governance actors was the existence of strong interest coalitions between the national state level and private actors. These coalitions are seen as the most relevant engine of urban development in Santiago. According to the stakeholders, however, private actor requirements are determinant here, as state regulation is widely regarded as weak. The results of the online survey suggest that the assessment of these coalitions, e.g., in the context of public–private partnerships (PPP) or privatized service provision, is a highly contested issue. Despite an overall critical evaluation of private sector influence, the majority of respondents approved of public–private collaboration. They pointed to the potential of PPP projects to facilitate public services and goods provision, e.g., in the maintenance of public parks. On the other hand, stakeholders who were skeptical of PPP criticized that private partners reaped considerably more profit from this type of cooperation than public entities, and only invested when large profits were guaranteed. Further points of criticism were the perceived lack of transparency in PPP projects and, more generally, lack of democratic legitimacy. Additionally, the suspicion was raised that PPP projects tended to neglect social and environmental concerns, e.g., the ecological effects of road construction. The National Commission for Environment (CONAMA) – replaced by the Ministry of Environment in October 2010 – responsible for approving projects that require an environmental impact assessment was criticized for not performing its duty.

15.2.3 Characteristics of the Institutional Framework

With regard to the shaping of governance in Santiago, the stakeholders in our research pointed to a broad range of formal institutions such as legislation, constitutional mechanisms of political control, and policy instruments. They furthermore cited informal institutions that have an impact on urban development, i.e., values, cultural principles, tacit knowledge and informal rules. The stakeholders confirmed the comparatively low significance of informality for the economy and urban development of Santiago (cf. Chap. 5). Quite a number of them, however, mentioned the problem of corruption. Although considered to have been of only moderate importance in the past, the perception of corruption has recently undergone change according to the participants in the online survey and is increasingly seen as a burden on public sector performance in Santiago. A second issue raised by a number of stakeholders, in particular in the workshops, is the observation that political culture in Chile is seen as heavily shaped by a market-oriented ideology and goes hand in hand with the predominance of a technocratic rationale in urban planning and development.

Three major issues with regard to formal institutions were touched upon. Firstly, stakeholders underlined the paramount influence of the central government. They pointed to such diverse fields as spatial planning, policies on air quality and energy policy. It became abundantly clear in our research that stakeholders saw decentralization as by no means accomplished in Chile. Accordingly, we find several complaints about the structure of the legal framework cementing the high degree of centralization of urban governance in Chile. Stakeholders also signalized that negative impacts at local level of decisions taken at a higher level are often not recognized there as such.

Secondly, stakeholders referred to inadequate public regulation in several policy fields and the absence of public sector instruments, e.g., lack of influence on infrastructure companies. Deficits in public regulation, however, are likewise attributed to failure to implement existing instruments appropriately. A few stakeholders at the round table discussion even argued that urban governance (in the sense of effective political and institutional guidance of urban development processes) was nonexistent in Chile. A significant number furthermore contended that planning instruments are frequently ignored. More specifically, they commented on the relevance of individual instruments of urban development and planning. Land use in the Metropolitan Region is supposed to be driven first and foremost by the Santiago Metropolitan Regulatory Plan (PRMS), the primary normative planning instrument formulated by the MINVU. Correspondingly, stakeholders regard it as the most important plan. With the Municipal Regulatory Plan (PRC), the municipal governments set norms for spatial development at the local level. By and large, however, the stakeholders were inclined to dismiss these and other indicative instruments as unfit for the proper steering of urban development processes. They referred to the lack of determination to implement these plans rather than to the plans themselves. As a result of their long implementation phase, for instance, the PRCs are perceived as toothless tigers, whereas building densities undergo rapid change.

Thirdly, the stakeholders emphasized the strong coherence deficit in the formal institutional framework. They see low linkages between the institutions in different policy areas, e.g., zoning and land-use plans (PRMS and PRC), and the normative Urban Transport Plan (PTUS) provided by the Ministry of Transport and Telecommunication (MTT). In the context of institutional deficiencies, stakeholders also hinted at overlapping areas of responsibility, in particular between the above-mentioned ministries (MOP, MINVU, MTT) and in the context of land-use management, waste management, transport and socio-spatial differentiation.

15.2.4 The Public Decision-Making Process

In order to understand the stakeholder perception of urban governance and the key issues, we asked them about current decision-making processes and outcomes of governance in Santiago. Vertical and horizontal coordination, i.e., the coordination

15 Dealing with Risks: A Governance Perspective on Santiago de Chile

Vertical coordination

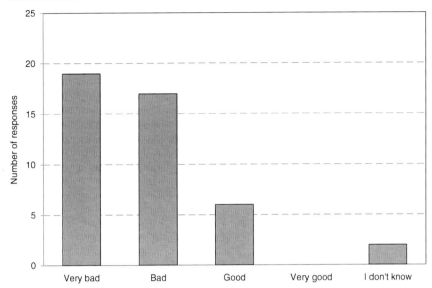

Horizontal coordination at national, regional and local government level

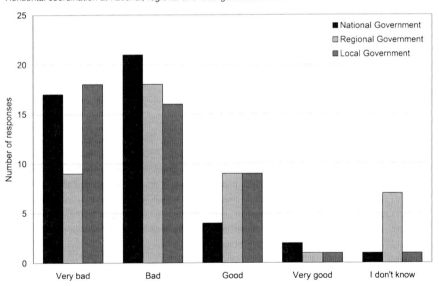

Fig. 15.2 Evaluation of decision-making process coordination (Source: Hölzl and Nuissl 2010)

and collaboration of actors representing different sectors and levels of policy-making, was another critical issue that was clearly a subject of discontent. In order to determine stakeholder attitudes to this issue, interviewees in the online survey were asked to evaluate vertical and horizontal coordination in Santiago on a scale ranging from 'very bad' to 'very good', and gave their general opinion on the challenges of coordination (see Fig. 15.2).

The figures indicate severe coordination deficits. Most stakeholders evaluated coordination issues negatively at all levels, with vertical coordination valued even less than horizontal coordination. Many complained about the lack of cooperation between the municipalities, which was thought to result from the absence at local level of resources and policy instruments to facilitate intermunicipal collaboration. Similarly, several stakeholders complained about the prevalence of sector-based approaches to planning and infrastructure policy. One participant in the online survey epitomized the stakeholder condemnation of most decision-makers as power seekers by referring to the public apparatus as "a monster with many heads that jealously guards its prerogatives" ("Un monstruo de muchas cabezas y celoso de sus prerrogativas").

Disintegration tendencies due to lack of cooperation are closely linked to the handling of information and knowledge and the issue of citizen participation. In the course of our research, stakeholders hinted several times at the lack of public awareness of critical issues as a result of inadequate public information and the corresponding low citizen engagement. On the issue of civil society participation, stakeholders declared that in Chile opportunities for citizen participation were still quite rare. The majority of stakeholders welcomed the progress made in some areas of urban development as the result of democratization at local level and growing civil society engagement. Diverging interests, particularly between the private sector and civil society, have provoked strong opposition in some parts of the city, and protests, citizen initiatives and even referenda are emerging (cf. SUR 2009). Stakeholders corroborated this observation, taking the example of the movement against construction of the urban motorway *Costanera Norte*, which was consequently built underground and led to the establishment of the community organization *Ciudad Viva* (see also Ducci 2000). Most of the stakeholders saw fresh opportunities for civic participation in the urban development process. In contrast to this optimistic outlook, some expressed pessimism with regard to current and future levels of civil society involvement. They pointed out that citizen participation in some policy fields is non-existent. Conflicts concerning water tariffs, for instance, are resolved by the concessionary company and the superintendence of sanitary services without civil society representation. Some stakeholders adopted a more critical stance on the quest for (more) participation. These interviewees complained about the apathy of citizens, who in their view react only when immediately affected by specific decisions. Similarly, some shrugged off the participatory efforts of civil society movements as "NIMBY" (not in my backyard) strategies.

15.2.5 Stakeholder Recommendations

The stakeholders involved in our research provided numerous recommendations on how to improve governance in Santiago. We will return to their detailed proposals in Sect. 15.4. This section describes the principal trends, which pertain to four specific areas: the regulatory function of the state, decentralization and coordination of public authorities, development of a common vision of city development, and encouragement of citizen participation.

Half of the recommendations made by the stakeholders in the online survey and in the policy-field workshops concentrated on the issue of state power. According to the stakeholders, the challenge of enhancing public regulation refers to the policy fields of spatial planning and land-use, transport, energy, water, social affairs, green space and environmental protection. On the one hand, they demanded a more resolute implementation of norms, planning instruments and control mechanisms already in place. On the other hand, they proposed measures to increase the regulative power of public authorities, such as law reforms and the introduction of new laws.

As results of the stakeholder survey indicate, the recommendations on state regulation are linked to a strong desire for more determined devolution of competencies and tasks from the national level to the regions and municipalities. There is almost no consensus, however, on how this goal is to be achieved. While stakeholders welcome further democratization efforts at the regional level, opinions on the desirability and likelihood of a reform that would increase the power of regional government are divided. Stakeholders from the regional government and academia in particular called for the establishment of a metropolitan government for the Santiago agglomeration, with a directly elected mayor. Other stakeholders were hesitant to endorse the idea of concentrating the responsibility for urban development in the hands of one person. They also believed that neither the central government nor the municipalities would allow the establishment of an *Alcalde Mayor* for the Greater Metropolitan Area. Concerning the local level, several stakeholders suggested supporting local authorities by transferring more competencies to this level.

With regard to coordination and cooperation, many participants in the online survey wished to see mutual recognition and the arrangement of respective self-interests (e.g., land use). This could lead to the development of an integrated approach to such pressing challenges as segregation, poverty, public transport and green space. Several stakeholders suggested improving the equalization transfer scheme, i.e., the Common Municipal Fund (FCM), in order to provide poor communities with the necessary resources to pursue their policies (see also Letelier 2004) and to create incentives for cooperation between the municipalities. Furthermore, in their view local authorities should be provided with the necessary information to raise their voice on issues of metropolitan development. In the same vein, a couple of stakeholders proposed new regulations to facilitate the engagement of local authorities in urban development with a regional scope. These should clearly define local discretionary powers, provide for fair negotiations between different authorities at all spatial scales and support intermunicipal cooperation with new instruments.

Several stakeholders supported the idea of a common *Leitbild* – a normative idea or model – with a shared vision of the Metropolitan Area that would contribute to improving coordination between urban development actors. They saw it as an incentive for urban governance actors and stakeholders to pull together, and believed that a joint vision required mid- or even long-term guidelines and clear policy goals to replace the predominance of political interests.

Finally, despite diverging ideas on the role of citizen participation in urban policy, the prevailing mood was one of support for more civil society involvement. Most stakeholders are of the opinion that increased citizen engagement presupposes opportunities for participation within the institutional framework and greater transparency of the decision-making process in general. Some had the view that citizen involvement would be strengthened by the establishment of a metropolitan government and direct elections.

15.3 Governance in Santiago: A Discussion of Risks

In the following discussion of governance-driven risks we address stakeholder views of governance problems in Santiago and link them to governance-related issues and urban policy fields elucidated in this volume. The stakeholder statements confirm the major importance of the governance matters of decentralization, privatization, participation and informality, all of which were discussed in Chap. 5. In order to identify crucial governance-driven risks, we apply our cumulative findings to the specific challenges of governance in Santiago. Their characteristics result from the interplay of decentralization, participation and privatization issues as well as of the key tasks of urban policy fields.

15.3.1 Extensive Power of the Private Sector and Clientelism

The strength of the private sector and its position, estimated by many of the interviewed stakeholders as too powerful, bear veritable risks for the future development of the urban system. Due to the uneven distribution of power, public authorities lack sufficient control, and social or environmental reforms often fail as a result of powerful private lobbies and the weak participation of civil society. In this context, we emphasize five major risks deriving from this predominance of private actors in urban policy.

Firstly, while many tasks of importance to the city as a whole, e.g. basic infrastructure, are delegated to the private sector, regulations to safeguard compliance with required standards and policy goals are inadequate. Responsibility for the collection of waste, for instance, was transferred to the municipalities; these, in turn, largely outsource public services to private contractors. There are, however, no official requirements or regulations for the recycling of waste. Planning standards associated with green space and the street network, for instance, appear to be lacking, while implementation of planning regulations at the local level seems weak. This frequently means that environmental impact studies, e.g., on hydroelectric power plants and urban megaprojects, are not carried out, making social, economic, environmental and spatial impacts an unknown quantity (cf. Romero and Vásquez 2005; Ramírez 2009). As a result, the development process lacks transparency and the examination of projects in terms of how they fit into the bigger picture is rendered impossible.

Yet another issue is associated with the profit and return logic incumbent on public–private partnerships. As the concession model of motorway funding demonstrates, for example, private investment in infrastructure is confined to areas of the city where profits are guaranteed or, at the least, leads to varying infrastructure quality across the city. Infrastructure investments with little indication of large profits are rare, a reality repeatedly pointed out in the stakeholder workshops. Profit-orientation and economies of scale by PPPs implies the preference for large-scale projects. This scenario often leads to conflicts of interest between public bodies and service providers, and ultimately to sweeping inequalities in the accessibility to basic services such as (quality) education and spatial mobility (cf. Rodríguez and Rodríguez 2010).

A major risk stemming from the strong position of the private sector is the oligopolistic market structure that characterizes the water, energy and waste sectors. *Aguas Andinas* serves approximately 1.5 million households with water (see Chap. 12). The Central Interconnected System (SIC), the major grid system in Chile, provides electricity for more than 92% of the population. Ninety percent of its capacity is dominated by three large holdings: Endesa, AES Gener and Colbún (see Chap. 9). Waste collection in Santiago is carried out by 18 separate enterprises, two of which alone control 42% of the market (see Chap. 13). Windows of opportunity for competition are thus few and far between, and the scope for small and middle-sized enterprises is extremely limited.

Long-established, closed elite networks pose another major governance-related risk, since formal public–private partnership arrangements tend to be based on informal networks and negotiations. The public and private actors involved in these cooperations belong for the most part to the MINVU, the MOP, municipal authorities, infrastructure companies and the real estate sector. In addition, close relationships, especially in the land-use management sector, exist between powerful politicians and private actors – as evidenced in the planning processes of peri-urban megaprojects (cf. Chap. 5 in this volume; Poduje and Yañez 2000; Ducci 2004; Heinrichs et al. 2011).

Although stakeholders showed growing concern about corruption, they stressed their overall assumption that informality is of only minor significance in Santiago. More research on informal facets would be useful, nevertheless, as informality refers to more than the informal economy and land occupation by the urban poor (cf. Roy 2005). Regarding real estate development, it might be interesting to know whether the planning system in place prompts 'unplanned' informal cooperation. Clientelism largely survived the regime change after the Pinochet dictatorship and has put down roots in municipal politics in Santiago (and many other Latin American cities) (cf. Massun 2006; Irazábal 2009). In Chile this is first of all due to a local political system set up around the figure of a pre-eminent mayor and to the clientelistic ties of other local politicians with community organizations. Within the managerial elitist model of local governance that is typical of many municipalities in Santiago, clientelism contributes to disempowerment of the urban poor (cf. Rivera-Ottenberger 2007). The drawback here is the combination of decentralization and lack of participation, which ultimately leads to socio-political exclusion. Finally, clientelism in addition to the extensive power of the private sector puts decision-making beyond the reach of ordinary citizens. The result is frequently the neglect of concerns and on-site realities, as well as of issues of common welfare and equity.

15.3.2 Over-Centralization and Coordination Deficits

The stakeholders in our research are convinced that insufficient decentralization and the lack of coordination among public actors are a major source of governance-driven risks in Santiago. The challenge of addressing these problems combines with that of generating cooperative multi-level governance. There is a universal observation that the dynamics of globalization and the re-scaling of economic and political processes has fundamentally altered the interrelation of the various levels of decision-making in society (e.g., Brenner 2003; Uitermark 2002). Unlike the classic welfare state and its extensive public services, which were founded on and legitimized a vast hierarchical apparatus, the normative understanding of multi-level governance as a new paradigm for policy and polity tries to adapt to contemporary challenges and to integrate increasingly diverse actors (cf. Marks and Hooghe 2004). The reorganization of state authorities and their responsibilities across the different levels of decision-making, however, has hitherto failed to establish a new system of governance.

This is evident at the regional level in Santiago, where there is no metropolitan government and the regional government is weak in terms of competencies and legitimacy. Similar to other megacities, the central government continues to have a strong direct influence on urban affairs. Inhabitants and numerous local decision-makers further identify with the municipality they live and work in rather than with the city (Schiappacasse and Müller 2004). They do not relate to a super-municipal entity, i.e., a regional government, and are thus unaware of a larger entity to which, and through which, cooperation could be fostered.

Governance in Santiago shows evidence of a problematic gap between the decision-making level and the level that has to cope with decisions made. Road planning and transport management, for instance, are notable examples of how the hands of the municipalities are tied, since scope for autonomous decision-making in this area is severely curtailed (cf. Zegras and Gakenheimer 2000). The problems of multi-level government, however, are at their most conspicuous at the local level. On the one hand, the municipalities find themselves in a weak position when it comes to interaction with the national level. Yet, on the other hand, they rarely manage to unite and cooperatively pursue their municipal interests vis-à-vis the central state or other actors. Here, the provision of adequate resources is no less important than the transfer of authority (cf. Siavelis et al. 2002; Schiappacasse and Müller 2004). Insufficient equipment, professionalization levels and municipal staff expertise coupled with low budgets and the ineffective redistribution of income through the FCM prevent municipalities from pursuing their interests (cf. Letelier 2004; Orellana 2009). Lack of resources is likewise a strong incentive for local authorities to compete for private investments rather than to cooperate with each other. This explains why a number of stakeholders in our study seemed hesitant about the (swift) transfer of responsibilities to the local level. It also points to the importance of recognizing the risk of setting up 'local traps' (Purcell 2006) when decentralization takes place without shifting sufficient resources. Thus, the dynamics of larger structures need to be taken into account, since increasing municipality responsibilities can generate greater injustice (cf. Miraftab 2008). On the whole, the most difficult task to resolve appears to be the strong disincentives for municipalities to engage in mutual cooperation.

The shortcomings of vertical and horizontal coordination in the municipalities, as well as of those at all urban policy levels and sectors in Santiago aggravate the problems caused by an over-centralized institutional framework, a common phenomenon in many megacities (cf. Siemens 2007; Borja and Castells 2003; UN-Habitat 2009; Ward 1996). The integration into the institutional framework of all sectors and levels relevant to urban development and the cooperation between various actors are clearly crucial to the success of urban governance. Failure to accomplish these tasks renders urban institutions – including the strategies, programmes and instruments that depend on them – less effective. Regarding town planning, for instance, weak institutional integration and contradictory planning norms seriously complicate application of the planning framework (cf. Sierra 2006; Chuaqui and Valdivieso 2004).

15.3.3 Predominance of Technocratic and Neoliberal Thinking

Beyond the framework of formal institutions, it is informal institutions and in particular predominant leitmotifs of proper policy-making that influence the structure of governance. The stakeholders pointed out that governance processes in Chile since the 1970s have been strongly influenced by a line of reasoning that manages to combine neoliberal and technocratic thinking (see also De Mattos 2005; Silva 2008; Zunino 2006). The latter is characterized by an attempt to professionalize public administration, while breaking the traditional influence of party politics. In this model, professionals and experts – referred to as technocrats – who work in public administration emerge as the new actors in the policy game. They are for the most part trained as engineers, economists, financial experts and managers, whereby administrative experience is not necessarily a condition. Chile is considered to be a model case of technocracy, at least since the era of Pinochet's military regime. Technocrats were key actors in changing the political system, both under the military government and after the return to democracy. Policy processes have therefore maintained an elitist character, since participation in decision-making presupposes an advanced degree or technical expertise. In the same vein, many public programs still draw on the idea of professionals working for the public good on behalf of a benevolent state, albeit in a pronounced top–down fashion (cf. Silva 2008). It is true that MINVU has recently begun to incorporate programmes such as *Quiero mi Barrio* or *Ciudad Parque Bicentenario*, aimed at improving housing quality, socially mixed housing, and integration, which take the issues of social capital and participation more seriously (see Chap. 8); the success of these programmes in overcoming the predominance of technocratic and market-oriented thinking in urban development policies, however, has still to be proven.

On the other hand, these values and informal rules for matters of governance create several risks. First of all, the absence of a common vision for the city as a whole beyond technical details and economic interests intensifies disintegration of the different actors, tasks and efforts in Santiago urban policy. If we take another look at the planning process, we find an absence of national guidelines, i.e., a national urban policy. Then, urban policy decisions based on a combination of technocratic reasoning and privatization goals often come at the expense of social and environmental considerations (see also Rodríguez and Rodríguez 2010) and make urgent reforms in these fields even more difficult. When technical and economic issues supersede social criteria and the involvement of citizens, as in the case of the public transport reform project Transantiago, efficiency itself is at risk. In general, if decision-making follows a technical rather than political rationale, the task of increasing citizen participation becomes an even greater challenge, since there is little scope for non-experts to participate.

15.3.4 Low Civil Society Engagement

As pointed out in the previous section, a substantial number of stakeholders were distinctly pessimistic about the future of citizen participation; referring to the 'NIMBY' rationale, some even voiced opposition to the notion of encouraging participation. Both attitudes should be taken seriously (cf. Cleuren 2008). It is nonetheless important to account for the fact that stakeholders from the private sector sometimes speak of citizens' apathy and the 'NIMBY' rationale as irreversible in order to legitimize their opposition to the quest for more participation.

Citizens tend to be perceived as consumers in Chile's largely neoliberal conception of governance and as participants are not seen as worthy of attention in the political decision-making process. The weak position of civil society emerged in the aftermath of the coup d'état in 1973. The once strong organization of citizens ceased to exist almost entirely in the 1970s and was at least partly 'replaced' by a culture of clientelism (Taylor 1998). This lack of citizen participation, however, is a major cause of many of the problems discussed thus far. The recent substantial rise in the number of urban conflicts poses new problems for public authorities, as projects meet with strong resistance at a late stage and consequently at tremendous social and economic cost – a genuine lose–lose situation (Poduje 2008). On the other hand, some stakeholders rightly see recent civil society initiatives born of these conflicts as positive signs. Indeed, the huge problems associated with the implementation of *Transantiago* – caused by lack of passenger information – led to protest from angry citizens. Today, information on changes to the public transport system is available and a certain level of participation exists – albeit informal and to a very small degree.

In general, however, the involvement of civil society remains low and in our view two main obstacles are crucial here. Firstly, there is little effort to stimulate discussion on urban policy challenges or to support certain measures with active campaigns. *Ciudad Parque Bicentenario*, for instance, is severely lacking in public communication strategies. In a country where segregation is frequently understood by the vast majority as a rule, this should have been a prerequisite in order to reduce prejudice and win over citizens for socially diverse housing projects (cf. Zunino 2006). Promising initiatives do not attract sufficient public attention, so that the respective draft proposals have no possibility of becoming institutionalized; this was the case with the proposed law to set a 5% rate of social housing for every residential development project. For similar reasons, environmental issues are likewise strongly underrepresented.

A second obstacle to citizen participation is the lack of fully developed instruments for participation. Mayors have been elected democratically since 1992, and in 2009 a law for the direct election of the Regional Council was passed (Law 20.390). However the change to the Organic Constitutional Law (19.175) for it to come into force is still awaited. The shrinking voter turnout, however, cannot be ignored. At the recent presidential elections it dropped to about 59%, with primarily young people failing to register their votes (cf. Quiroga 2010). Beyond elections, the set of participative instruments at the local level is still rudimentary in Chile. According to Irazábal (2005), many of the current forms of citizen participation in planning processes in Latin American cities fall into the category of 'non-participation' on Arnstein's 'Ladder of Citizen Participation' (1969). Citizen participation in the elaboration of Municipal Regulatory Plans (PRC), for instance, provides for information rather than for the concrete involvement of local people (cf. Zenteno 2007; Poduje 2008). Likewise, when it comes to infrastructure investments, uneven power distribution and public tendering behind closed doors strongly affect decision-making processes and outcomes, culminating in planning decisions that cater to private interests.

15.4 Governance in Santiago: Opportunities and Recommendations

Having identified a set of potential governance risks to urban development, it should be noted that opportunities and chances are also inherent in the contemporary evolution of governance in Santiago. A highly encouraging trait is the willingness and general ability of stakeholders to stimulate urban development and enhance urban governance. This potential must clearly be built on if governance in Santiago is to improve. In this spirit, we wish to provide some recommendations. They address and are arranged according to aspects discussed in the preceding section and refer to concrete policy field issues and their fundamentals. They are layered onto existing opportunities and constitute an entry point for the discussion of recommendations made by stakeholders in the course of our study. The viability of their application to policy fields will also be considered.

15.4.1 Restricting Private Sector Power Through State Regulation

Strengthening public control of urban development processes seems indispensable to the achievement of more sustainable urban development in Santiago. Accordingly, the stakeholders in our study first of all addressed the issue of the regulative power and efficiency of the state when asked to formulate recommendations for future urban policies. Two aspects are of particular interest here: the creation or improvement of regulations and the strengthening of the regulatory power of public authorities to implement existing rules and procedures.

15.4.1.1 Improving the Regulatory Framework

Proceeding from stakeholder suggestions on how to improve the regulatory framework, we see the following recommendations pertaining to specific urban policy fields as vital elements of a sustainable urban development strategy. At this point, our primary interest is to use these proposals to illustrate the kinds of rules, incentives and control instruments we deem crucial to regaining some control over the activities of private actors, thereby widening the scope and opportunity to implement public decisions:

- To enable low-income households to live in a consolidated area of the city with higher land values, it is recommended that small-scale housing programmes be implemented rather than vast private housing projects. This could be effected by subsidies differentiated according to the land values in the respective municipality, which should also be legally stipulated, e.g., in the form of quotas in the Municipal Regulatory Plan (PRC). As witnessed in the context of conditional planning however, these quota schemes are difficult to implement. Furthermore, in the stakeholder workshop on socio-spatial differentiation, the strengthening of Santiago's sub-centres was strongly recommended, especially the development of social and technical infrastructure.
- Regulations for transport management and the improvement of air quality should include the introduction of emission standards, a congestion charge for the inner city and restrictive parking management. In addition, sustainable road construction should follow a strategic plan rather than the proposals of private construction companies.
- Similarly, appropriate laws are required for the promotion of renewable energies and energy-saving practices. This also holds true for the regulation of renewable energy feed-in and energy saving (e.g., standards for domestic appliances). The participants in the stakeholder energy workshop regarded a 5% share of energy production from renewable energy as too low, not least because the law applies

to new plants only. Moreover, subsidies for producers and households could help to increase the distribution of solar thermal technologies.
- With respect to water consumption, stakeholders proposed the introduction of a differentiated price system based on specific water uses (domestic, agricultural, industrial and energetic). Other measures proposed were the definition of maximum amounts and adjustment of the water code. Investment in new technologies such as recycling or downcycling water and rainwater are other possibilities. Yet another demand – appropriate to all infrastructure provision – is to increase stipulations for infrastructure companies (e.g., to render tariffs transparent and to avoid costs shifting to consumers and the creation of monopolistic structures).
- With regard to the challenges of waste management, a differentiated charging system could contribute to waste reduction and be implemented in the context of a tariff system reform. New norms are required to enforce the pre-treatment of waste. Stakeholders also called for incentives, e.g., for small enterprises to promote new technologies for waste treatment, composting and energetic use. This applies in particular to the energetic use of biomass (instead of payments for biogas burning). Finally, aid programmes could help to organize and support workers ('cartoneros') in the informal sector at the municipal level (see also Irazábal 2009).

15.4.1.2 Enforcement of Regulations and Plans

Beyond the legal framework, the concrete implementation of rules and instruments, regardless of whether they are new, improved, or already in place, is equally decisive. Land-use policy and planning instruments are the most critical in this context. Firstly, more determined implementation of existing instruments is necessary, if control over new development schemes is to be maintained; support should, if necessary, come from the controlling authorities. Many of the stakeholders made a number of recommendations on the shortcomings in planning practices. They called for more environmental impact studies and the integration of environmental aspects into all plans; consideration of different timelines in planning; the elaboration of problem-specific plans (e.g., air quality); the design of land-use plans that account for neighbourhood scales. The elaboration of thematic strategic plans was another frequently mentioned proposal (for a broad discussion of strategic planning, see Healey 2007). To be effective, these plans would have to be integrated into the regulatory planning system framework and impact on land rights; frequently they are either modified or not complied with. Similarly, some stakeholders called for the use or improvement of evaluation systems and the establishment of expert committees for urban and regional planning (for a detailed discussion of planning innovations, see UN-Habitat 2009).

Regarding the improvement and reform of existing planning instruments, compensation measures for environmental and other infringements are the subject of heated debate in Chile (cf. Zegras and Gakenheimer 2000; Poduje 2006). Unless they receive more political support, however, local authorities will not be in a position to demand the introduction of appropriate compensation rights for damage caused by developments carried out on their territory. More generally, public requirements for the implementation of new developments must be clarified and put into concrete terms. Furthermore, it is essential to extend the regulative competences of land-use planning beyond the urbanized areas. Implementing these propositions would require amendments to existing laws or the introduction of new (planning) laws.

Although the amount of public control over public affairs could undoubtedly be increased, it is vital that obstacles to its implementation by the public authorities be taken seriously. On the one hand, the strong conflict of interests, especially in the fields of land use and planning, makes powerful resistance from the private sector inevitable. On the other hand, public authority deficits clearly obstruct the regulation of activity in this sector.

15.4.2 Enabling Multi-level Governance

The exchange with stakeholders revealed that unless the challenges of coordination and multi-level governance are tackled successfully, implementation of an integrated sustainable urban development strategy is unlikely. The key task in this context is decentralization. The participants in our study highlighted the difficult role of the municipalities in the Santiago Metropolitan Area. The absence of common objectives emerged from comments in the online survey as a major problem, i.e., the inability of local elites to actually pursue the shared interests of the municipalities in the region. Regardless of the unwillingness of local authorities to coordinate their activities, the lack of resources and instruments (such as common investments or joint development projects) to support intermunicipal cooperation must be addressed. The systematic inclusion of regional and local authorities in planning decisions is crucial to achieving this goal. Furthermore, considering the aforementioned limited resources and low level of professionalization at the municipal level, it is paramount that all efforts to strengthen municipalities be accompanied by capacity building. Finally, a reform of the hegemonic position of mayors in Chile should be discussed in order to enhance the democratic character of local governments (cf. Rivera-Ottenberger 2007; Cleuren 2007).

Many of the problems at local level can only be solved if the regional level is strengthened at the same time. The latter has the potential to facilitate both, since it is in a position to mediate competency transfers to local authorities, on the one hand, and intermunicipal cooperation, on the other (cf. Borja and Castells 2003). Hence, it might be useful to look more closely at possible arrangements for the governing of a Metropolitan Region. It might furthermore help to contextualize diverging stakeholder views on this question. Some arrangements of metropolitan governance described in the far-reaching research on this issue fit into the 'metropolitan reform tradition' that advocates governmental consolidation, whereas others correspond to the 'public choice perspective' that considers the autonomy of cooperating municipalities the most efficient model (for broader discussion of these and other models, see Heinelt and Kübler 2005a). Correspondingly, in Santiago we might distinguish three lines of discourse on how to overcome the disparities and coordination deficits within the public sector in the future. Interestingly enough, although the stakeholders are familiar with these discourses, few efforts have as yet been made to elaborate on them further with scientific studies or policy papers.

- A primary course of action would be to increase incentives for intermunicipal cooperation. However, only a few stakeholders advocate this 'soft' model of metropolitan governance.
- A second option consists in the transfer of certain responsibilities and decision-making competencies to the regional or metropolitan authority. This would include, first and foremost, land-use planning, social policies and the entire domain of transport management. The introduction of a metropolitan transport authority for Greater Santiago (made up of 34 municipalities) has frequently been discussed.
- A third arrangement is the establishment of a metropolitan government for the Greater Metropolitan Area of Santiago (comprised of between 34 and 39 municipalities) with a directly elected mayor. It could hold either extensive or basic capacities for strategic, operational and administrative planning. Since some stakeholders had reservations about such a solution, an alternative arrangement would be a metropolitan government executed by a council of representatives from different sections of society.

According to Schiappacasse and Müller (2004), five tasks are crucial to metropolitan sustainability in Santiago: identity, democratic representation, social equilibrium, intermunicipal collaboration and strategic planning. For two fundamental reasons, the creation of a directly elected metropolitan government is in our view the most promising solution (cf. Chuaqui and Valdivieso 2004, p. 106). Firstly, the problems Santiago is experiencing can only be solved if the bulk of citizens is involved. They in turn will only be willing to accept compromises if they have a more active role in the governance of their city. Santiago's citizens must be convinced to identify more strongly with the metropolis and its economic and cultural potential. Establishing a common and directly elected metropolitan political entity could bolster this connection. Secondly, against the backdrop of the deficits revealed in this research, it might be more effective to create a new metropolitan authority and abolish existing entities with cross-municipal responsibility for certain policy fields, rather than to add yet another entity, e.g., to serve transport concerns.

Before embarking on the path to achieve new arrangements of metropolitan governance, communication and collaboration procedures need first of all to be facilitated, e.g., through mediation processes. Only then can established negotiation practices and deadlocks be overcome. In this context, intermunicipal cooperation could be deepened by focusing on crucial topics. The regional government and the municipalities in the Metropolitan Area could initiate joint activities, e.g., on the mitigation of air pollution, disaster protection and the synergetic potential of intermunicipal cooperation in conducting revitalization programmes. Furthermore, the current regional government could pave the way for the establishment of a stronger, democratically legitimized regional authority by taking an active role in metropolitan affairs. Similar to the process of drawing up the PROT, it could design an integrated development strategy for Santiago, one that is urgently required. In terms of citizen participation at the regional level, the new law on the direct election of Regional Councillors is a vital step towards more democracy. The same would be required for the *Intendente*, the head of the regional authority. In conclusion, the creation of a division for local and citizen empowerment would also help to meet the requirements for increased participation and account for different spatial scales.

Undoubtedly the question of enabling decentralization and multi-level governance primarily concerns the central state. Regional, metropolitan and local authorities will ultimately gain significance only if the national state is prepared to waive some of its competencies. This, in turn, will not occur until such time as civil society, i.e., the voter, makes a strong plea for this course. Encouraging such public demand at the national level is one reason why public awareness and support for the reawakening of civil society have become major tasks in enhancing governance in Santiago.

15.4.3 Supporting the Reawakening of Civil Society

The establishment of a metropolitan government and the direct election of a mayor would as a matter of course constitute a strengthening of citizen participation in the form of elections. It is no less important, however, to reinforce existing instruments at the local level, where citizen participation is principally organized.

Although various stakeholder recommendations focused on civil society participation in urban governance, stakeholders tended to be unspecific in their suggestions, which contained little or no concrete ideas on what should be done to increase participation. One stakeholder proposed the introduction of citizen budgeting at the regional and local level, an instrument already in use, for instance, in the municipalities of Cerro Navia and San Joaquín (cf. Montecinos 2006). With regard to land-use planning, a number of stakeholders pointed out that citizen participation should not be confined to the official obligation to inform local residents. Instead it should call for an early and extended involvement of citizens in processes of problem diagnosis and plan drafting (e.g., with respect to the Municipal Regulatory Plan). The suggestion was also made to bring key urban development actors together, including representatives of civil society, such as community organizations (*juntas de vecinos*) and NGOs. The introduction of a round table is one method in this context. The combined efforts of the NGO *Cordillera* and the municipality of La Florida to curb urbanization processes in the Cordillera should be mentioned as an interesting example in this regard. Indeed, the various examples stakeholders referred to during the study are encouraging. The above-mentioned community organization *Ciudad Viva* is now cooperating with the public sector for the design of bicycle lanes throughout Santiago. In the area of housing policies there is gradual interest in a stronger involvement of future residents in the planning process, so as to adapt housing to people's needs and make use of social capital (cf. Chap. 8). These examples illustrate the opportunities for institutional and cultural change through innovative governance approaches, which leave a lasting impression on people's minds.

Nevertheless, to achieve profound progress in civic participation, these initial attempts must be institutionalized and a comprehensive body of formal rules and regulations laid down. The Citizens' Councils (CESCO) and Local Development Strategies (PLADECO), the official instruments for more direct citizen participation at local level, must be saved from the fate of being ignored or manipulated in the current practice of local decision-making (cf. Greaves 2004). For some policy fields (e.g., water and energy) a more reliable dissemination of and access to information would be a step towards participation. As discussed, initiating public awareness and discussion is a prerequisite for plans to reform.

15.4.4 Facilitating the Transformation of Political Culture

Finally, we will briefly address an issue repeatedly brought up by the stakeholders themselves; it could be referred to as the political culture issue. A large number of comments in the online survey pertained to implementing democratic procedures, citizen participation and transparency of the political process. Many of the stakeholders seem convinced of the need for change in the long-established traditions of the political system and certain cultural values described earlier as a combination of neoliberal and technocratic reasoning and clientelism. To achieve this, the decision-making elite would have to radically rethink their role and their actions (see also Heinelt and Kübler 2005b). The findings of the analysis suggest moreover that such a change in political culture could lead to more reliable, successful and accountable political leadership in the long term.

It is worth mentioning in this context, the call for a common vision. Initial signs of this guiding principle are found in the *Transantiago* project. Despite the surfeit of problems associated with the project, the reform of the public transport sector is in itself encouraging, even though it was motivated by inter-urban competition and the race to become a "global city". This discourse should be exploited for more fundamental and ambitious reform challenges. By strengthening the burgeoning use of bicycles, for example, Santiago could become an international showcase example for innovative transport solutions in Latin America.

Yet many stakeholders clearly had a broader concept in mind when they referred to a *Leitbild* for Santiago. The terms they used to describe what is meant by such a concept – "territorial vision" ("una visión de territorio"), "city lineaments" ("lineamiento ciudad"), "city policies" ("políticas de ciudad") – and what it should address, e.g., "heritage" ("patrimonio"), "environment" ("medioambiente"), "public space" ("espacio público"), indicate that at the heart of this debate is the attempt to address Santiago as a whole and to define common goals. As already seen, stakeholder views on fundamental issues were quite similar, allowing for hope that a common guiding principle might evolve in the future. Of course, the various stakeholders equate such a vision with varying urban policy goals, with particular stress on quality of life and welfare issues. What we want to emphasize here, however, is the significance of such a common vision in itself. Developing, discussing and advancing a shared vision beyond technical requirements and economic targets, and embedding it into public consciousness seems vital to addressing the challenges of urban governance identified in this research: to encourage citizen participation, to enhance cooperation and integrated approaches by the various actors, and to implement public decisions and planning.

15.5 Conclusion

This chapter focused on governance-driven risks in Santiago de Chile and aimed at formulating recommendations that might facilitate the implementation of programmes and strategies for sustainable urban development. In this endeavour we included the observations, views and proposals of stakeholders in Santiago urban policy obtained through empirical research.

In our view the main challenges of urban governance for sustainability consist, firstly, in restricting the extensive power and influence of the private sector by strengthening state regulation, especially with regard to land use and planning. This seems indispensable to increasing the scope and opportunity for making and implementing public decisions. Furthermore, this could contribute to overcoming a high degree of pessimism and severe lack of trust in public urban policy and its instruments on the part of the stakeholders, as the interviews revealed. Secondly, to enable multi-level governance capable of addressing current challenges, it is vital to improve coordination among public authorities throughout all levels and sectors, and to tackle the long-standing problem of over-centralization. In the long run, the establishment of a metropolitan government for the Santiago agglomeration with a directly elected mayor seems to be required. Thirdly, the task of increasing citizen participation must be recognized as a core issue in augmenting the legitimacy of urban policy and avoiding the growing and costly disputes over urban development projects. A comprehensive body of formal rules and regulations is vital to encouraging civil society involvement. Finally, it is crucial to facilitate the transformation of the prevailing political culture in the field of (urban) governance, which is often shaped by clientelism and a predominance of technocratic and neoliberal rationality. Here, an important contribution could be to support the development of a common vision for the city of Santiago de Chile, beyond technical details and economic interests.

These changes, all of which are closely connected, could reduce the risk of fragmentation of the city in social, political and physical terms and, instead, help to embark on a path of more sustainable urban development and pursue a common, integrated approach to the solution of pressing problems. The findings of this chapter indicate that, despite some encouraging developments of late, promising efforts and intentions will soon be confronted with the limitations of current governance structures and the distribution of power. To transform the ideas of sustainable urban development into action and to take advantage of existing opportunities and initiatives, change is nonetheless paramount.

References

Arnstein, S. (1969). A ladder of citizen participation. *AIP Journal, 35*(4), 216–224.
Borja, J., & Castells, M. (2003). *Local and global. The management of cities in the information age* (4th ed.). London: Earthscan.
Brenner, N. (2003). Metropolitan institutional reform and the rescaling of state space in contemporary Western Europe. *European Urban and Regional Studies, 10*(4), 297–324.
Chuaqui, T., & Valdivieso, P. (2004). Una ciudad en busca de un gobierno: Una propuesta para Santiago. *Revista de Ciencia Política, 24*(1), 104–127.
Cleuren, H. (2007). Local democracy and participation in post-authoritarian Chile. *European Review of Latin American and Caribbean Studies, 83*, 3–18.
De Mattos, C. (2005). Santiago de Chile: Metamorfosis bajo un nuevo impulso de modernización capitalista. In C. De Mattos, M. E. Ducci, A. Rodríguez, & G. Yánez (Eds.), *Santiago en la globalización: ¿una nueva ciudad?* (pp. 17–46). Santiago: SUR-EURE libros.
Dillman, D. A. (2007). *Mail and internet surveys: The tailored design method* (2nd ed.). New York: Wiley.
Ducci, M. E. (2000). *Governance, urban environment, and the growing role of civil society.* Washington, DC: Woodrow Wilson International Centre for Scholars.
Ducci, M. E. (2004). Las batallas urbanas de principios del tercer milenio. In C. De Mattos, M. E. Ducci, A. Rodríguez, & G. Yánez (Eds.), *Santiago en la globalización: ¿una nueva ciudad?* (pp. 137–166). Santiago: SUR-EURE libros.
Greaves, E. (2004). Municipality and community in Chile: Building imagined civic communities and its impact on the political. *Politics & Society, 32*(2), 203–230.
Häder, M., & Häder, S. (Eds.). (2000). *Die Delphi-Technik in den Sozialwissenschaften. Methodische Forschungen und innovative Anwendungen.* Wiesbaden: Westdeutscher Verlag.
Healey, P. (2007). *Urban complexity and spatial strategies: Towards a relational planning for our times. The RTPI Library Series.* London/New York: Routledge.
Heinelt, H., & Kübler, D. (2005a). *Metropolitan governance: Capacity, democracy and the dynamics of place.* New York: Routledge.
Heinelt, H., & Kübler, D. (2005b). Conclusion. In H. Heinelt & D. Kübler (Eds.), *Metropolitan governance: Capacity, democracy and the dynamics of place* (pp. 188–201). New York: Routledge.
Heinrichs, D., Lukas, M., & Nuissl, H. (2011). Privatisation of the fringes – a Latin American version of post-suburbia? The case of Santiago de Chile. In N. Phelps & F. Wu (Eds.), *International perspectives on suburbanization: A post-suburban world?* London: Palgrave-MacMillan (forthcoming).
Hölzl, C., & Nuissl, H. (2010). *Governance in Santiago de Chile – Stakeholder prospects for the future* (UFZ discussion paper 3/2010) Leipzig. http://www.ufz.de/data/Disk_Papiere_2010_3_Nuissl12823.pdf. Accessed 17 Sept 2010.
Irazábal, C. (2005). *City making and urban governance in the Americas. Curitiba and Portland.* Aldershot: Ashgate.
Irazábal, C. (2009). *Revisiting urban planning in Latin America and the Caribbean.* http://www.unhabitat.org/downloads/docs/GRHS2009RegionalLatinAmericaandtheCaribbean.pdf. Accessed 17 Sept 2010.
Letelier, S. L. (2004). *Alcances y Desafíos de la Descentralización Fiscal en Chile.* http://www.fes.cl/documentos/descent/alcancesydesafios.pdf. Accessed 17 Sept 2010.
Marks, G., & Hooghe, L. (2004). Contrasting visions of multi-level governance. In I. Bache & M. Flinders (Eds.), *Multi-level governance* (pp. 15–30). Oxford: University Press.

Massun, I. (2006). *Clientelismo político*. Moreno: Métodos.
Miraftab, F. (2008). Decentralization and entrepreneurial planning. In V. Beard, F. Miraftab, & C. Silver (Eds.), *Planning and decentralization: Contested space for public action in the global south* (pp. 21–35). New York: Routledge.
Montecinos, E. (2006). Descentralización y democracia en Chile: Análisis sobre la participación en el presupuesto participativo y el plan de desarrollo comunal. *Revista de Ciencia Política, 26* (2), 191–208.
Orellana, A. (2009). La gobernabilidad metropolitana de Santiago: La dispar relación de poder de los municipios. *Revista EURE, 35*(104), 101–120.
Poduje, I. (2006). El globo y el acordeón: planificación urbana en Santiago. In A. Galetovic (Ed.), *Santiago: ¿Dónde estamos? ¿Hacia dónde vamos?* (pp. 231–276). Santiago de Chile: Centro de Estudios Públicos.
Poduje, I. (2008) *Participación ciudadana en proyectos de infraestructura y planes reguladores* (Serie Temas de la Agenda Pública, No. 22) Santiago de Chile: Pontificia Universidad Católica de Chile.
Poduje, I., Yáñez, G. (2000). Planificando la ciudad virtual: megaproyectos urbanos estatales y privados. In: *Seminario Internacional Las regiones metropolitanas del Mercosur y México: entre la competitividad y la complementariedad*. Buenos Aires, Programa de Investigación Internacional Grandes Regiones Metropolitanas del Mercosur y México, Mexiko.
Purcell, M. (2006). Urban democracy and the local trap. *Urban Studies, 43*(11), 1921–1941.
Quiroga, Y. (2010). *Politischer Wechsel in Chile: Nach einem halben Jahrhundert gewinnt die Rechte demokratische Wahlen*. FES Kurzbericht. http://library.fes.de/pdf. Accessed 17 Sept 2010.
Ramírez, P. (2009). *Los errores del MOP en Vespucio Sur que costarán 25 millones de dólares*. In Ciperchile, February 20, 2009, http://ciperchile.cl/2009/02/20/los-errores-del-mop-en-vespucio-sur-que-costaran-25-millones-de-dolares/. Accessed 4 Dec 2010.
Rivera-Ottenberger, A. (2007). Decentralization and local democracy in Chile: Two active communities and two models of local governance. In V. Beard, F. Miraftab, & C. Silver (Eds.), *Planning and decentralization: Contested space for public action in the global south* (pp. 119–134). New York: Routledge.
Rodríguez, A., & Rodríguez, P. (2010). *Santiago, a Neoliberal City* (working paper). http://www.socialpolis.eu/index.php?option=com_docman&Itemid=199&task=doc_download&gid=272. Accessed 17 Sept 2010.
Romero, H., & Vásquez, A. (2005). La Comodificación de los territorios urbanizables y la degradación ambiental en Santiago de Chile. *Scripta Nova – Revista Electronica de Geografía y ciencias sociales, 9*(194), 68.
Roy, A. (2005). Urban informality. Towards an epistemology of planning. *Journal of the American Planning Association, 71*(2), 147–158.
Schiappacasse, P., & Müller, B. (2004). Desarrollo metropolitano integrado: El caso de Santiago de Chile. *Urbano, 7*(10), 68–74.
Siavelis, P. M., Valenzuela Van Treek, E., & Martelli, G. (2002). Santiago: Municipal decentralization in a centralized political system. In D. Myers & H. Dietz (Eds.), *Capital city politics in Latin America: Democratization and empowerment* (pp. 265–295). London: Lynne Rienner Publishers.
Siemens AG (2007). *Megacity challenges. A stakeholder perspective*. München: Siemens AG. http://w1.siemens.com/entry/cc/features/urbanization_development/all/en/pdf/study_megacities_en.pdf. Accessed 17 Sept 2010.
Sierra, L. (2006). Urbanismo por decreto: centralismo y confusión institucional en la ciudad chilena. In A. Galetovic (Ed.), *Santiago: ¿Dónde estamos? ¿Hacia dónde vamos?* (pp. 299–328). Santiago de Chile: Centro de Estudios Públicos.
Silva, E. (2008). *In the name of reason: Technocrats and politics in Chile*. University Park: Pennsylvania State University Press.

SUR – Corporación de Estudios Sociales y Educación. (2009). Conflictos urbanos en Santiago de Chile. Mapa de conflictos urbanos. *Notas Digitales*, 1, 1–6. http://constructoresdeciudad.sitiosur.cl/wp-content/uploads/2009/12/Nota1-final-conflictos-urbanos.pdf. Accessed 17 Sept 2010.

Taylor, L. (1998). *Citizenship, participation and democracy: Changing dynamics in Chile and Argentina*. Basingstoke/Hampshire/London: Macmillan Press/St. Martin's Press.

Uitermark, J. (2002). Re-scaling, 'scale fragmentation' and the regulation of antagonistic relationships. *Progress in Human Geography, 26*(6), 743–765.

UN-Habitat. (2009). *Planning sustainable cities: Global report on human settlements 2009*. London: United Nations Human Settlements Programme. Earthscan.

Ward, P. (1996). Contemporary issues in the government and administration of Latin American mega-cities. In A. Gilbert (Ed.), *The mega-city in Latin-America*. Tokyo/New York/Paris: United Nations University Press.

Zegras, C. & Gakenheimer, R. (2000). *Urban growth management for mobility: The case of the Santiago, Chile Metropolitan Region* (Report prepared for the Lincoln Institute of Land Policy and the MIT Cooperative Mobility Program). http://web.mit.edu/czegras/www/Zegras_Gakenheimer_Stgo_growth_mgmt.pdf. Accessed 17 Sept 2010

Zenteno, J. (2007). *Planificación urbana, planes reguladores comunales y reacción ciudadana* (Report prepared for Sur Corporación de Estudios Sociales y Educación). http://www.sitiosur.cl/descargadocumentos.php?PID=71. Accessed 17 Sept 2010.

Zunino, H. (2006). Power relations in urban decision-making: neo-liberalism, 'techno-politicians' and authoritarian redevelopment in Santiago, Chile. *Urban Studies, 43*(10), 1825–1846.

Chapter 16
Synthesis: An Integrative Perspective on Risks in Megacities

Dirk Heinrichs, Kerstin Krellenberg, Bernd Hansjürgens, and Francisco Martínez

Abstract The central motivation for this book was the observation that megacities are places of inherent opportunity and risk. The book combines the empirical study of urban sectors in the Metropolitan Region of Santiago de Chile with theoretical considerations of risk and explores several common risk concepts. This final chapter provides a discussion and some conclusions to key questions. Firstly, it provides a systematic review of the extent to which different approaches to risk help to understand the *Risk Habitat Megacity*. Secondly, it considers what research methods and approaches are appropriate to understand the complex system that the Megacity is and provides orientation and action knowledge. Finally, and with special reference to the case of the Metropolitan Area of Santiago, it discusses what forms and strategies of governance constitute an adequate response to these challenges.

Keywords Complexity • Metropolitan Region of Santiago de Chile • Research methodology • Risk

16.1 Megacities as Risk Habitats: A Challenge for Integrative Research

Urbanization is *the* phenomenon of the twenty-first century. Since 2007 cities have been the habitat of more than half of the world's population. Megacities in particular have attracted increasing attention. They are characterized by vast populations (in some cases more than ten million inhabitants) with high population densities, infrastructure and resources, but also by the velocity of change and the complexity

---

D. Heinrichs (✉) • K. Krellenberg • B. Hansjürgens • F. Martínez
German Aerospace Center (DLR), Institute of Transport Research, Rutherfordstr. 2, 12489 Berlin, Germany
e-mail: dirk.heinrichs@dlr.de

of their structures. The latter refers to the high division of labour that prevails in megacities and leads to an extreme of specialization, an intense relationship between the city and its hinterland in terms of resource and energy flows, a unique concentration of innovation and high-level education, and a diversity of interrelated formal and informal decision-making structures (including decisions by private households, companies, various levels of governmental authorities, interest groups), all of which shape the behavioural patterns of their inhabitants. We see these interrelated elements – size (or scale), speed (or velocity of change) and complexity – as the three decisive features of today's mega-urban agglomerations, with high explanatory powers for the analysis of megacity processes and challenges. Furthermore, processes of global change take place in a highly condensed form first and foremost in megacities, allowing the underlying forces to be studied and analysed in a nutshell (megacities as 'laboratories of global change').

The central motivation for this book was the observation that megacities are places of inherent opportunity and risk. As opportunity they are places where people can generate income and establish a basis for economic survival under conditions that are liveable. They are also places of advanced innovation, however, where social and human capital can be accumulated. The per capita resource use in megacities could be lower than in rural areas if efficient and effective resource use methods (e.g., synergies, economies of scale) were to be employed. Low unitary production costs lead to high productivity and economic opportunities, and ultimately to competitive advantages. This explains the tendency to agglomerate economic activities and production in urban areas (Bettencourt et al. 2007). Apart from being a source of opportunity, megacities are places of risk. Risks can take many different forms. The following risks are addressed in several chapters of this volume, most of which are related to natural resources and the environment: natural risks stemming from earthquakes, flood risks related to land-use change, social risks associated with socio-spatial segregation, risks linked to energy supplies and transport system quality, health risks as a result of poor air quality, water quantity- and water quality-related risks, and risks associated with waste management.

The book addressed these issues along two main questions, focusing on a selected case study: the Metropolitan Region of Santiago de Chile. The first question is how to define, analyse and reduce the wide variety of risks. The second question is, how to foster opportunities. By exploring risks and opportunities in one city, this book provided a unique opportunity to take into account the side effects of sectoral policies and inter-regional aspects (affecting different parts of the urban agglomeration or the entire metropolitan region). It likewise sought to make the complexity of urban systems visible, addressing the challenge of complexity and integration with a multi-sector and multi-disciplinary approach. Against this background the objectives of the book were threefold:

- To systematically examine risk trends in mega-urban agglomerations by analysing diverse risks elements;
- To evaluate the extent and severity of risks, and

16 Synthesis: An Integrative Perspective on Risks in Megacities

– To develop instruments, measures and governance strategies to cope with adverse risks in an integrated and coherent manner as a guideline for urban development.

Tackling these research questions required a solid integrative perspective that not only explores the risks in the fields concerned, but also takes unintended side effects, the interconnectedness of risks, and the backward slopes into account. The underlying notion of integration in this book is a broad one, rendering the several dimensions of integration explicit. They include the integration of scientific disciplines (interdisciplinarity), cognitive integration (developing a common understanding through learning), the integration of methods and results (e.g., with respect to spatial and temporal scales), and stakeholder integration (transdisciplinarity).

The remainder of this final chapter is devoted to summarizing the major findings of the book, emphasizing the various forms of integration. Section 16.2 focuses on the use of risk concepts and addresses cognitive integration. Section 16.3 looks at methodological and conceptual issues, touching on aspects of spatial and temporal scale, data provision and conceptual frames. Section 16.4 deals with the concrete situation in Santiago de Chile, outlining several key findings on risk and governance responses. Section 16.5 concludes with an outlook.

16.2 Different Risk Concepts: Achievements and Open Questions

The definition of risk used in this book was not all-encompassing. Instead, three separate risk concepts were chosen for the analysis of the Risk Habitat Megacity:

– The first risk concept was derived from the hazard community. Risks were understood here as hazardous events characterized by statistical measures (expected damage and probability of damage occurrence). People and infrastructure as 'elements at risk' face exposure to these hazards and have a specific vulnerability, understood as the result of exposure to risk and the capacity to cope with adverse hazardous effects. This concept forms the basis for the risk analysis of earthquakes (Chap. 6), as well as of land use and floods (Chap. 7).
– The second risk concept was developed in cooperation with the integrative sustainability concept of the Helmholtz Association. The Helmholtz Integrative Sustainability Concept serves as a formulation of the desired targets against which risks can be measured. The concept was put into operation with the definition of specific sustainability indicators. Risks are measured in terms of how far indicators depart from set targets. This perspective guided the risk analysis on energy (Chap. 8), transportation (Chap. 10), air quality (Chap. 11), water (Chap. 12), and waste management (Chap. 13).
– The third risk concept, which was applied to social-spatial differentiation and social exclusion (Chap. 8), is that of systemic risks. The causes of this kind of

risk are not manifested in external shocks, but exist as an integral part of the way in which economic, social and technical systems function in modern society. Risks of this type emerge particularly in complex societies defined by interdependencies, complex linkages and backward loops, as is the case in mega-urban agglomerations.

Although far from suggesting that the adopted concepts in the various chapters are the preferred options for examining risk in the cases under review, we argue that the idea of multiple perspectives is a more adequate approach to risk in complex situations than the single perspective. A single risk concept would fail to satisfy the analysis requirements for the broad range of risks concerned. As shown in Part II, Chaps. 6–13, each concept has its own emphasis and explanatory powers when it comes to investigating how risk is generated, measured and managed. It is not surprising that the hazard-oriented concept, which has its origins in the hazard community of the natural sciences, was applied to earthquakes in Chap. 6 and to the impact of land-use change on flood risk in Chap. 7. Neither of these chapters, however, represent the hazard risk concept in its pure form. Rather, they apply an 'extended' risk concept. In addition to the analysis of hazards in the chapter on earthquakes, the aspect of vulnerability is taken into account and includes the exposure of elements at risk (people, infrastructure, economic values). In the chapter on land use, an even broader perspective is chosen, defining risk as the combined product of hazard, elements at risk and vulnerability.

The distance-to-target risk concept was applied in Chaps. 9–13. The distance to a pre-defined sustainability target is adopted as a measure of assessing the extent and severity of the risk. The considerable advantage of this risk concept lies in the facility with which future developments can be evaluated by applying scenario techniques. If 'pictures of the future' are developed, a risk analysis on the basis of the distance-to-target concept can reveal information on the quality of future developments, i.e., whether such developments will lead to higher or lower risks. This in turn allows for adjusting policies towards a more sustainable direction. The systemic risk approach is applied in Chap. 8. Here, this approach points to the fact that socio-spatial segregation trends spring from complex human social interaction. The subsequent risk of social exclusion or inclusion is not externally driven, but caused by internal developments.

What has megacity research gained by applying these different risk concepts? First of all, this approach refines and extends the knowledge of complex risks in megacities and creates added value to risk discourses. Secondly, it deepens our understanding of the emergence of risks and their severity. The approach furthermore opens up a new perspective for researchers and stakeholders embedded in 'traditional thinking'. To give one example: the widespread perspective on natural hazards among natural scientists could be overcome by demonstrating that risks are rarely the result of 'natural' processes. This holds true even in the field of earthquakes, where the extent and severity of risks is also determined by the extent of the exposure of elements at risk and their capacity to cope with adverse impacts. While earthquakes cannot be prevented, people can respond in many ways: in an area prone to earthquakes they decide precisely where to build houses and infrastructure; they decide how to build and what materials should be used; they also decide on precautionary measures with regard to earthquakes (e.g., rescue plans). All of these factors illustrate that human beings have several options when it comes to determining the degree and severity of risks.

16.3 Methodological and Conceptual Insights

The studies presented in this book touch on a wide range of methodological and conceptual issues and challenges: how to measure (indicators), how to incorporate different temporal and spatial scales, and how to integrate the perspectives of diverse stakeholders. These aspects are key elements of integration, since comprehensive approaches and appropriate risk strategies – as outlined in Sect. 16.2 above – require common frameworks. Complexity and multiple risks in megacities likewise call for a variety of methods and approaches. Consequently, the sectoral risk analyses undertaken in Chaps. 6–13 representing a broad spectrum of disciplines from the natural and social sciences are based on multiple methods that range from qualitative interviews and analysis of statistical and remote sensing data to geospatial analysis and mathematical forecast development and coupling. They also have common points of reference in terms of indicators and spatial and temporal scales.

The sectoral studies were based on the analysis of specific factors that drive urban development in Santiago de Chile (see Chap. 4). In the context of topics that applied the distance-to-target approach to (future) risk analysis, where the distance to a more desirable (sustainable) development was evaluated, working with (sustainability) *indicators* was crucial. Indicators were additionally used to analyse flood and earthquake risks as a function of the tripartite product of natural hazards, elements at risks and vulnerability. In all cases the use of indicators facilitated the analysis of changing conditions over time, including historical trends as well as future scenario techniques, and led to comprehensive conclusions. The same holds true for the comparative study of several Latin American megacities presented in Chap. 2 and their overall sustainability performance discussed in Chap. 14. Furthermore, the work with indicators and other analytical tools, such as common concepts and scenarios, played an important role in confronting the interrelation of processes and the interdependencies of, e.g., the use of resources. Single risks may overlap across thematic fields, and in some cases even refer to *risk chains*, as seen in the case of transport emissions, their consequences on air quality and the related health impacts.

In other words, providing conceptual knowledge on a range of risk concepts and indicators, and applying this knowledge to various topics is seen as a highly useful approach to the integrative and comparative perspective on multiple risks and opportunities in megacities.

With respect to *spatial scale*, the Metropolitan Region of Santiago de Chile, with its 52 urban and rural municipalities, served as the common reference area for all of the studies. This unit could not be used exclusively, however, since content and availability of data in some cases dictated whether research was conducted on smaller or larger spatial scales. In other instances, e.g., energy, the topic needed a wider, national spatial perspective. Several topics analysed different levels in order to highlight certain trends. Since national census data in Santiago de Chile was a primary data source, analysis in several studies included one or more of the four levels: municipal, district, block ('manzana') and household, as shown in the following examples:

- The analysis of socio-spatial differentiation processes (Chap. 8) focuses on 39 municipalities out of the total 52 that make up the Metropolitan Region. These 39 municipalities are considered urban municipalities or the Greater Metropolitan Area of Santiago. On the one hand, each of the municipalities was analysed in itself, while on the other hand, municipalities were grouped into five municipal clusters to highlight some of the broader trends. In short, the research content determines the selection of the spatial scale and the extent of the analysis.
- Research on flood risk in Santiago de Chile (Chap. 7) was primarily determined by the availability of remote sensing data that more or less covered the Metropolitan Area of Santiago de Chile (MAS, 34 municipalities). Hence associated population data was analysed at the MAS level to allow for an overlay with remote sensing data. Due to restrictions on the spatial coverage of hazard maps, in-depth research on flood hazard concentrated on two selected municipalities.
- Other sectors such as air quality (Chap. 11) showed no direct connection to administrative sub-units but rather analysed the distribution of emissions across the Metropolitan Area of Santiago. This reference area was selected for its close linkage with models used in the transport sector (Chap. 10).
- The study on energy supply (Chap. 9) accommodated the pronounced interrelations between the city and the national level. The powerful role of the national level is also evident in terms of data availability. As the chapter shows, most information exists on a national scale, with considerably less filtering in at the level of the Metropolitan Region.

The examples indicate that spatial scale is a vital issue for the integration and comparison of the various sectors. Taking the Metropolitan Region as the main reference point gave (limited) support to the comparison and integration of the results.

In multi-sectoral and comparative studies, *temporal scale* is another significant element of integration, albeit one that is heavily defined or restricted by the time periods of the available data. Census data as input for a wide range of analyses presented in this volume (e.g., socio-spatial processes, flood risk, transportation, health) is collected every 10 years (...1992, 2002 ...). Other surveys covering intermediate time periods, e.g., the Encuesta CASEN, follow the same procedure in data collection to a limited degree only, making comparison with census data a challenging task.

The inclusion of local *stakeholders* in participatory workshops and surveys was a further indispensable element of integration, in this case of perspectives. Consensus exists on the ground rule that no action or orientation knowledge should be generated without considering the opinions and needs of decision-makers, civil society and the business community. The integration of stakeholders nevertheless posed a huge challenge. On the one hand, analysis and integration through methods and concepts was crucial to opening up a structured discussion with stakeholders on the topic of feasible *action and orientation knowledge* in the context of risks in megacities. On the other hand, transforming scientific conclusions into appropriate information for orientation knowledge and into action is far more challenging. The incorporation of multi-stakeholder perspectives undoubtedly enhances the research work. This form of *ground-truthing* likewise helped to increase stakeholder acceptance of the research findings.

In conclusion, these various elements of data integration and the analysis of risks and opportunities for one vast study object such as a megacity are a major advance in terms of understanding system dynamics and interrelations. There are, however, obvious limitations that remain difficult to overcome.

In addition to methodological and conceptual integration, the integration of research conclusions from the individual fields under consideration is vital. It can best be achieved with a concrete case study, in this case the Metropolitan Area of Santiago de Chile. The comprehensive perspective on the state of existing risks, their trends and the future challenges in the Metropolitan Region of Santiago that emerge from the studies presented in Chaps. 6–13 of this volume is summarized in the following section.

16.4 Lessons for Santiago de Chile: Emerging Risks and Governance Responses

Overall the analysis of the current state of and future trends in Santiago de Chile is the account of an urban development model characterized by remarkable economic progress and an increase in the well-being of the population. Many of the 'generic' risks associated with megacities, in particular those linked to the lack of access to basic services such as water or energy, do not apply in Santiago. Poverty levels in the Metropolitan Region are in decline (Chap. 14). Access to drinking water and sewage infrastructure has increased, and hygienic standards have improved (Chap. 12). Household access to electricity supplies, with low system interruption, is universal (Chap. 9). Social housing policy has successfully reduced the housing deficit (Chap. 8). These positive trends go hand in hand with significant technological advancement and improved standards of living, as documented in the case of declining concentrations of air pollutants (Chap. 11).

The chapters related to the consumption of resources and/or the generation and release of emissions (including waste) into the environment, however, also highlight the immense resource intensity, as well as the persistence of certain 'unsustainable' development paths. This side of the mega-urban system contains several inherent risks. The level of land consumption for built-up areas, for example, and the loss of agricultural land forces a transition from natural hazard to risk and exposure as a result of increased settlement in flood- and landslide-prone areas (Chap. 7). The exploitation of available water resources in the water catchment leads to water stress and supply deficits in dry years, which are expected to increase (Chap. 12). For three reasons, environmental risks of this kind are likely to intensify in the future:

- Firstly, the population in Santiago is projected to increase substantially from currently 6.5 million inhabitants to approximately 8 million in 2030. The likely consequences are increasing resource use and the associated emissions. This leads to the exposure of a growing number of people and amount of capital to extreme events and greater health risks.
- Secondly, resource-intense forms of consumption will most likely continue to grow under the current development model. One example is the projected surge in the generation of waste, which will potentially exhaust landfill capacities and increase landfill gas emissions relevant to climate (Chap. 13). Another example is the transportation system (Chap. 10). The predicted rise in motorization rates and individual car use that will inevitably culminate in greater congestion and longer travel times could lead to reduced quality of life.
- Thirdly, these trends may be aggravated by climate change in some sectors. Growing water demands, for instance, are accompanied by a decrease in water supplies due to the impact of climate change (higher temperatures, more droughts, a drop in water quantities from glaciers).

While the nexus between population growth, economic growth and greater resource use, on the one hand, and critical environmental consequences and risks, on the other, is easily perceived, the implication of social risk is perhaps less obvious. While there is substantial debate in the literature on risks related to social exclusion and the lack of social cohesion, the situation in Santiago is ambiguous. As already pointed out, the tremendous achievement of providing universal access to basic services and shelter for the inhabitants of the Metropolitan Region has pruned the deficits. Yet, significant socio-economic differences continue to exist: socio-spatial differentiation processes, for example, have led to large-scale concentration of lower socio-economic status households in massive social housing estates on the urban periphery, a move associated with a host of negative consequences such as stigmatization, unemployment and crime. This puts residents at the risk of being drawn into a vicious cycle of exclusion (Chaps. 3 and 8). These areas constitute hot spots of social risk and disintegration. At the same time, however, the city centre and several peripheral municipalities with a traditionally lower-class population show evidence of a trend towards gentrification, brought about first and foremost by the immigration of young people, and of medium and higher-income groups.

Should pollution and daunting resource deficits begin to have serious effects on certain groups, e.g., in the form of rising prices for water, energy or mobility, disadvantaged areas could become a social risk. Under these circumstances policies and regulations will be crucial to mitigating the negative effects, especially on lower-income groups. Taking the example of water, the government currently subsidizes social groups at the lowest end of the scale, enabling them to cover their water bills. The situation is likely to spiral in subsequent decades as population numbers rise, water resources dwindle, and new infrastructure provisions, such as channelling water between river basins and desalinization plants, prove costly.

This opens the debate on the role of 'governance', understood broadly as the regulation of publicly relevant affairs (Chap. 5). One of the contributions in this volume explores how governance itself can pose or aggravate risks and even reinforce dangerous trends (Chap. 15). In fact, many of the chapters highlight a form of governance characterized by a short-term, one-sided, 'reductionist' perspective rather than one that sets sustainable (understood as long-term and comprehensive) solutions in motion as a response to complex challenges. The risk trends caused or exacerbated by governance can be illustrated along three main concerns:

- The first concern relates to the temporal implications of the current development paradigm. The topics discussed in the chapters reveal that little consideration is given to the long-term environmental risks contained in the current practice of resource exploitation and resource use. Referring back to the issue of water, current water rights are purchased in accordance with the 1980 Water Code, with no prospect of strategic water management planning or consideration of long-term ecological impacts (Chaps. 12 and 14).
- The second governance deficit is the lack of spatial comprehensiveness and coverage. This is particularly evident in the context of binding land-use regulations. Competencies are essentially limited to the urban municipalities in the metropolitan region (Chaps. 5, 7, 15) and thus neglect the overall need to steer land use and building activities away from a metropolitan-wide perspective. This factor is compounded by weak capacities at the regional level to mediate and direct spatial development, and to adjust to the contemporary challenges of a growing metropolis and the integration of increasingly diverse actors.
- A third consideration relates to the so-called 'silo effect': the regulation of each development sector is confined to its own concerns, making no provision for unintended side effects of policies, projects or measures. One example is the interaction between energy supply options and emissions, on the one hand, and air quality and health impacts, on the other. Other prominent issues are transportation, housing and social-spatial segregation. Perhaps one of the key lessons of this book is understanding the urgent need to advance the task of sectoral integration.

Yet another valuable lesson from this study is the realization that certain risks related to, for example, water and energy supplies or the levels of pollution, can be alleviated by the introduction of appropriate technology. Other risks are intrinsic to megacities and their remarkable tendency to agglomerate the economic, cultural and social activities mentioned above. The research presented in this book highlights two cases from these sectors, which are relevant to the dilemma of suggesting appropriate governance: transport and socio-spatial differentiation.

Common to the risks involved in these two areas is the danger of their growth as development advances; at least in the period that draws developing cities close to the developed world. Economic forces dominate risk growth, although investment of increasing wealth in policies and technology could help to stunt this growth. What tentative suggestions could be made in the interests of innovative contemporary governance structures and processes that have a tendency to favour short-term 'reductionist' solutions over those that are sustainable? How can 'metropolitan governance' cope more effectively with and mediate the significant challenges ahead? The concerns highlighted above provide a 'mega-challenge' to governance in the Metropolitan Region of Santiago de Chile. They likewise point to the possible direction of attempts to innovate current practice. We refer to four crucial elements:

- The first aspect is to reinforce the perspective on the 'long-term' social and environmental consequences of current trends. Beginning today to address the resource constraints of tomorrow is of the utmost importance. It would open opportunities to discuss technological, social, economic and ecological options in an integrative manner. Long-term orientation, however, is likewise relevant in the context of infrastructure investment, since decisions determine the future of the city in a way that is difficult and costly to reverse. This is extremely relevant in cities such as Santiago de Chile, where the pace of economic development implies a high annual rate of infrastructure development. The how of planning determines the sustainability of the city and its future. As this book demonstrates, introducing long-term sustainability indicators is a useful tool to identify and monitor distant risks.
- The second aspect is to move towards a kind of 'metropolitan governance' that fully embraces all urban and rural locations and the relation of the region to outside locations. It touches on the difficult question of how to allocate competencies, functions and capacities between the metropolitan area and the municipalities. Although detailed analysis of such angles would exceed the scope of this book, it can be said that the current division of competencies between the metropolitan area and the municipalities is clearly inadequate.
- The third aspect is to demand systematic cross-sectoral analysis in order to comprehend more fully the trade-offs between critical trends. This includes strengthening coordination between public authorities at all levels and sectors, and financing instruments such as common and multi-year investments or joint development projects.
- A fourth requirement is that metropolitan governance safeguards inclusiveness in the sense of ensuring citizen involvement and participation. As established in various contributions in this volume, public consent must be acknowledged as a core issue in augmenting the legitimacy of urban policy and preventing costly disputes over urban development projects. Given the anticipated risks ahead, this point may prove to be even more vital in the future than it is today.

Although the challenges involved in enhancing governance cannot be solved here, these aspects are of the essence if governance decisions are to embrace greater sustainability and long-term social and environmental risks to be considered more systematically.

16.5 Perspectives

Risk Habitat Megacity, with its analyses of different sectoral fields, conceptual discussions and scientific conclusions, is the result of interdisciplinary and transdisciplinary research. The research endeavour adopted the perspective of the megacity as a complex system. Following the ideas of complexity research, the work presented here approaches various 'sub-systems' from different angles and with different research methodologies. Yet, while attempting to discover the state of and development in individual realms of the urban agglomeration, it likewise gained a more thorough understanding of the entire system and its constitutive elements. It maintains an integrative perspective on the megacity, attuning analysis to the mutual interdependence of the processes involved and providing a basis for modelling and scenario techniques. With its problem perspective and context-specific investigation, this approach links the generation of orientation knowledge with action-oriented knowledge and the implementation of solutions.

Given that the number and size of mega-large and mega-complex systems will continue to grow in the future, there is significant potential for further research in this direction. It is hoped that the approach presented here and its intention of shifting governance in megacities 'from response to action' and from 'sectors to systems' will make a valuable contribution.

Reference

Bettencourt, L. M. A., Lobo, J., Helbing, D., Kühnert, C., & West, G. B. (2007). Growth, innovation, scaling and the pace of life in cities. *Proceedings of the National Academy of Sciences USA (PNAS), 104*, 7301–7306.

Printed by Books on Demand, Germany